PASSIVE COOLING OF BUILDINGS

PASSIVE COOLING
OF BUILDINGS

EDITORS
M. SANTAMOURIS AND
D. ASIMAKOPOULOS

from Routledge

First published by James & James (Science Publishers) Ltd. in 1996

This edition published 2013 by Earthscan

For a full list of publications please contact:

Earthscan
2 Park Square, Milton Park, Abingdon, Oxon OX14 4RN
Simultaneously published in the USA and Canada by Earthscan
52 Vanderbilt Avenue, New York, NY 10017, USA

First issued in paperback 2020

Earthscan is an imprint of the Taylor & Francis Group, an informa business

A catalogue record for this book is available from the British Library

ISBN 13: 978-0-367-57963-0 (pbk)
ISBN 13: 978-1-873936-47-4 (hbk)

Typeset by Edgerton Publishing Services, Huddersfield, UK

Table of contents

Preface

The increase in household income in Europe and the relatively low cost of electricity have helped air conditioning to become highly popular. This has resulted in a significant increase in building energy consumption in southern Europe. Sales of air-conditioning equipment in this region have increased considerably over the past few years and are now worth close to 1.7 billion ECU per year. In Greece, for example, while the annual sales of packaged air conditioners was close to 2000 units in 1986, the market jumped to over 100,000 units in 1988.

The specific energy consumption of buildings for cooling needs is dictated by the climate, the type of building and the installed equipment. A comparison of the annual specific energy consumption for cooling in large air-conditioned office buildings in Denmark, Greece, The Netherlands, Norway, Sweden and the UK, shows a variation of between 15 to 110 kWh/m^2. It should be noted that, despite the Mediterranean climate, Greek office buildings do not use more electricity than Swedish or British offices for heating, ventilating and air-conditioning (HVAC) systems during summer. Clearly, the problem of cooling in buildings is not peculiar to southern climates alone. It may be just as important in northern climatic regions, depending on the building type and construction.

The impact of the use of air conditioners on electricity demand is a serious problem for almost all southern European countries. Peak electricity loads force utilities to build additional power plants in order to satisfy the demand, thus increasing the average cost of electricity.

Environmental problems associated with the use of ozone-depleting CFC refrigerants used in conventional air conditioners present an additional argument for minimizing the use of HVAC systems for energy savings in the cooling sector. Problems of indoor air quality associated with the use of air conditioning should also be taken into account. Recent studies of air-conditioned buildings have shown that illness indices are relatively higher for these than those for non-air-conditioned buildings.

Alternative passive cooling techniques, based on improved thermal protection of the building envelope and on the dissipation of building's thermal load to a lower-temperature heat sink, have been proved to be very effective. These strategies and techniques have already reached a certain level of acceptance in architectural and industrial circles. Passive techniques as alternatives to air conditioning can bring important energy, environmental, financial, operational and qualitative benefits.

The study and application of passive cooling is a multilayered and multidisciplinary process. It is important to treat the subject in conjunction with other aspects of architectural design; it should not be considered in isolation. A useful framework for considering passive and hybrid cooling in the context of environmental design can be summarized as follows: prevention of heat gain, modulation of heat gain and heat dissipation. Protection from heat gain may involve:

- landscaping and the use of outdoor and semi-outdoor spaces,
- building form, layout and external finishing,
- solar control and shading of building surfaces,
- thermal insulation, and
- control of internal heat gain.

This book has three main objectives:

- to report on basic knowledge to date in the field of passive cooling, as well as on new tools developed during the past few years;
- to present the recent progress in the field that has been achieved through prominent research programmes; and
- to identify priorities for future research into passive cooling.

The book provides information on all available passive cooling methods and techniques, their potential effectiveness, their basic principles and the criteria needed to identify those most appropriate for specific types of buildings. It also includes presentations of several easy-to-use methods that are available for calculation of the cooling potential of the most important techniques, as well as of the overall thermal performance of buildings.

This book was prepared by the Central Institution for Energy Efficiency Education (CIENE), within the framework of the SAVE Programme 'Short Educational Structure on Energy Efficiency in Buildings' of the European Commission, Directorate General XVII for Energy.

The main goal of the SAVE Programme is to undertake actions that promote the efficient use of energy in all member states of the EU. In this way, the Programme contributes to achieving the aims of the EU energy policy for rational use of natural resources and reduction of carbon dioxide emissions in the atmosphere.

CIENE was established in October 1992 as a SAVE Programme, under the auspices of the National and Kapodistrian University of Athens, Greece. Its operation is harmonized with the philosophy of the SAVE Programme of establishing a network of energy centres throughout Europe. The main role of CIENE is to organize and coordinate specialized training and continuing-education programmes on energy efficiency and energy management.

This book is addressed to anyone who is interested in energy conservation in buildings. The information provided can be used by building owners, occupants and those professionals directly or indirectly involved in the construction, operation and maintenance of buildings. In some respects, this book may be of more interest to those who have some background experience of energy-related topics and building applications.

We thank the various authors, who devoted a great deal of time to perfecting their chapters. We wish to acknowledge the help, encouragement and support of the DGXVII staff who work for the SAVE Programme, and especially that of Mr E. Dalamangas who has closely followed the overall project.

We would like to acknowledge the help of a group of individuals who have provided valuable contributions, comments and suggestions for this book : S. Alvarez of AICIA Seville, Spain; F. Allard of the University of La Rochelle, France; J. Goulding of University College Dublin, Ireland; M. Grosso of the Politecnico di Torino, Italy; G. Guaraccino of LASH/ENTPE, Lyon, France; O. Lewis of University College Dublin; E. Maldonado of the University of Porto; C.A. Roulet of EPF Lausanne, Switzerland; A. Tombazis of Meletitiki Ltd; P. Wouters and L. Vandaele at BBRI; and S. Yannas of the Architectural Association of London.

Any comments with regard to the contents and structure of this book are welcome.

M. I. Santamouris
D. N. Asimakopoulos
Editors

Athens, January 1996

1

Cooling in buildings

Ever since humans have moved into shelters, in search of a more stable environment, they have looked for ways to improve indoor conditions. Inevitably, however, the indoor environment is influenced by prevailing outdoor conditions, daily and seasonal changes in climate and varying occupant requirements due to the type and operation of the building. Depending on the location and season, emphasis is given to either cooling or heating of indoor spaces, in an attempt to counterbalance the unfavourable outdoor conditions and achieve indoor comfort by controlling the indoor temperature, humidity, light availability and air quality.

Historically, for practical reasons and owing to natural laws, humans have been most successful in controlling their environment in situations requiring heat. Maintaining a warm environment has always been considered necessary during the cold season. However, in modern society, maintaining a cool environment during the warm months has proven to be as important for the optimum utilization of human resources and productivity.

The greatest advance has been in the change from rather artful design practices of indoor spaces to the detailed analytical methods that are necessary in order to handle complex building structures. Modern buildings, taking full advantage of currently available state-of-the-art technology, provide an indoor environment with high living and working standards.

Air-conditioning (A/C) systems can be used for year-round environmental control, in terms of temperature, moisture content, and air quality. However, like all mechanical systems, A/C systems consume valuable electrical energy for their operation. The shortage of conventional energy sources and escalating energy costs have caused reexamination of the general design practices and applications of A/C systems and the development of new technologies and processes for achieving comfort conditions in buildings by natural means. Indoor thermal comfort implies that humans are satisfied with the prevailing air conditions in the space – neither too warm nor too cool.

The historical development of the various processes and systems for mechanical or passive cooling, along with current trends and practices in the field, are reviewed in the following sections.

HISTORICAL DEVELOPMENT

Cooling is the transfer of energy from the space or the air supplied to the space, in order to achieve a lower temperature and/or humidity level than those of the natural surroundings. The development of cooling processes has passed through several stages, starting from simple intuitive applications of natural cooling techniques, such as shading, evaporative cooling and air circulation for enhancing the comfort sensation, to mechanical cooling systems, known as air conditioners, based on mechanical refrigeration cycles.

In fact, it appears that there has been a return to the utilization of several well-known techniques and processes that were used successfully even in the early periods of civilization. The principles of passive cooling are the same, but they are now enhanced with the available technological know-how and they are optimized so that they can be successfully incorporated into the building design and operation, in a suitable form for providing the best results.

In the early stages of history earth shelters were used by humans as a readily and naturally available living space, which provided protection from high and low temperatures, as well as from other unfavourable weather conditions. Building architecture quickly developed as an art, as the needs and demands of humans were changing with time, along with the appropriate know-how and availability of tools.

Well before the development of mechanical systems, though, several techniques for providing cooling and comfortable indoor conditions were applied in building architecture. The use of these techniques was not based on the understanding of the physical processes involved, but rather on conceptual experience. The majority were simple applications, like air movement through open spaces, external and internal shading, appropriate arrangement of the immediate surrounding spaces (vegetation, open pools and ponds) and use of proper building materials ('cold' marble and light surface colours). In addition, human mobility in indoor spaces was used extensively in order to avoid spaces with uncomfortable conditions during the day, as a result of direct solar gains.

Building design incorporated various fundamental and simple, but effective, principles. The large openings of the buildings, allowed for ample cross air movement, which can have a significant cooling effect. Even if the outdoor air is not at the desired temperature, air movement creates a cooling sensation as it moves around the human body.

The building itself provided sufficient protection to the occupants by properly shading the living spaces from direct solar gains. The landscaping around the buildings was primarily designed for aesthetic reasons, but at the same time improved the microclimate around the building, by providing shading and evaporative cooling. Extensive use of vegetation around the buildings provided the necessary shading, while absorbing large amounts of incident solar radiation and maintaining lower air temperature, which is further reduced by evapotranspiration from the trees. Open pools, fountains, ponds and running water were quite popular in the historical development of architecture, especially in southern dry climates. The phase change during water evaporation can decrease the dry bulb temperature of the air, though at the same time, there is an increase of the water content of the air.

Light coloured outer surfaces have been used extensively in traditional Greek architecture. The picturesque white villages in the Greek islands provide more than an aesthetically pleasing feeling. White surfaces reflect much of the incident solar radiation, thus reducing the heat transferred into interior spaces.

One of the most effective ways, however, of dealing with the problem of high temperatures during the day has also been the behavioural response of people. Use of different, cooler indoor spaces during the day or changes in building use altogether during summer, were very common behavioural actions. Gatherings in cooler open spaces, during the day or at the midday break, and outdoor sleeping during the night were also some simple ways of dealing with high temperatures and uncomfortable conditions.

Progress in science and technology has introduced tremendous changes in all fields, including the management of indoor conditions. In particular, advances in the fields of thermal sciences (thermodynamics, heat transfer, fluids) had as a result the design, development and production of mechanical systems capable of satisfying practically all needs in the field of cooling and refrigeration. The development of a practical cooling and refrigerating machine dates back to the middle of the nineteenth century, although there is evidence of the use of evaporative effects and ice for cooling in very early times [1]. Mechanical cooling for comfort got its start early in the twentieth century, but advances in this area have been rapid.

Mechanical cooling is achieved by several means [2], including:

- Vapour compression systems;
- Gas compression systems involving expansion of the compressed gas to produce work;
- Gas compression systems involving throttling or unrestrained expansion of the compressed gas; and
- thermoelectric systems.

Vapour compression is the most commonly used method in air-conditioning systems. The first equipment dates back to a British patent application by Jacob Perkins in 1835 [3]. The early systems were driven by steam engines and used ether as the working fluid. Compressors and overall equipment size became smaller as electric motors were substituted for the original steam engines and as vapour compression systems were applied to space air conditioning and commercial refrigeration.

Compression is accomplished mechanically or through absorption methods. The simple (theoretical) single-stage vapour compression refrigeration cycle is shown in Figure 1.1.

Cooling is accomplished by evaporation of the working fluid (a liquid refrigerant) under reduced pressure and temperature, in the evaporator, as a result of heat transfer from a high-temperature space. The refrigerant then enters the compressor as a slightly superheated vapour at a low pressure. It leaves the compressor and enters the condenser as vapour at some elevated pressure, where it is condensed as a result

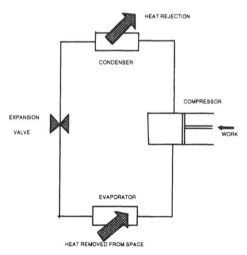

Figure 1.1 A simple vapour compression cycle

of heat rejection to ordinary cooling water or atmospheric air. The relatively high-pressure liquid is then throttled as it flows through the expansion valve. The thermo-dynamic cycle is completed as the remaining low-pressure liquid again enters the evaporator.

The most common refrigerants used in the 1920s and 1930s, were ammonia, car-bon dioxide, methyl, ethyl, and methylene chloride, and isobutane, but they exhib-ited major disadvantages. Finally, the development of nontoxic, nonflammable working fluids, with acceptable operating temperatures and pressures and high effi-ciencies solved the problem for many decades. Eventually, CFC-12 was accepted as a standard working fluid for air-conditioning applications and refrigerators, HCFC-22 for residential air conditioning, CFC-11 for most commercial air-conditioning applications, and HCFC-502 for low temperature refrigeration.

Gas compression systems involving expansion of the compressed gas to produce work have found commercial applications in air refrigeration systems used for cooling aircraft spaces. The two gas compression systems are also used in the lique-faction of various gases.

Finally, there are some systems that produce cooling by thermoelectric means [4]. The thermoelectric device, such as a conventional thermocouple, utilizes two dis-similar materials. One of the junctions is located in the space under cooling and the other in ambient air, as shown in Figure 1.2. When a potential difference is applied, the temperature of the junction located in the cooled space decreases and the tem-perature of the other junction increases. Under steady-state operating conditions, heat is transferred from the cooled space to the cold junction. The other junction reaches a higher temperature than the ambient and, as a result, heat will be trans-ferred from the junction to the surroundings.

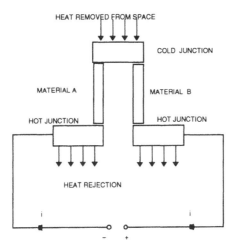

Figure 1.2 A thermoelectric device

Since in most cases there is a need both to cool and to heat a space, depending on the season, it appears that a system which can be used for both cooling and heating would be most attractive. This can be achieved with a heat engine, which is also known as a heat pump. The principle of the heat pump was introduced in the late mid-nineteenth century by Lord Kelvin. The system, however, received widespread application in the USA after World War II [5].

Technology advances that have provided systems with a high efficiency and performance and lower initial cost are described later in this chapter. However, the apparent concern for an overall reduction of energy consumption, including the increased cost of energy, and inherent problems of mechanical compression systems that operate with refrigerants which have been linked to atmospheric pollution, have caused a re-examination of alternative technologies and systems. Many of the previously described techniques of passive cooling, which have been applied successfully in the past, are now being revitalized, enhanced with new research findings, as well as the application of current technology and advanced system design know-how.

Published research has shown that a return to alternative energy sources, techniques and systems can be used to satisfy a major portion of the cooling needs in buildings. The trend, which started following the energy crisis in the mid-1970s, has stimulated a very active area of research with significant findings and great success in commercial applications.

This area has been named natural and passive cooling. The terms cover all naturally occurring processes and techniques of heat dissipation and modulation, as well as overheat protection and related building design techniques. This means without any other form of energy input than renewable energy sources or the use of other major mechanical systems. Included are several well known methods, enhanced with new capabilities allowed by technology advances, better understanding of the physical processes involved and optimum utilization of their potential effectiveness, in both traditional and modern applications, through a more comprehensive coupling and co-ordination of the available techniques and systems with the architectural design of the building in its environment.

Passive cooling techniques are also closely linked to thermal comfort of occupants. Indeed, some of the techniques used for passive cooling do not remove the cooling load of the building itself, but rather extend the tolerance limits of humans for thermal comfort in a given space.

It is also possible to increase the effectiveness of passive cooling with mechanically assisted heat transfer techniques that enhance the natural cooling processes [6]. Such applications are called 'hybrid' cooling systems. Their energy consumption is maintained at very low levels, but the efficiency of the systems and their applicability is greatly improved.

Passive cooling refers to techniques which can be used to [6, 7]:

- prevent heat gains and
- modulate heat gains.

Protection from or prevention of heat gains involves the following design techniques:

- microclimate and site design
- solar control
- building form and layout
- thermal insulation
- behavioural and occupancy patterns and
- internal gain control

while heat gain modulation can be achieved by proper use of the building's thermal mass (thermal inertia), in order to absorb and store heat during daytime hours and return it to the space at a later time.

Natural cooling refers to the use of natural heat sinks for excess heat dissipation from interior spaces, including :

- ventilation
- ground cooling
- evaporative cooling and
- radiative cooling.

By combining different passive and natural cooling techniques, it is possible to prevent overheating problems, decrease cooling loads and improve comfort conditions in buildings.

The principles governing each of the above topics are briefly presented in the following discussion, for each major category. A more comprehensive treatment is presented in other chapters of this book. The majority of these methods were extensively investigated in the USA and Israel following the major oil crisis in the mid-1970s. In Europe, similar research was initiated in the mid-1980s. A comprehensive overview of the current state of passive cooling research is available in [8–10].

Passive cooling by heat gain prevention

The microclimate and proper site design can greatly influence the thermal behaviour of a building. The overall principle is that a building must be adapted to the climate of the region and the microclimate (that is, the immediate environment around a building). The site design is influenced by economic considerations, zoning regulations and adjacent developments, all of which can interfere with the design of a building with regard to the incident solar radiation and the available wind. Vegetation does not only result in pleasant outdoor spaces, but can also improve the microclimate around a building and reduce the cooling load. It has been estimated that a full size tree evaporates 1,460 kg of water on a sunny summer day, which is equivalent to 870 MJ of cooling capacity [11]. Areas with high vegetation may exhibit noticeably lower air temperatures (by 2 to 3°C).

Solar control is the primary design measure for heat gain protection. The use of various shading devices in attenuating the incident solar radiation as it enters into the building [12], can significantly reduce the cooling load. External shading of opaque walls with surrounding natural vegetation, external or internal shading and high performance glazing can result in satisfactory thermal and optical performances. This subject is treated in Chapter 10.

The building form and internal space layout determines the exposure of interior spaces to incident solar radiation, as well as to daylight and wind [6]. The building shape controls both heat losses and heat gains, by reducing or increasing the ratio of the exposed surface to the volume. It is determined by planning regulations, space availability, neighbouring site development, architectural styles, client preferences and cost constraints.

The building envelope determines the physical processes taking place between the outdoor environment and indoor spaces. The objective is to limit thermal gains during the summer, due to high outdoor air temperatures and incident solar radiation. Thermal insulation can reduce the heat conducted through the building materials. During summer it reduces thermal gains and during winter it reduces thermal losses. The level of thermal insulation in buildings is determined by national codes and is mandatory in most countries. An increase of insulation above recommended values results in a very small decrease in the cooling load.

Behavioural and occupancy patterns in a building can be properly adjusted in order to achieve thermal comfort and consequently reduce the energy consumption for cooling. Dressing according to the prevailing weather conditions, adjusting the levels of physical activity, moving to cooler spaces in the building, adjusting thermal controls (opening and shutting windows, use of blinds and curtains for shading) are simple but yet effective actions [13].

The control of internal gains can be achieved by proper design and operation of internal heat sources, such as artificial lighting, equipment and occupants. For example, use of energy-efficient lamps can reduce heat dissipation into the space, in addition to reducing electrical energy consumption. Human activity can greatly influence internal heat gains, but it is difficult to modify, other than by appropriate zoning within the building's spaces. Accordingly, groups of people with similar activities should be placed in the same area in order to be able to best satisfy their needs. Equipment which requires special air conditions, such as computers or other electronic equipment, and spaces with high equipment concentration must be taken into account when calculating the cooling load. Similarly, a large number of occupants in a given space will increase the latent heat of the space and the amount of air ventilation required. In general, internal heat loads must be carefully accounted for, especially in large buildings that extensively utilize artificial lighting, in spaces that are usually heavily equipped and in spaces that are occupied by a large number of people. In fact, large office buildings require some cooling almost all year round. The methodology for cooling load calculations is presented in Chapters 7 and 14.

Passive cooling – heat gain modulation

The thermal mass of a building (typically contained in walls, floors and partitions, constructed of material with high heat capacity) absorbs heat during the day and regulates the magnitude of indoor temperature swings, reduces peak cooling load [14] and transfers a part of the absorbed heat into the night hours [15]. The cooling load can then be covered by passive cooling techniques, since the outdoor conditions are more favourable. An unoccupied building can also be pre-cooled by ventilation during the night and this stored coolness can be transferred into the early hours of the following day, thus reducing energy consumption for cooling by close to 20% [16]. The role of thermal mass in reducing and regulating the cooling load and indoor air temperatures is treated in more detail in Chapter 8.

Natural cooling – heat sinks

Natural or forced ventilation is one of the primary means of reducing the cooling load in buildings (removing heat from indoor spaces) and of extending indoor thermal comfort conditions for humans when the outdoor conditions (temperature and humidity) are favourable. Ventilation is also necessary in all indoor spaces, in order to introduce the required levels of fresh air and to control odours and indoor pollutants. Quantified ventilation research has been conducted continuously since the 1930s [17].

Natural ventilation is caused by pressure differences at the inlets and outlets of a building envelope, as a result of wind velocity and/or stack effects. To enhance its effectiveness, the wind can be properly channelled and moved through the building. Night ventilation, wind towers and solar chimneys are the main natural ventilation techniques. The subject of passive cooling with natural ventilation is also addressed in the penultimate section of this chapter and treated in more detail in Chapter 9.

Forced ventilation is achieved by mechanical means, using fans to achieve and control the appropriate airspeed. Ceiling, attic or simple portable room fans can be used to control the indoor air movement and achieve the appropriate air changes. Ceiling fans allow for higher indoor air temperatures, since increased indoor air movement enhances occupants' thermal comfort conditions (Chapter 6). As a result, cooling load requirements are reduced, with reductions ranging between 28% and 40% of the cooling costs [18], depending on the climate. Ceiling fans have been proved to be a viable technology and their low capital, operational and maintenance costs make them especially attractive.

Ground temperatures remain almost constant during the day at depths exceeding 1 m, while there are small seasonal changes which extend to a depth of 9 to 12 m. Even during hot summer days, ground temperatures remain significantly lower than ambient air temperatures, with small daily variations compared to the diurnal cycle of ambient air temperature and solar radiation [19]; this is due to the high thermal capacity of the soil. Accordingly, it is possible to couple the building with the ground for cooling purposes [20], either indirectly or directly. Indirect ground coupling involves a hybrid system. Indoor air is circulated through underground pipes, to dissipate heat to the lower-temperature heat sink and then the cooled air is returned to the building. A similar concept can be traced to the mid-16th century in Italy [21] and to ancient Greece on the Aegean island of Delos. The performance of buried pipes depends on the inlet air temperature, ground temperature, air circulation rate and pipe characteristics such as the thermal properties of the materials, dimensions and depth [22].

Another way to take advantage of the lower ground temperatures is by placing the building in direct contact with the ground [23]. In early human history, underground shelters were used extensively, utilizing the ground as the first insulating material. Today, partial or total underground construction is a viable alternative to conventional architectural styles, because of the resulting energy conservation associated with such a construction, especially in climates with extreme ambient conditions [19]. In Europe, there are a large number of one- or two-storey buildings partially buried in hillsides, with reported energy savings of 50 to 90% [10]. Earth-sheltered buildings can also provide isolation from noise and air pollution, although there are several design and operational difficulties that need to be taken into account.

Evaporative cooling applies to all processes in which the sensible heat in an air stream is exchanged for the latent heat of water droplets or wetted surfaces [24]. Warm outdoor air comes into contact with water droplets, which can be sprayed directly into the air stream, or it is passed through a wetted porous material. The moisture evaporates, thus extracting heat from the warm air and lowering its temperature. The majority of evaporative cooling applications use hybrid systems. Small mechanical systems, like pumps and fans, are necessary for moving the fluids (air and water) used during the process. When the water evaporation occurs in direct contact with the ambient air this is a direct process. The problem with such a process is that it increases the moisture content and consequently it cannot be used in regions with high humidity levels. In such a case, it is possible to use an indirect process, in which air is cooled without addition of moisture by passing through a heat exchanger, with a consequent reduction in efficiency. Evaporative cooling techniques were best demonstrated at the Seville Universal Exhibition (EXPO'92) in Spain [25–29].

Cool towers can also be used to move air masses through interior spaces and eliminate the use of blowers and its electrical cost. In a prototype application, a cool tower was fitted to a test house at the Environmental Research Laboratory, University of Arizona, Tuscon [8]. The air circulation achieved by the mechanism inside the house was 2.7 air changes per hour. Maximum air temperatures inside the house did not exceed 26°C, during the summer period. Traditionally, cool towers have been used in Middle Eastern architecture [30] and in natural-draught evaporative cool towers [31, 32], by moving ambient air through evaporative cooling pads, resulting in a lower air temperature at the exit of the tower. In a similar approach [33], the towers use fogging devices (atomizers) as the cooling source of the system, which allows for better regulation relative to the prevailing conditions and minimization of air-pressure losses.

Radiative cooling is a technique which can be used either passively or as a hybrid technique [34]. It is based on the fact that every object, being at a temperature higher than 0°K, emits energy in the form of electromagnetic radiation. If two elements at different temperatures are facing each other, a net radiant flux will occur. In the event that a low-temperature element is kept at a constant temperature, then the other element will radiate and cool down, in order to reach an equilibrium with the colder object. In the case of building radiative cooling, the building envelope (or another appropriate device such as a metallic flat-plate radiative air cooler) is cooled by dissipating infrared radiation to the sky, which acts as a low-temperature environmental heat sink. The amount of radiant exchange depends on the temperature difference between the sky temperature and the building element. Clear skies exhibit low sky temperatures, while cloud cover, air humidity and pollution decrease the effectiveness of radiative cooling processes [35].

The roof is the most important passive radiative cooling system in a building, because it continuously faces the sky dome. Roof colour influences the thermal performance of a building because it governs the absorption and reflection of incident solar radiation during daytime and the emission of long-wave radiation during night time, especially for light structures [36]. Measurements of the potential of this technique [37] give a cooling potential of 0.014 kWh m^{-2} per day. To enhance the performance of this technique, it is possible to protect the roof during the day with a movable insulation system and expose it during the night. Operable insulation can be in the form of horizontal movable panels or hinged panels, positioned vertically during the night. Measurements of the performance of this technique give a cooling potential of 0.266 kWh m^{-2} per day [37]. Alternatively, it is possible to use water bags placed on the roof, a system known as the Skytherm [38]. This is the only system that, with favourable climatic conditions, can provide both cooling and heating.

The potential of radiative cooling for 28 locations around southern European countries has been simulated and evaluated in [39] for a flat-plate radiator (hybrid system). A comparison has also been performed with the results from four localities in south-eastern USA. Accordingly, it has been concluded that southern Europe exhibits a promising potential for the use of this technique. The estimated mean daily useful cooling energy delivered by a flat plate radiative cooler ranges between 55 and 208 Wh m^{-2} for average sky conditions, and between 68 and 220 Wh m^{-2} for clear sky conditions, depending on the location. For areas like the southern USA, with high humidity levels, the potential effectiveness is limited, ranging between 41 and 136 Wh m^{-2} for average sky conditions and between 69 and 182 Wh m^{-2} for clear sky conditions. Some results on specific aspects of radiative cooling have also been reported in [40–42].

Comparative information regarding the thermal performance for a typical building of several passive and hybrid cooling techniques involving the use of natural cooling techniques, such as ground cooling, direct and indirect evaporative coolers and night ventilation techniques are available in [43]. The calculations were performed for a typical 80 m^2 one-storey dwelling located in Athens, Greece. Accordingly, the results have shown that :

- Night ventilation techniques can provide a part of the building's cooling load, being more effective during the months with lower night outdoor air temperature (June and August). The maximum depression of the peak indoor temperature does not exceed 1°C;
- For ground cooling, the earth-to-air heat exchangers have to be buried at a depth ranging between 3.5 and 5 m, with peak indoor air temperature reductions ranging between 2 and 5°C, respectively. The diameter and air velocity variations were found to affect the indoor air temperature significantly. For example, an increase of the diameter from 0.20 m to 0.22 m results in an indoor temperature decrease of 1.5°C.
- The use of a direct evaporative cooler, a parallel-plate-pad evaporative cooler with a 50 m^2 wetted area, can reduce the maximum peak indoor air

temperature by 4 to 6°C. The indoor temperature decrease is satisfactory for a fan rate higher than 1,500 r.p.m.

- The use of indirect evaporative coolers can lead to acceptable indoor temperature levels for an air speed through the cooler of 0.1 m s^{-1}. During the hot month of July, an air speed of 0.3 m s^{-1} is recommended.

BUILDINGS AND ENERGY CONSUMPTION FOR COOLING

The total primary energy consumption in the European Community in 1990 was 725 million tons of oil equivalent (mtoe) [44]. Transportation and industry each consumed 220 mtoe, while for domestic and tertiary uses the total is 275 mtoe. The final energy consumption in the domestic/tertiary sector, for each member state, is given in Table 1.1. Oil with a 44.3% share, was the primary energy source, while solids represented 23.8%, natural gas 17.5%, nuclear 12.9% and all other technologies, including renewable energy sources, were only utilized to satisfy 1.6%.

Table 1.1 Final energy consumption in the domestic/tertiary sector in the EEC member states for 1990

Member state	Energy consumption [mtoe]
Belgium	12.0
Denmark	5.8
France	51.0
Germany	68.0
Greece	4.0
Ireland	3.0
Italy	38.2
Luxembourg	0.7
Netherlands	19.0
Portugal	2.3
Spain	13.5
UK	57.5

During the last decade, Belgium, Denmark, Germany and The Netherlands have achieved a small energy consumption decrease, France, Luxembourg and the UK have maintained a constant average, while there has been an increase in Greece, Ireland, Italy, Portugal and Spain. Clearly, further actions are necessary. In 1989 the European Commission adopted a programme of urgent measures to reinforce and expand efforts in energy-efficient improvements, in energy conservation, combined with the use of more non-fossil-fuel energy sources, and in the protection of the environment.

Overall, energy consumption in commercial and residential buildings represents approximately 40% of Europe's energy budget [45]. This is allocated for satisfying the buildings' energy needs for heating, cooling, lighting and the other electrical appliances and equipment.

Passive solar design reduces fossil and nuclear fuel consumption for cooling in buildings. According to a study on the current and future use of passive solar energy in buildings in the European Community [46], there is a large potential for considerable energy savings in cooling of buildings, especially in southern regions. In Italy, for example, during 1990 the solar energy contribution to non-domestic cooling was 2.3 mtoe and 5.2 mtoe for domestic cooling purposes. In 2010, there is a technical potential to reach 5 mtoe and 12 mtoe, respectively. The consequent reduction in atmospheric pollution in 2010 is estimated to reach 3 million tones of CO_2 for domestic cooling and 0.01 million tones of SO_2. Similar information for other south European countries is shown in Figure 1.3.

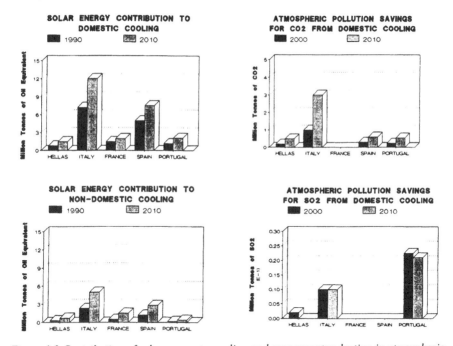

Figure 1.3 Contribution of solar energy to cooling and consequent reduction in atmospheric pollution in southern European countries [46]

Energy consumption for cooling of building in southern countries is of primary concern. A recent study carried out in Greece [47], a country with a representative south European climate, has provided a comprehensive picture of the energy consumption in buildings. These energy audits were performed in over 1,000 public and commercial buildings, including air-conditioned and naturally ventilated office, commercial, school, hospital and hotel buildings, around the country. Each type of building exhibits individual characteristics and problems due to specific operational needs, and therefore requires independent treatment and analysis.

Accordingly, the average annual final energy consumption for all the audited buildings in the five categories is 187 kWh m^{-2} in offices [45], 152 kWh m^{-2} in commercial buildings [48], 93 kWh m^{-2} in school buildings [49], 406.8 kWh m^{-2} in hospitals [50] and 273 kWh/m^2 in hotels [48].

To appreciate the impact of mechanical air-conditioning systems on energy consumption, one should refer to Table 1.2. The total consumption and the annual average final energy consumption for cooling in air-conditioned (A/C) and naturally ventilated (N/V) buildings is given for each one of the five building categories. In offices cooling represents 12.6% of the total average annual final energy consumption, in commercial buildings 12.2%, in schools only 2%, in hospitals less than 1% and in hotels about 4%. As shown in Table 1.2, air-conditioned buildings consume considerably more energy than naturally ventilated buildings, ranging between 4% to over 54%, depending on the type of building.

Table 1.2 Annual average values of total final energy consumption in buildings in Greece

Building Category	A/C Buildings		N/V Buildings	
	Number of Buildings	Total Energy [kWh/m^2]	Number of Buildings	Total Energy [kWh/m^2]
Offices	153	226	33	179
Commercial	261	215	254	206
Schools	16	183	150	119
Hospitals	26	352	6	236
Hotels	78	285	62	270

In comparison, typical primary energy values for offices in northern climates range between 270 to 350 kWh m^{-2} [51]. The electrical energy consumption in mechanically air-conditioned buildings ranges between 93 to 115 kWh m^{-2} for school buildings [52], while for a newly constructed 25,000 m^2 office building (with laboratory facilities and heavily equipped spaces) the electric energy use amounted to 214 kWh m^{-2} [53].

Information from a study in France has provided similar results [54]. For offices, the annual average total energy consumption ranges between 145 kWh m^{-2} in small communities (population less than 2,000 inhabitants) and 165 kWh m^{-2} in large cities (population greater than 50,000 inhabitants). For schools, the annual average total energy consumption ranges between 147 kWh m^{-2} in large cities and 175 kWh m^{-2} in small communities. For social and cultural buildings the corresponding energy consumption ranges between 77 kWh m^{-2} in small communities and 188 kWh m^{-2} in large cities.

The cumulative distribution of the total final energy consumption in each category of the audited buildings in Greece, is shown in Figure 1.4.

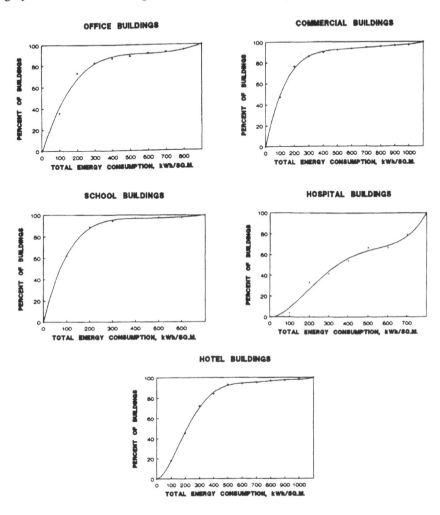

Figure 1.4 Cumulative distribution of the total energy consumption in buildings

Office buildings exhibit an annual average total energy consumption of 187 kWh m^{-2}. The minimum value from the collected data was 6 kWh m^{-2} and the maximum value 2,100 kWh m^{-2}. Approximately 39% of the office buildings have an energy consumption less than 100 kWh m^{-2}, while for 36% of the buildings their energy consumption ranges between 100 and 200 kWh m^{-2}.

Commercial buildings have an annual average energy consumption of 204 kWh m^{-2} (with for the audited buildings a minimum value of 3 kWh m^{-2} and a maximum value of 6,400 kWh m^{-2}). About 48% of the buildings consume less than 100 kWh m^{-2}, while for 32% the values range from 100 to 200 kWh m^{-2}.

Schools operate for nine months and remain closed during the summer months. Some schools also operate only during the morning hours and, as a result, the total energy consumption has an annual average value of 110 kWh m^{-2} (a minimum value of 6 and a maximum value of 900 kWh m^{-2}). Accordingly, 27% of the buildings consume less than 100 kWh m^{-2}, while 47% have an energy consumption ranging between 100 and 200 kWh m^{-2}.

Hospitals exhibit the highest annual average total energy consumption (407 kWh m^{-2}), as a result of their continuous 24-hour operation and the use of a large amount of health-care equipment. The actual data ranged between 100 and 1,400 kWh m^{-2}. About 33% of the buildings consume less than 200 kWh m^{-2}, while 21% of the buildings are in the range of 200 to 400 kWh m^{-2}.

Hotels exhibit the second highest annual average total energy consumption, equal to a value of 273 kWh m^{-2}. Among the audited buildings, the minimum value was 20 kWh m^{-2} and the maximum value was 1,500 kWh m^{-2}. Approximately 19% of the hotels have an energy consumption less than 100 kWh m^{-2}, while for 33% of the buildings the consumption ranges between 100 and 200 kWh m^{-2}.

The impact of air conditioning on the total energy consumption of a building can be significant. For example, in N/V and A/C office buildings the annual average total energy consumption is 179 kWh m^{-2} and 226 kWh m^{-2}, respectively. Consequently, the use of A/C in office buildings increases the annual energy consumption by an average value of 40 to 50 kWh m^{-2} [45]. The distribution of electrical consumption of the N/V and A/C buildings, in each one of the five categories, is shown in Figure 1.5.

For offices, 42% of the N/V and 38% of the A/C buildings exhibit an annual average electrical energy consumption lower than 100 kWh m^{-2}. In addition, 87% of the N/V and 84.7% of the A/C buildings present an electrical consumption lower than 300 kWh m^{-2}.

For commercial buildings, the annual average electrical energy consumption is less than 100 kWh m^{-2} for 58% of the N/V and 38% of the A/C buildings, while for 25% of the N/V and 35% of the A/C buildings the consumption is between 100 and 200 kWh m^{-2}.

For schools, 27% of the N/V buildings and 29% of the A/C buildings consume, on an average annual basis, less than 100 kWh m^{-2} of electrical energy, while 49% of the N/V and 35% of the A/C buildings have a consumption between 100 and 200 kWh m^{-2}.

For hospitals, over 50% of the N/V and only 5% of the A/C buildings have a consumption less than 100 kWh m^{-2}, while 25% of the N/V and 29% of the A/C exhibit a consumption ranging between 100 and 200 kWh m^{-2}.

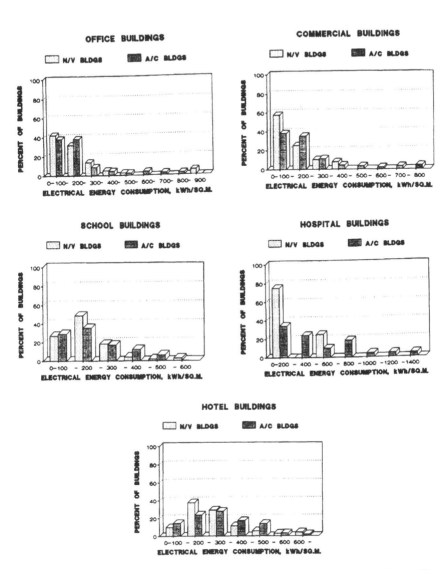

Figure 1.5 Distribution of annual average electrical energy consumption of naturally ventilated (N/V) and air-conditioned (A/C) buildings

For hotels, 47.2% of the N/V and 38% of the A/C buildings have an annual average electrical energy consumption less than 200 kWh m^{-2}, while 40.3% of the N/V and 45% of the A/C buildings have a consumption between 200 and 400 kWh m^{-2}.

Overall, this information, along with other related guidelines [55], can prove very useful for efficient energy planning and building design. Following a series of simulations and analysis, it has been shown in [47] that it is possible to reach an overall 20% energy conservation. In particular, several interventions in the buildings, to reduce the energy consumption for cooling, were investigated and assessed. Accordingly, it is possible to achieve the following energy savings for cooling by:

- reducing external loads (proper shading can result to a 30% reduction);
- reducing internal loads from lighting by using fluorescent lamps (80 lm/W) can reduce the cooling load by an average of 10%;
- using natural cooling techniques (like ground, evaporative cooling), where possible;
- using night ventilation techniques, with six air changes per hour, can result in substantial energy savings for cooling, up to 80%, depending on the type of building; and
- using ceiling fans can result in a very high percentage of energy savings for cooling, up to 95%, depending on the type and function of the building.

Significant energy savings for cooling, but also for heating, purposes can be achieved by using control strategies of the indoor air temperature [56]. Indoor air temperature is controlled in such a way as to drift from the optimum set-point to conditions which are still acceptable for most occupants. In office buildings this can result to energy savings up to 10% [57]. In another recently completed study of a large office building [58], indoor air temperature was allowed to increase in the afternoon hours (after 3 pm). Simulation results have shown that it is possible to achieve a reduction of seasonal consumption of chilled water of between 34 and 40% and a reduction in the energy budget for heating, ventilating and air-conditioning systems of 11%.

Finally, it is important to consider an additional parameter for energy savings in existing mechanical cooling systems. Proper maintenance of air-conditioners can reduce electrical peak demand, save energy and consequently reduce operating costs, according to a pilot project recently completed in the USA [59]. Savings and peak reduction estimates for individual repair measures such as correcting low airflow rates (due to clogged-up filters), repairing overcharging and undercharging of the refrigerant and repairing duct leakage can achieve 8 to 18% savings in cooling.

CURRENT TECHNOLOGY OF AIR-CONDITIONING SYSTEMS

Air-conditioning systems are used to control the temperature, moisture content, circulation and purity of the air within a space, in order to achieve the desired effects on the occupants of the space or on the products and equipment manufactured or stored there. Year-round A/C includes summer cooling (maintaining the indoor temperature below that of the outdoor air), winter heating (maintaining the indoor temperature above that of the outdoor air), dehumidification (lowering the water content of the ambient air to acceptable levels, usually during the cooling season), and humidification (addition of water vapour to the indoor air, usually during the heating season).

Air-conditioning systems are grouped into two main categories, namely central units and independent units.

Central air-conditioning systems

High-capacity central A/C systems are primarily used in large buildings. The main unit is usually placed in a mechanical room, generally at a distance from the conditioned spaces. The central unit is connected, through a duct/piping system, to individual units which are installed in each air-conditioned space. The outdoor air is introduced into the main A/C unit and is mixed with an amount of recirculated air. The mixture then passes through air filters, in order to remove any dust or other foreign particles.

The air is then conditioned, according to the operating mode (cooling or heating) of the system. For cooling purposes, the air is cooled and if necessary is also dehumidified. For heating purposes, the air is pre-heated by passing through a system supplied with steam or hot water, if necessary water vapour is also added passing through a humidifier, and finally the air is heated again by using steam or hot water. The air is then transported with fans (at speeds ranging between 5 and 15 m s^{-1}) through a duct system, usually at ceiling height, and is then diffused and circulated inside the conditioned spaces. There are various types and designs of inlet–outlet air diffusers, which are usually placed at ceiling height, floor level or near the windows, in order to ensure the best possible air circulation in the space.

For applications where there are zones in the building which may require warm and cool air at the same time, it is possible to use two independent air-duct systems, one for each air stream. In this way the air supplied into each zone of the building can be independently controlled to meet specific requirements. A far more costly design of several air duct systems is also necessary in some buildings, in order to supply air in significantly different conditions. For example, computer rooms require a year-round constant temperature of 20°C, while working environments may require temperatures during summer of 25–26°C.

Finally, depending on the type of building, an appropriate percentage of indoor air is transferred back to the central unit, for mixing with fresh but not conditioned outdoor air. This results in considerable energy savings compared to using and conditioning 100% outdoor air. Sometimes this is mandated by the type of building, as in the case of hospitals. Central A/C systems can be sized accordingly, in order to satisfy the cooling and heating loads of the building.

Rather than use all-air systems, it is also possible to use air–water systems. One possible way is to partially condition the air at the central unit, transfer it to the individual spaces, where the air is treated in a heat exchanger by using either a hot or

cold water stream, in order to achieve the desired temperature. These systems require, in addition to the air ducts, a pipe system for the circulation of water and also
a system for the heating or refrigerating of the water. These systems allow the occupant to control and easily achieve the desired conditions inside the space.

To reduce the cost of the air–water units, it is possible to exclude the air circulation system from the central unit. Instead, the local system in each conditioned space
is supplied with hot or cold water, while the indoor air is circulated by a fan in the
system, and passed through a heat exchanger where it is heated or cooled. These
systems are known as fan-coil units and can be placed on the floor or at ceiling
height.

Some innovative elements which have proved successful in the operation of central systems include the use of cogeneration concepts for refrigeration prime movers
and waste energy recovery using absorption refrigeration for subcooling to boost the
coefficient of performance of the refrigeration plant [60].

For large scale applications, chilled water storage may also be used in order to increase the performance, reliability and energy efficiency and to lower the operational cost of a central cooling system. A partially buried 10,000 m^3 chilled water
storage reservoir was coupled with a 4200 ton central chiller plant, in order to reduce the annual cooling electricity usage for a 102,000 m^2 defence electronics manufacturing facility [61]. The thermal energy storage system reduced net annual cooling electricity usage at the facility by 970 MWh (representing 9% of the total energy
consumption) during the first year of operation and by 3 GWh (about 28.3%) during
its second year of operation, in comparison to a conventional cooling system.

Independent air-conditioning systems

Smaller, independent A/C systems can be placed in any space without the need of a
central system. They are mostly used in buildings where occupants desire to install
air conditioning easily and quickly in selected spaces only, or in older buildings
where the renovation cost for installing central systems is not justified. Usually, independent A/C systems are small units, since they are only used for satisfying the
loads of small spaces, typically ranging between 9000 and 25,000 Btu/hr.

There are primarily two types of systems, namely monoblock and split units.

Monoblock units are independent air-conditioning systems that use older design
and technology. All the components are placed in the same housing. They are usually placed inside the structural material of an exterior wall of the space (in a hole of
the size of the unit) or in some cases they are placed at a window that has been properly modified.

Split units offer easier installation, since they do not require any major structural
changes of the space's construction material. One part of the unit is placed outside
(housing the heavy mechanical parts – compressor and a fan), while another part,
containing the evaporating unit and a blower, is placed anywhere in the indoor

space. The two units are connected by two small-diameter pipes for transferring the working fluid. Indoor air is circulated through the indoor unit, passes through the heat exchanger and then through a filter, before it is returned into the space. Operating noise levels are very low (since the main mechanical parts are placed outdoors), while the indoor units are designed to be aesthetically pleasing, in various sizes and with various controls (thermostats, operating timers, fan speeds, telecontrols). The indoor units can be easily mounted on the wall or ceiling, or placed on the floor, depending on the needs of the occupant.

Heat pumps

The most popular current technology for small to medium cooling/heating loads is the heat pump. Air conditioners and heat pumps were first used extensively in Japan, but they have now penetrated the European and US markets. The characteristics of these systems are analysed in the following discussion.

Any refrigeration system acts, to an extent, as a heat pump. Heat is transferred from a low-temperature space and is rejected at a higher temperature level. If the system is only to produce a cooling effect, then the rejected heat is usually wasted. In a heat pump, the rejected heat in the condenser is utilized to heat a space during the periods that heat is required.

A simple schematic diagram representing the operation of a heat pump is shown in Figure 1.6.

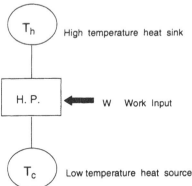

Figure 1.6 Basic heat pump operation

Heat pumps are categorized according to the type of operation of the compressor, as electrically driven or as an internal combustion. Depending on the location of the various components of the heat pump system, they are identified as monoblock units (when all components are placed in the same housing) or split units (when the condenser is separated from the remaining system). Depending on the type of the heat source for the evaporator (air, water, ground) and the heat sink (air, water), they are identified as air-to-air, air-to-water, water-to-water, water-to-air and ground-to-water heat pumps. Of these, outdoor air and well water are the most commonly used, working as a reversed Carnot cycle (smaller size and economical operation).

Heat pumps operate most efficiently with small temperature differences between the heat source and heat sink $(T_h - T_c)$. Consequently, one must look for a heat source with the highest possible temperature, which has to be lower than the desired one; otherwise there is no need for the heat pump. Heat pumps are available in various sizes, small (6000 to 25,000 Btu/hr), medium size (25,000 to 100,000 Btu/hr) and large units (100,000 to 600,000 Btu/hr).

Small, split-system heat pumps with variable speed drives are becoming quite popular throughout the world, owing to their ease of installation, quiet operation and energy savings. During 1990, over 30% of all home air conditioners sold had a variable speed feature [62]. The electrical energy savings of variable-speed systems can be significant, resulting in reductions ranging between 10 and 30%. The variable speed permits the air-conditioner to maintain high efficiency while operating at part load. To achieve this, systems utilize an electronic chip called an inverter.

The inverter converts the mains frequency (i.e. 50 or 60 cycles per second) alternating current into direct current and then back to alternating current with a frequency of 30–120 cycles per second, depending on the cooling load. The motors can then vary the speed over a wide range of heat removal rates. Because of the power consumption of the inverter, however, a power loss ranging between 7 and 10% has been reported. At maximum cooling load conditions, this can even reach 35%.

Mechanical compression systems using vapour refrigerants are predominant among the cooling methods. All the mechanical cooling systems operate in a cycle that requires work and accomplishes the objective of transferring heat from a low-temperature system to a high-temperature system. In order to assess the effectiveness of the various thermodynamic systems, it is possible to use a dimensionless parameter, called coefficient of performance (COP), which is defined as the ratio of the useful effect to the net energy supplied from the external source. For a heat pump that operates in the cooling mode, the useful cooling effect is (Q_L), the heat transferred from the cooled space, while the energy that costs is the work (W) and is supplied from external sources. Accordingly, for the theoretical single-stage cycle of Figure 1.6, the coefficient of performance is given by:

$$\text{COP} = Q_L / W = Q_L / (Q_H - Q_L) = 1 / (Q_H / Q_L - 1)$$

The COP of the various systems is a determinant factor on the final energy consumption of the unit. An increase, for example, of the coefficient of performance (for example by using a heat pump with a COP of 2), can reduce the electrical energy consumed by the cooling system, by as much as 25%.

Problems and prospects in the A/C industry

During recent years, the use of mechanical air conditioners in southern European countries has increased significantly. This is primarily due to an increase in the living standards in these countries and a reduction in the cost of A/C units. Figure 1.7 shows that there is a clear trend of increasing sales with gross national product (GNP) in European Community member states [63]. In Greece, the sales of A/C units showed an unprecedented increase during the late 1980s due to a series of heat waves over a period of three years (Figure 1.8). Sales of A/C units in Greece, have increased by 900% during this period [46].

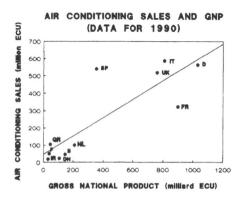

Figure 1.7 Sales of air conditioners as a function of Gross National Product in European Community member states [63]

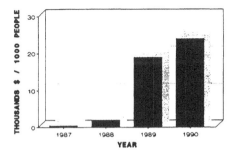

Figure 1.8 Sales of air-conditioners in Greece during the period of 1987 to 1990

Unfortunately, as a result of the sudden increase in demand for air-conditioning systems and primarily for small split units, the Greek market was invaded by a number of low-quality, low-efficiency, but also low-priced, units. The systems were primarily old-technology air-conditioners, not designed for the prevailing climatological conditions in Greece. However, customers looking for a low-price bargain purchased and installed thousands of these systems. The fact that there were no official standards or relavant legislation meant that there was no protection for the consumer.

The negative impact was immediate. Bad performance, high operating costs and numerous malfunctions of the systems started a domino effect. People lost confidence in air-conditioning systems and the market exhibited a decline. The positive result was that most of the market invaders were eventually forced out of business. The negative overall result, though, was that people who had made an investment, in their attempt to justify it, continued using the systems. In addition to all this, the properly operating units and the continuous promotion of sales by the well established major air-conditioning companies created a more serious, and, it appears permanent, negative effect.

Clearly, the impact of this increase on electrical energy consumption has been alarming. During this period, the peak electric energy load in Greece occurred, for the first time, during the summer [64]. Similar trends have also been observed in most southern regions in Europe, the USA [65], Japan and the Middle East. During the summer of 1990, Tokyo Electric Power Company, the world's largest private utility, experienced a summer peak only 1% below its maximum available generation capacity [62], due primarily to the increasing popularity of air-conditioner systems. In Kuwait, air-conditioning systems consume 75% of the total electrical energy consumption during the summer months [66].

This can potentially create severe problems in the availability of electric energy, especially in countries where the electrical energy production is primarily based on hydroelectric power plants, as in Greece. Hydroelectric power plants do not operate up to their potential during the dry summer months. Consequently, in order to avoid possible energy shortages, if the same trends in energy consumption continue, power companies must build new power generation plants to satisfy these short seasonal needs. This will clearly require significant financial investment, which is difficult to justify if a plant is to be used only during the summer period. Furthermore, if thermal plants are considered, this will have a direct impact on atmospheric pollution. All these aspects constitute a very grim scenario, mandating immediate action to reverse it.

Ice thermal storage systems are receiving greater acceptance throughout the industry as an effective means of shifting summer peak power in commercial office buildings and providing lower operating costs [67]. Baseload power for freezing ice is taken off-peak at night. This also has a positive effect on power-generating plant efficiency, because of the higher load factor and constant operation, with minimum on and off daily cycles [68]. In addition, with a totally integrated off-peak ice storage design, it is possible to offset the relatively high kW/ton required to make ice by reducing pump and fan horsepower [69].

Off-peak air conditioning (OPAC) is already a major energy saver [68]. Partial storage with chiller priority is the most common application. Ice is produced only as a back-up which, in most climates, is only 10–15% of the time. Even during these times, the chiller operates during the day at 5–7% more efficiency than with no storage. This is because the chiller, being upstream of the ice storage, is handling the higher half of cooling the return brine temperature (i.e. 16 to 11°C). When the chiller is operating to build ice, it is at full constant load (no unloading, part loading or on/off operation).

Taking advantage of 1–3°C chilled water available with ice storage, equipment cost can be minimized throughout the water and air distribution systems. The installation of a full storage, dynamic ice harvester shifting a daily load of 300 tons, a 340 m^3 ice storage tank, a heat exchanger, and a low-temperature variable-air-volume (VAV) air distribution system in a 17,000 m^2 commercial office building [67] has proved the applicability of such systems. Substantial energy savings and utility costs result from reduced electrical demand and use of less expensive off-peak energy. Additional benefits include improved air quality levels, since low water temperatures at the cooling coil and low air-temperature distribution cause reduction in the growth of bacteria and fungus in the air ducts. Overall, advances in ice storage technology are expected to reduce the energy used for mechanical cooling of buildings by about 15% in the 1990s [68].

Despite these positive prospects, in recent years the A/C industry has been facing a major shake-up, possibly the worst in its history. Ratification of international agreements such as the 1987 Montreal Protocol on the phase out of the commonly used chlorofluorocarbon (CFC) refrigerants in A/C systems have created a major problem for the A/C and related industries as they try to identify and use new working fluids. Certain CFCs have been related to the depletion of the ozone layer and the greenhouse effect. Guidelines have been set for a steep reduction of fully halogenated CFCs, such as R-11, 12, 113, 114, 115 and halons 1211, 1301 and 2402.

The members of the European Union agreed in 1990 on an immediate 85% cut in the production of CFCs and the elimination of their use by the end of the century. In the USA, the Environmental Protection Agency has adopted a regulation on the Protection of Stratospheric Ozone, based on contributions from the industry and its representatives, various engineering societies, the Department of Energy and other interested parties. The freeze on fully halogenated CFCs took effect in the summer of 1989 by setting production to 1986 levels, while 20 and 30% reductions will follow over a ten-year period. Efforts are being made to shorten this time frame.

Consequently, other refrigerants with good thermophysical properties which satisfy the new environmental restrictions must be made available to replace those phased out. Identification and testing of substitute refrigerants has become a top priority to the air-conditioning and refrigeration industry. Partially halogenated refrigerants (hydrogenated chlorofluorocarbons), such as HCFC-22, are considered safer. Refrigerants not containing chlorine (hydrogenated fluorocarbons), such as HFC-152a and 134a are safest. Furthermore, the concept of employing a multi-component refrigerant (mixtures) introduces a desirable degree of freedom in developing substitute refrigerants.

The selection of alternative working fluids for vapour compression processes depends on the particular application and on some distinctive properties, according to the chemical nature of the refrigerant, its thermophysical properties and on their compliance with health, safety and environmental restrictions. A comprehensive approach for the selection and screening of novel refrigerants is described in [70].

Three of the most promising candidates among the existing refrigerants for future use are HCFC-123, 124 and HFC-134a. These alternative refrigerants satisfy our environmental concerns, since they have about 98% less potential to deplete ozone than R-11 (commonly used in industrial refrigeration systems).

For an acceptable capacity of a vapour compression cycle, the cooling effect of the substitute refrigerants must not be much lower, and the specific volume must not be much higher, than the corresponding properties of the original equipment refrigerant. The COP cannot be much lower without also degrading the ability of the system to, at least temporarily, meet condenser and evaporator temperatures. The efficiency of the compressor should not be much degraded if an otherwise acceptable refrigerant is employed.

CURRENT TECHNOLOGY FOR NATURAL VENTILATION IN BUILDINGS

Natural ventilation is the predominant passive cooling technique. In general, air ventilation (passive, hybrid or mechanical) of indoor environments is also necessary in order to maintain the required levels of oxygen in the space and the maintenance of acceptable indoor air quality levels.

Traditionally, ventilation requirements were achieved by natural means. In the majority of older buildings, infiltration levels were such as to allow considerable amounts of outdoor air to enter the building, while additional requirements were satisfied by simply opening the windows.

Modern architecture and energy-conscious design of buildings has reduced to a minimum the air infiltration inside buildings, in an attempt to reduce its impact on the cooling or heating load. Better construction of buildings has resulted in actually sealing the building from the outdoor environment. In particular, the construction of large glass office buildings, where the opening of windows is not allowed, has practically eliminated the possibility of using natural ventilation techniques for supplying fresh air to indoor spaces.

It is clear that building design approaches have to be re-evaluated and adapted to the new requirements, in order to facilitate the application of natural ventilation techniques for cooling purposes. The following discussion addresses the principles of the most up-to-date techniques for natural cooling.

The use of natural ventilation, provided that outdoor climatological conditions are favourable, can provide for the following:

- a reduction of the cooling load (Chapter 8);
- enhancement of thermal comfort conditions (Chapter 6); and
- maintenance of proper indoor air quality levels [71, 72].

The effectiveness of natural ventilation techniques is determined by:

- prevailing outdoor conditions – microclimate (wind speed, temperature, humidity and surrounding topography), and
- the building itself (orientation, windows – number, size, location).

Current trends of natural ventilation

Current trends for the successful implementation of natural ventilation techniques for cooling purposes are presented in the following discussion. A detailed analysis of the various concepts involved along with the information needed by building designers for their proper application, are analysed in Chapter 8.

New buildings allow for 0.2–0.5 air changes per hour (ACH) by infiltration, while with the windows wide open during summer it is possible to achieve 15–20 ACH. Even larger air changes, around 30 ACH, can be achieved by natural means, but there is a need for a large number of window openings and careful placement within the space. The overall problem of energy conservation in a building becomes more complex since a large number of openings will increase the thermal load during winter, by increasing thermal losses, since a glass window can not compare in its thermal performance with a well insulated wall. It is also important to link the necessary number of windows and their placement with the requirements for natural lighting purposes.

Natural ventilation is used to create a volumetric flow, for renewal of indoor air quantities and transfer of heat, resulting from the outdoor wind speed and/or stack effects, in order to:

- remove heat stored into the building materials; and
- enhance heat dissipation from the human body to the environment.

The savings from natural ventilation depend on:

- the number of air changes;
- the construction of the building (heavy or light construction);
- the microclimate; and
- the temperature and humidity in the space.

Outdoor temperature, humidity and wind velocity are determinant factors for the successful application of natural ventilation techniques. For cooling purposes, the incoming air should be at a lower temperature than the indoor air temperature. However, even at higher temperatures, the resulting air flow inside the space can cause a positive effect on the thermal comfort conditions of the occupants, since it increases heat dissipation from the human body and enhances evaporative and convective heat losses. An increase of air velocity by 0.15 m s^{-1} compensates an increase of $1°C$ at a relative humidity of 75%. There is a direct link between thermal comfort and ventilation rates, which is explained in detail in Chapter 6.

Natural ventilation techniques for cooling purposes are also very effective during the night hours, when outdoor air temperatures are usually lower than the indoor ones. As a result the cooling load is reduced and the peak indoor air temperatures can be reduced by 1 to $3°C$.

Air humidity is the most important limiting factor for the application of natural ventilation techniques. High levels of humidity have a negative influence on thermal comfort. As a result, in regions with high relative humidity levels during summer, the use of conventional air-conditioning systems is necessary in order to remove water vapour from indoor air (dehumidification). Under such circumstances, natural ventilation during the day- or night-time hours should be avoided.

Overall, natural ventilation research has concentrated on the design and optimization of window openings, for enhancing natural ventilation in relation to daylighting, solar gains, and thermal losses during winter. Dealing with the prediction of the wind potential in urban environments and air movement within the various spaces inside a building are areas which require additional research. The European Commission research programme for passive cooling of buildings, PASCOOL [73], has addressed most of these problems and reported significant advances.

In addition to the use of windows and other openings for natural ventilation, there are some additional means of enhancing the air movement. Wind towers utilize the kinetic energy of wind, which is properly channelled within the building in order to generate air movement within a space. They have been successfully integrated in many architectural designs [74, 75]. Solar chimneys are constructions used to promote air movement throughout the building using solar gains. They are positioned on the sunward side of the building to make the best possible use of direct solar gains. Performance information is provided in [76, 77].

To increase the cooling effectiveness of natural ventilation techniques, especially at locations with low outdoor air velocity and variable wind directions, it is possible to incorporate wing-walls into the building's design [78]. Wing-walls simply project outward next to a window and even a slight breeze against the wall creates a high-pressure zone on one side and a low one on the other. The pressure differential draws the outdoor air in through one open window and out through the adjacent one. This technique is most suitable for providing increased ventilation rates in rooms with windows on only one side, as is the case in most schools. Classrooms are usually next to long hall-ways allowing windows on one side only, which reduces the possibility for cross ventilation. As a result, a number of schools in southern USA have been built with wing-walls, with very satisfactory results. Especially in Florida, where the law requires that new schools be designed with a non-mechanical method of venting classrooms, this concept has received wide acceptance.

Ceiling fans can also be used to enhance indoor thermal comfort conditions, by controlling indoor air circulation. Ceiling fans allow for higher indoor air temperatures and result in reduced cooling requirements by enhancing comfort conditions. They have been proved to be a viable technology and are especially attractive because of their low capital, operational and maintenance costs.

For a detailed overview of current practice in the use of natural ventilation in passive cooling, the reader should also refer to [9, 10, 79] and Chapter 9.

CONCLUSIONS

Natural and passive cooling techniques in buildings have been proved successful through numerous demonstration and pilot projects. They exhibit a high potential for reducing the energy consumption for cooling in commercial and residential buildings, which represents a significant percentage of the total energy consumption in buildings.

Technology advances have provided air-conditioning systems with high efficiency, good overall performance and lower initial cost. However, the current concerns about reducing energy consumption, the increased cost of energy and inherent problems of mechanical compression systems, which operate with refrigerants that have been linked to atmospheric pollution, have caused a re-examination of alternative technologies and systems. Natural and passive cooling techniques, which have been applied successfully in the past, are being revitalized, enhanced with new research findings and applied with current technology and advanced system-design know-how.

Passive cooling refers to techniques which can be used to prevent and modulate heat gains. Protection from heat gains involves the use of microclimate and site design, solar control, building form and layout, thermal insulation, behavioural and occupancy patterns and internal gain control. Heat-gain modulation can be achieved by properly using the thermal mass of the building itself (thermal inertia) in order to absorb and store heat during the daytime hours and return it to the space at a later time.

Natural cooling refers to the use of natural heat sinks for excess heat dissipation from interior spaces, including ventilation, ground cooling, evaporative cooling and radiative cooling.

Combining different passive and natural cooling techniques, it is possible to prevent overheating problems, decrease cooling load and improve indoor thermal comfort conditions.

In particular, it is possible to achieve an overall 20% energy conservation in new and existing buildings. Significant energy savings for cooling can be achieved, for example by reducing external loads using proper shading of the building facade, by reducing internal loads from lighting with the use of energy-efficient fluorescent lamps, by using natural cooling techniques like ground, evaporative and radiative cooling and by night ventilation.

Substantial energy savings can also be achieved in existing mechanical cooling systems. Proper maintenance of air-conditioners can reduce electrical peak demand, save energy and consequently reduce operating costs. Individual repair measures may include, for example, correcting low air flow rates, repairing overcharging and undercharging of refrigerant and repairing duct leakage. Heat pumps and other high performance mechanical systems can replace older technology equipment in order to achieve considerable energy savings for applications where air-conditioning systems are necessary.

REFERENCES

1 McQuiston, F.C. and J.D. Parker (1982). *Heating, Ventilating and Air Conditioning. Analysis and Design*, 2nd edition, John Wiley & Sons, New York.

2 Threlkeld, J.L. (1970). *Thermal Environmental Engineering*, 2nd edition, Prentice Hall, Englewood Cliffs, NJ.

3 Fischer, S. (1991).'Energy use impact of CFC alternatives', *Energy Engineering*, Vol. 88, No. 3, pp. 6–21.

4 Van Wylen, G.J. and R.E. Sonntag (1978). *Fundamentals of Classical Thermodynamics*, 2nd edition, John Wiley & Sons, New York.

5 Sauer, H. and R.H. Howell (1983). *Heat Pump Systems*, John Wiley & Sons, New York.

6 Santamouris, M. (ed.) (1990). 'Horizontal study on passive cooling', *CEC-Building 2000 Project*, Chapter 1, pp. 1–7, organized by the Directorate General 12, Commission of the European Communities.

7 Goulding, J.R., J. Owen Lewis and T.C. Steemers (eds) (1992). *Energy in Architecture, The European Passive Solar Handbook*, Commission of the European Communities, Brussels.

8 Cook, J. (1990). 'The state of passive cooling research', *Passive Cooling*, ed. J. Cook, Chapter 7, pp. 539–569. MIT Press, Cambridge, MA.

9 Antinucci, M., B. Fleury, J. Lopez d'Asiain, E. Maldonado, M. Santamouris, A. Tombazis and S. Yannas (1992).'Passive and hybrid cooling of buildings. State of the art', *International Journal of Solar Energy*, Vol. 11, pp. 251–272.

10 Steemers, T.C. (1991). 'The state of the art in passive cooling', *International Journal of Solar Energy*, Vol. 10, pp. 5–14.

11 Moffat, A. and M. Schiller (1981). *Landscape Design. How to Save Energy*. William and Narrow Company, New York,

12 Yannas, S. (1990). 'Solar control techniques', *Proc. Workshop on Passive Cooling*, eds E. Aranovich, E. Oliveira Fernandes and T.C. Steemers, Ispra, Italy, 2–4 April, pp. 75–97.

13 Markus, T.A. and E.N. Morris (1980). *Building, Climate and Energy*. Pitman, London.

14 Antinucci, M. (1990). 'Heat attenuation as a means to limit cooling load and improve comfort', *Workshop on Passive Cooling*, eds E. Aranovitch, E. Fernades and T. Steemers, Ispra, Italy, pp. 111–119.

15 Braun, J. (1990). 'Reducing energy costs and peak electrical demand through pptimal control of building thermal storage', *ASHRAE Transactions*, Vol. 96 No. 2, pp. 876–888.

16 Brown, M. (1990). 'The thermal mass of buildings in reducing energy consumption', *ASME Journal of Solar Energy Engineering*, Vol. 112, p. 273.

17 Chandra, S. (1990). 'Ventilative cooling', *Passive Cooling*, ed. J. Cook, Chapter 2, pp. 42–84. MIT Press, Cambridge, MA.

18 Chandra, S., P. Fairey and M. Houston, (1986). 'Cooling with ventilation', Florida Solar Energy Center, SERI/SP-273-2966, DE86010701.

19 Labs, K. (1979). 'Underground building climate', *Solar Age*, Vol. 4, pp. 44–50.

20 Labs, K. (1989). 'Earth coupling', *Passive Cooling*, ed. J. Cook, Chapter 5, pp. 197–346. MIT Press, Cambridge, MA.

21 Fanchiotti, A. and G. Scudo (1981). 'Large scale underground cooling system in Italian 16th century Palladian villa', American Solar Energy Society Passive Cooling Conference, Miami.

22 Tzaferis, A., D. Liparakis, M. Santamouris and A. Argiriou, (1992). 'Analysis of the accuracy and sensitivity of eight models to predict the performance of earth-to-air heat exchangers', *Energy and Buildings*, Vol. 18, pp. 35–43.

23 Givoni, B. and L. Kats, (1985). 'Earth temperature and underground buildings', *Energy and Buildings*, Vol 8, pp. 15–25.

24 Yellott, J.I. (1989). 'Evaporative cooling', *Passive Cooling*, ed. J. Cook, Chapter 3, pp. 85–137. MIT Press, Cambridge, MA.

25 Velazquez, R., J. Guerra, S. Alvarez and J.M. Cejudo (1991). 'Case Study of Outdoor Climatic Comfort: The Palenque at EXPO'92', PLEA'91, Seville, Spain.

26 Guerra, J., J.L. Molina and J.M. Cejudo (1991). 'Thermal performance of water ponds: modelling and cooling applications', PLEA'91, Seville, Spain.

27 Alvarez, S., E.A. Rodriguez and J.L. Molina (1991). 'The Avenue of Europe at EXPO'92: Application of cool towers in architecture and urban spaces', PLEA'91, Seville, Spain.

28 Rodriguez, E., S. Alvarez and R. Martin, (1991). 'Direct air cooling from water drop evaporation', PLEA'91, Seville, Spain.

29 Rodriguez, E.A., S. Alvarez and R. Martin, (1991). 'Water drops as a natural cooling resource. physical principles', PLEA'91, Seville, Spain.

30 Bahadori, M.N. (1986). 'Natural air-conditioning systems', *Advances in Solar Energy*, Vol. 3, pp. 283–356, Plenum Press, New York.

31 Cook, J. (1989). 'Phoenix Solar Oasis: cooling design of urban spaces', *14th National Passive Solar Conference*, Denver, pp. 350–355.

32 Kent K. and T. Lewis (1990). 'Natural Draft Evaporative Cooling', *15th National Passive Solar Conference*, pp. 157–160, Austin.

33 Alvarez, S., E.A. Rodriguez and J.L. Molina (1991). 'The Avenue of Europe at EXPO'92: Application of cool towers', PLEA'91, Seville, Spain.

34 Martin, M. (1989). 'Radiative cooling', *Passive Cooling*, ed. J. Cook, Chapter 4, pp. 138–196. MIT Press, Cambridge, MA.

35 McCathren, J.R. and J.M. Akridge, (1982). 'Radiant cooling to the night sky in hot, humid climates', *Progress in Passive Solar Energy Systems*, American Solar Energy Society, pp. 877-882,

36 Santamouris, M. (1990). 'Natural cooling techniques', *Proceedings of Passive Cooling Workshop*, eds E. Aranovitch, E. de Oliveira Fernades, T.C. Steemers, Ispra, pp. 143–153.

37 Givoni, B. (1981). *Proceedings of the. International Passive Hybrid Cooling Conference*, eds A. Bowen, E. Clark and K. Labs, Miami Beach, p. 279.

38 Hay, H.R. and J.I. Yellott (1969). 'Natural air conditioning with roof ponds and movable insulation', *ASHRAE Transactions*, Vol. 75, No. 1, pp. 165–177.

39 Argiriou, A., M. Santamouris, C.A. Balaras and S.M. Jeter (1993). 'Potential of radiative cooling in southern Europe', *International Journal of Solar Energy*, Vol. 13, pp. 189–203.

40 Catalanotti, S., V. Cuomo, G. Piro, D. Ruggi, V. Silvestrini and G. Troise (1975). 'The radiative cooling of selective surfaces', *Solar Energy*, Vol. 17, pp. 83–89.

41 Bartoli, B., B. Catalanotti, B. Coluzzi, V. Cuomo, V. Silvestrini and G. Troise, (1977). 'Noctural and diurnal performances of selective radiators', *Applied Energy*, Vol. 3, pp. 267–286.

42 Papadakis, G., G. Voulgaris, A. Frangoudakis and S. Kyritsis (1988). 'Night sky radiation in Athens during the summer. influence of city pollutants', *International Journal of Solar Energy*, Vol. 6, pp. 279–289.

43 Agas, G., T. Matsaggos, M. Santamouris and A. Argiriou (1991). 'On the use of the atmospheric heat sinks for heat dissipation', *Energy and Buildings*, Vol. 17, pp. 321–329.

44 *Energy in Europe* (1992). Directorate General for Energy, DG XVII, Commission of the European Communities, Brussels.

45 Santamouris, M., A. Argiriou, E. Dascalaki, C.A. Balaras and A. Gaglia, (1994). 'Energy characteristics and savings potential in office buildings', *Solar Energy*, Vol. 52, pp. 59–66.

46 *Passive Solar Energy as a Fuel* (1990). The Commission of the European Communities, Directorate General XII for Science Research and Development, Brussels.

47 Santamouris, M., A. Argiriou, E. Dascalaki, M. Vallindras, A. Gaglia and J. Sigalas (1992). 'Energy conservation in office buildings', Final Report, Greek Productivity Centre and Ministry of Industry, Research and Technology.

48 Santamouris, M., E. Dascalaki, C.A. Balaras, A. Argiriou and A. Gaglia (1993). 'Performance assessment and the potential for energy conservation and the use of alternative energy sources in buildings', *Proceedings of 3rd European Conference on Architecture, Solar Energy in Architecture and Urban Planning*, 17–21 May, Florence.

49 Santamouris, M., E. Dascalaki, C.A. Balaras, A. Argiriou and A. Gaglia (1994). 'Energy consumption and the potential for energy conservation in school buildings, in Greece', *Energy, The International Jourmal*, Vol. 19, pp. 653–660.

50 Santamouris, M., E. Dascalaki, C.A. Balaras, A. Argiriou and A. Gaglia (1994). 'Energy performance and energy conservation in health care buildings in Greece', *Energy Conversion and Management*, Vol. 35, pp. 293–305.

51 Campbell, J. (1988). 'Use of Passive Solar Energy in Offices', *Passive Solar Energy in Buildings*, ed. P. O'Sullivan. The Watt Committee on Energy, Elsevier Applied Science Publishers, New York.

52 Lefebvre, R.R. (1993). 'New HVAC system reduces operating costs', *ASHRAE Journal*, Vol. 35, No. 4 (April), pp. 20–23.

53 Boldt, J.G. (1993). 'Separate HVAC systems maximize energy efficiency', *ASHRAE Journal*, Vol. 35, No. 4 (April), pp. 16–19.

54 *Communes, Bilan Energetique du Patrimoine et des Services, Dépenses & Consommations 1990* (1990). Ministère de l'Interieur et de la Securité Publique, Agence de l'Environnement et de la Maîtrise de l'Energie.

55 Balaras, C.A. (1993). 'Energy Efficient Building in Greece', in *The European Directory of Energy-Efficient Building*, ed. B. Cross. James & James Science Publishers, London.

56 McNall, P., E. Pierce and J. Barnett, (1978). 'Control Strategies for Energy Conservation', in *Energy Conservation Strategies in Buildings*, ed. J.A. Stolwijk. J.B. Pierce Foundation.

57 Fleming, W.S. (1979). 'Energy conservation: an investigation of the thermal comfort alternative', *ASHRAE Transactions*, Vol. 85, No. 2, pp. 813–824.

58 Zmeureanu, R. and A. Doramajian (1992). 'Thermally acceptable temperature drifts can reduce the energy consumption for cooling in office buildings', *Building & Environment*, Vol. 27, No. 4, pp. 469–481.

59 Proctor, J. (1991). 'An ounce of prevention: residential cooling repairs', *Home Energy*, Vol. 8, No. 3, pp. 23–28.

60 Amberger, R.F. and J.A. DeFrees (1993). 'retrofit cogeneration system increases refrigeration capacity', *ASHRAE Journal*, Vol. 35, No. 4 (April), pp. 24–27.

61 Fiorino, D.P. (1993). 'Chilled water storage system reduces energy costs', *ASHRAE J.*, Vol. 35, No. 4 (April), pp. 30–33.

62 Meier, A. (1991). 'Variable-speed drive: Energy winner, power loser', *Home Energy*, Vol. 8, No. 3, pp. 8–9.

63 King, A. '(1993). The growing market for air conditioning', *Proceedings: Advanced Systems of Passive and Active Climatisation*, 3–5 June, Barcelona.

64 Balaras, C.A., M. Santamouris, M. Vafiadis, P. Vovolis, O. Voumdas, D. Ragousis (1991). 'Correlation of bioclimatic indices with peak electric load, in the Athens area for the period of 1986–1989', Institute for Technological Applications, Greek Productivity Centre, Athens.

65 United States Department of Energy (1985). *Non-residential Buildings Energy Conservation Survey: Characteristics of Commercial Buildings*, Washington, DC, pp. ix, 21, 105.

66 Al-Jamal, K., O. Alameddine, H. Al-Shami and N. Shaban (1988). 'Passive cooling evaluation of roof pond systems', *Solar & Wind Technology*, Vol. 5, pp. 55–65.

67 Landry, C.M. and C.D. Noble (1991). 'Making ice thermal storage first-cost competitive', *ASHRAE Journal*, Vol. 33, No. 5, pp. 19–22.

68 MacCracken, C.D. (1991). 'Off peak air-conditioning: A major energy saver', *ASHRAE Journal*, Vol. 33, No. 12, pp. 12–23.

69 Harmon, J.J. and H. Chung Yu (1991). 'Centrifugal chillers and glycol ice thermal storage units', *ASHRAE Journal*, Vol. 33, No. 12, pp. 25–31.

70 Balaras, C.A. and S.M. Jeter (1991). 'A methodology for selecting and screening novel refrigerants for use as alternative working fluids, *Energy Conversion and Management*, Vol. 31, No. 4, pp. 389–398.

71 Argiriou, A., C.A. Balaras, E. Dascalaki, A. Gaglia, G. Gountelas, K. Moustris, M. Santamouris and M. Vallindras, (1992). 'A survey of indoor air quality in office buildings in Athens, Greece', *Proceedings of the International Conference on Indoor Air Quality and Ventilation*, Athens.

72 Balaras, C.A., M. Santamouris, E. Dascalaki, A. Argiriou, D. Assimakopoulos, I. Tselepidaki and M. Loizidou (1993). 'indoor air quality and health symptoms in town halls in Athens, Hellas', *Proceedings of Indoor Air '93, 6th International Conference on Indoor Air Quality and Climate*, Helsinki, 4–8 July.

73 Santamouris, M. and A. Argiriou (1993). 'The CEC Project PASCOOL', *Proceedings of 3rd European Conference on Architecture, Solar Energy in Architecture and Urban Planning*, 17–21 May, Florence.

74 Bahadori, M. (1988). 'A passive cooling heating system for hot arid regions', *Proceedings of 13th National Passive Solar Conference*, Cambridge, MA, June, pp. 364–367.

75 Karakatsanis, C., M. Bahadori and B. Vickery (1986). 'Evaluation of pressure coefficients and estimation of air flow rates in buildings employing wind towers', *Solar Energy*, Vol. 37, pp. 363–374.

76 Bouchair, A. (1988). 'Moving air using stored solar energy', *Proceedings of 13th National Passive Solar Conference*, Cambridge, MA, June.

77 Dimoudi, A. (1989). 'An evaluation of the use of solar chimneys to promote the natural ventilation of buildings', M.Sc. Thesis, Cranfield Institute of Technology.

78 Melody, I. (1987). 'Solar cooling research', *Solar Today*, January, pp. 23–24.

79 Fleury, B. (1990). 'Ventilative cooling: State of the art', *Proceedings of Workshop on Passive Cooling*, Ispra, Italy, 2–4 April, pp. 123–133.

2

Passive cooling of buildings

Buildings, although static in space, are dynamically related to time. In addition to offering a shelter and fulfilling aesthetic criteria, they should ensure means of comfort (thermal, visual, acoustic) for their inhabitants.

The thermal behaviour of buildings is affected by various parameters. These include the climatological ones, which are environmental variables and are not subject to human control. The other type of parameters is the design variables, which are under control at the design stage. Insufficient attention to the aspect of a building's thermal behaviour at the first stages of its design can lead to an inhospitable internal environment. During summer, especially in climates with hot weather, buildings are exposed to high intensities of solar radiation and high temperatures. This may result in overheating conditions that exceed the threshold of thermal comfort in the interior of buildings. Under these conditions, cooling of buildings is of great importance.

Traditional architecture can show examples of harmonization with local climate. In hot and dry climates the buildings were of massive construction with few and small openings and light colours on the external surface. Fountains, pools, water streams and vegetation accomplished the cooling effect through evaporation. In hot and humid climates, where ventilation is desirable, we find lightweight structures with large openings and large overhangs. Modern buildings, in many cases, fail to follow the examples set by tradition. Part of the blame for the failure, from the climatic point of view, has been attributed to the so called 'international style', that brought science and technology to its design, adapting design ideas and features regardless of different climatic regions. This was connected with the separation of the envelope's design, which was the task of the architect, and the interior operation, which was entirely left to service engineers. This approach led to a total dependence on mechanical equipment to support the energy needs of buildings. With the present state of the art of air conditioning, the cooling needs of any building can be met, but at the cost of using unnecessary amounts of energy.

BENEFITS OF PASSIVE COOLING

The energy required for heating and cooling of buildings is approximately 6.7% of the total world energy consumption. By proper environmental design, at least 2.35% of the world energy output can be saved [1]. In hot climate countries, energy needs for cooling can amount to two or three times those for heating, on an annual basis [2]. Utilization of the basic principles of heat transfer, coupled to the local climate, and exploitation of the physical properties of the construction materials, could make possible the control of the comfort conditions in the interior of buildings. Even in areas with average maximum ambient temperature around 31.7°C, comfortable conditions inside buildings can be achieved by means of proper building design [3] that frequently makes the use of air-conditioning units in dwellings unjustified. It is estimated [4] that an increase of 9 mtoe (million tons of oil equivalent) per annum in the total technical potential solar contribution (all potential solar usage is exploited) is possible in all the EU countries by the year 2010, compared to 1990, if passive cooling is applied in dwellings.

Passive cooling strategies in the design of buildings should be considered, since the extensive use of air-conditioning units is associated with the following problems:

Environmental

- Wide use of air-conditioning units has caused a shift in electrical energy consumption to the summer season and an increased peak electricity demand. Peak electric loads impose an additional strain on national grids, which can only be covered by development of extra new power plants. It is estimated [5] that in the USA the total electric peak load induced by air-conditioning units is about 38% of the non-coincident peak load. In addition to environmental implications, increased electrical peak loads result in an increase of the average cost of electricity to cover the construction of new power stations.

- Increased electrical energy production contributes to exploitation of the finite fossil fuels, to atmospheric pollution and to climatological changes. During the production process (fuel conversion), CO_2, which is one of the main causes of the greenhouse effect, is released. Coal-fired power plants burn and emit approximately 0.5 kg of carbon in the form of CO_2 for each kWh generated [6]. From the pre-industrial period to 1987, the ever-increasing use of fossil fuel and deforestation together have raised atmospheric CO_2 concentration to some 24% [7].

- Heat rejection during the production process (for electrical energy and air-conditioning units) and from the operation of air-conditioning units themselves increases the phenomenon of the 'urban heat island' (the climate modification due to urban development, which produces generally warmer air in cities than the surrounding countryside). Studies [8] show that summer heat islands are of average daily intensity of 3–5°C, resulting in discomfort and increased air-conditioning loads. It is estimated [9] that 5 to 10% of the urban American electrical demand is for additional air conditioning to compensate for heat islands.

- Ozone-layer depletion can be caused by CFCs and HFCs (the most common refrigerants of currently used air-conditioning units) from possible leakage during manufacture, system maintenance or unit failure. It has been estimated [10] that in the UK the average annual rate of refrigerant leakage from building air-conditioning units is 20% of the total machine charge and that around 75% of CFC consumption in 1991 by the refrigeration systems for building air conditioning was for servicing existing systems. Although long-term alternatives to CFCs and HFCs will have zero ozone-layer depletion potential (ODP), most of them will still be strong greenhouse gases.

Indoor air quality

- Increased indices of illness symptoms (lethargy, headache, blocked or runny nose, dry or sore eyes, dry throat and sometimes dry skin and asthma), known as 'sick building syndrome' [11] are reported in people working in air-conditioned buildings [12, 13].
- Occupants' dissatisfaction with indoor comfort conditions.

Economic

- Economic and political dependence of countries with limited natural resources on other countries, richer in natural resources.
- Installation of air-conditioning units presents an extra cost in the construction of a building, followed by an additional operation and maintenance cost. In EU countries 3.1 million ECU were spent on air-conditioning units in 1991, with Germany the largest contributor (22% of total) [14].
- Expenses for importation of A/C units. Countries with hot climates exhibit an increased rate of sales of air-conditioning units. In Greece, sales of packaged air conditioning have increased by 900 per cent over recent years [15], with 80% of them delivered to the residential sector [14].

PASSIVE COOLING STRATEGIES

Passive cooling can broadly cover all the measures and processes that contribute to the control and reduction of the cooling needs of buildings. It includes all the preventive measures to avoid overheating in the interior of buildings and strategies for rejection to the external environment of the internal heat, either generated in the interior or entering through the envelope of the building. Coupling of buildings with ambient cooling sinks that can absorb the indoor built-up heat has to be provided [16]. Natural sources that can be utilized as cooling sinks are:

- the upper atmosphere (sky)
- the ambient air
- the earth.

Heat rejection to the heat sinks can be achieved either by the natural processes of heat transfer: conduction, convection, radiation, evaporation (natural cooling) or enhanced mechanically with use of small power fans or pumps (hybrid cooling).

To avoid overheating and create thermal comfort conditions in the interior of buildings during summer, the cooling strategies should be designed at three levels:

1   prevention of heat gains in the building
2   modulation of heat gains
3   rejection of heat from the interior of the building to heat sinks (natural or hybrid cooling).

The first two measures aim to minimize the heat gains and the air temperature inside the building, while the third one attempts to lower the interior air temperature.

Prevention of heat gains

Prevention of heat gains is the first step towards improvement of the thermal comfort conditions in the interior of buildings and includes every measure that provides minimization of heat gains in it. The heat gains in a building can be classified as (Figure 2.1):

Figure 2.1 Sources of thermal heat gains in a building

- **External heat gains**, which originate from the association of the building with the outdoor environment. The main sources are:

     – *Solar radiation*, which is directly transmitted indoors through the glazing or is absorbed by the opaque elements and consequently conducted indoors.
     – *Ambient temperature*, which is conducted through the building fabric and convected by ventilation and air infiltration.
 • **Internal heat gains**. The main sources are:
     – *Metabolic heat*, the heat produced by occupants, which can be significant, especially in crowded spaces. (An average adult produces 110 W, which can rise to about 800 W when playing basketball.)
     – *Artificial lighting*, which is a significant heat source in office and commercial buildings, while its contribution to residential buildings is usually small.
     – *Appliances*, from which a significant amount of heat is produced, especially in modern offices.
     – *Cooking, bathing, etc.,* which add an extra heat input in the interior of buildings.

The contribution of each type of heat gains depends on the envelope and the use of the building.

Protection from heat gains could be achieved through (Figure 2.2):

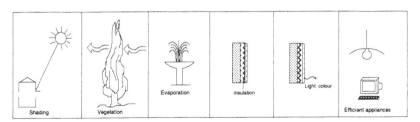

Figure 2.2 Ways of preventing overheating

 • **Modulation of the microclimate of the building**. Microclimate is the climate around the building and any modifications can improve indoor comfort conditions and reduce its cooling load. The microclimate can be modified through:
     – *Landscaping*. Trees around the building can protect it from the solar radiation, reduce the ambient temperature through evapotranspiration and offer at the same time protected spaces for outdoor activities. Evaporative cooling from pools and fountains, combined with wind channelling by vegetation can improve the indoor thermal environment of buildings.
     – *Site planning and siting of the building*. Proper site design and a building's siting can exploit existing topographical features, adjacent build-

ings and vegetation for natural solar protection, local breezes for summer ventilation or existing vegetation and water surfaces for natural cooling.

- **Form and morphology of the building**. Manipulation of the building's geometry and positioning and the sizing of its openings can regulate the exposure to incident solar radiation and wind. The arrangement of the internal spaces and the openings can influence the availability of natural daylighting and air flow in the interior of the building.

- **Solar protection of the envelope and shading of openings**. Preventing solar radiation from reaching the building and its interior is one of the fundamental measures for avoidance of overheating conditions in its interior. Although solar protection of the envelope is beneficial, most crucial is shading of the openings. This can be achieved by creating obstacles to the sunpath (shading devices) or manipulating the solar-optical properties of glazed surfaces (high reflectivity, thermochromic, electrochromic, holographic glasses).

- **Thermal insulation.** Thermal insulation in the building's envelope can function as a means to reduce heat transmittance through its fabric. In addition, through a special arrangement of the insulation layer, it can act as a radiant barrier [17]. Positioning a low-emissivity material (like an aluminium foil) next to an air gap in a multi-layer element (Figure 2.3) reflects the radiation, reducing in this way the inner-layer temperature and the radiant temperature of the internal spaces, while during the night it blocks radiant exchange, thus reducing the required cooling load.

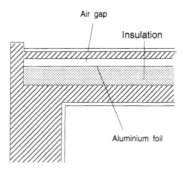

Figure 2.3 Schematic representation of a radiation barrier on the roof

- **Surface properties.** The colour of the external surface determines the amount of solar radiation received on the external surface of the building's envelope, as light colours are associated with reduced solar heat absorption. Irregularities of the texture of the external surface can modulate the heat absorbed and radiated back to the environment. The textured surface, because of its irregularities, has an absorbing area less than its emitting surface and also presents an increased surface for convective heat transfer. This permits a textured surface to cool down faster than a flat surface.

- **Control of internal gains**. Internal gains can be significant, especially in specific types of buildings, like offices and commercial buildings. Maximum use of natural daylighting and high-efficiency artificial lighting can significantly reduce the cooling load. Use of appliances of high efficiency and the location of them in positions that allow easy rejection of the heat produced can improve the internal thermal environment. Also allocation of the spaces according to the occupation densities and types of activities should be considered.

Modulation of heat gains

Modulation of heat gains can be achieved through the use of materials with high thermal storage (or thermal mass, as it is also called) in the building's structure. High-thermal-mass materials, like brick and concrete, act as a storage for both heat and cold as they heat up and cool down relatively slowly.

This procedure provides an attenuation of peaks in internal temperatures, by delaying the discharge of heat until a later time, when outdoor ambient temperatures are lower. At the same time it reduces the heat flow reaching the interior of the building, as part of the stored heat in the envelope is reradiated and convected back to the external environment during the evening hours. During the evening, the thermal mass acts as a storage of cold that is gradually recovered during the day.

The distribution of thermal mass in the building's envelope and its size play an important role in the effectiveness of the cooling discharge operations (mainly night-time ventilation and radiative cooling) and the amount of stored coolness. For assessment of the effectiveness of indoors-positioned thermal mass, the concept of the '*diurnal heat capacity*' was introduced [18, 19]. This is defined as the amount of energy stored in the material and returned back to the indoor space during the diurnal cycle per unit surface area, for one degree diurnal swing in the surface temperature of the storage element (Wh m^{-2} K).

The larger the swings in outdoor temperature, the more important is the effect of the thermal mass. The role of thermal mass attains its highest importance for buildings in continuous occupation, such as residences. Some recommendations for buildings in Israel are presented by Shaviv [20] and for European conditions in a study by INSA [21].

Heat rejection

In many climates, the preventive and modulating measures related to heat gains are not sufficient to keep indoor temperatures at a comfortable level during the day. The design of the building should ensure means to reject the heat build-up in the interior of the building (natural or hybrid cooling). The main techniques of natural cooling, according to the mode of heat transfer and fluid flow, can be classified as (Figure 2.4):

- cooling with ventilation
- radiative cooling
- evaporative cooling
- earth cooling.

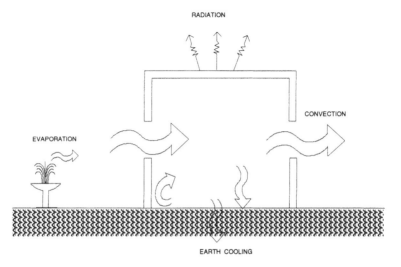

Figure 2.4 Modes of heat transfer

Some of the techniques provide a direct instantaneous cooling effect. In others, the coolness is collected during night-time and is released the next day, thus smoothing the effect of the accumulated heat inside the building. Materials suitable for storage, which can even store the coolness of the cold winter ambient air, are [22] (Figure 2.5):

- building mass, for storage for a few hours
- rock beds, which can store coolness for a few hours or months
- earth, for seasonal storage from winter to summer
- water, for a few hours' or a few months' storage
- phase-changing materials, for daily and seasonal cooling storage.

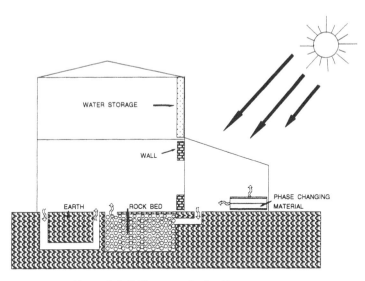

Figure 2.5. Different methods of heat storage

VENTILATION

Ventilation of buildings has been a major cooling technique throughout the world. It is based on the fundamental heat-transfer mode of convection, where the air flowing next to a surface carries away heat, provided it is at a lower temperature than the surface. When it passes over the human body, it increases the evaporation rate from the skin and enhances heat extraction.

Air movements through buildings result from the difference in pressure indoors and outdoors which can be achieved (Figure 2.6) by:

- **Natural forces**:
  - wind-induced pressure difference
  - pressure difference induced by temperature gradients between the inside and outside of the building (sack effect) (Figure 2.7) [23]
- **Mechanical forces**:
  - Pressure difference induced mechanically, like a fan.

Ventilation of buildings can be divided [24] into:

- **Comfort ventilation** (or daytime ventilation), which provides human comfort, mainly during daytime, through higher indoor air speeds. The air flow can provide a direct psychological cooling effect even at temperatures as high as 34°C. The higher air speed increases the rate of sweat evaporation

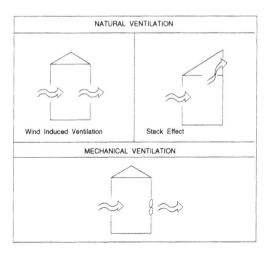

Figure 2.6 Ventilation strategies

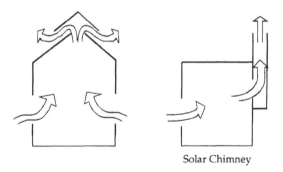

Figure 2.7 Stack-effect ventilation

from the skin, minimizing in this way the discomfort from the feeling of wet skin. Thus, it is especially beneficial when the relative humidity of the air is high [25].

Although daytime ventilation can raise the temperature of the internal surfaces and air, its overall effect on comfort can be positive, provided that indoor comfort conditions can be experienced for the outdoor air temperature, with an acceptable indoors air speed [25]. It is estimated that increasing the air velocity by 0.15 m s^{-1} compensates for a 1°C increase in air temperature at moderate humidity levels (less than 70% RH) [26].

The design objectives should be:
– to obtain a continuous air flow throughout the building
– to direct the air flow through the occupied zones
– to achieve high air velocities at the level of the occupants.

The structure materials should be of low storage.

- **Nocturnal convective cooling**. Air movement through the building during the night, when the outside air temperature is low, lowers the indoors ambient temperature and is also associated with storage of its coolness in a storage mass. The storage mass can be distributed in the structure itself, for example in the walls, floors, ceiling or in a specialized thermal store, such as a rock bed or water store mass. The building is kept cool during the next day, provided the windows are closed, as the cool structural mass is able to absorb the heat that penetrates through the envelope or is generated inside the building. When the storage mass is outside the building, it can be used to precool the ventilation supply air or to cool the building by a closed-circuit flow.

The effectiveness of the nocturnal ventilation is linked to:
- the ventilation rate
- the storage area
- the area that comes into contact with the flowing air
- the heat capacity and thermal conductivity of the storage material.

The design of the building should ensure a high ventilation rate through the building and especially over the storage surfaces.

This strategy is beneficial in climates with a large diurnal temperature swing, over approximately 15°C, where the night minimum temperature during summer is lower than about 20°C [24]. The indoor air-temperature pattern in a nocturnally ventilated building follows the outside one. In a high-mass, well insulated building, shaded and closed during the day, a drop of 35–45% in the internal air temperature, relative to the outdoor temperature, is possible through nocturnal ventilation [24].

RADIATIVE COOLING

Radiative cooling is based on the fundamental principle that any warm body emits thermal energy in the form of electromagnetic radiation to the facing colder ones. The sun is radiating heat (in the form of short-wave radiation) to the earth during the day and earth is radiating back heat (in the form of long-wave radiation) to the cool sky. The radiant heat loss takes place both day and night; especially in the northern hemisphere, the upper atmosphere layers are cool enough during the day to provide a useful heat sink [27]. During the day, the absorbed solar radiation counteracts the cooling effect of the long-wave emission.

The radiative cooling is stronger in clear night skies but it is reduced by the existence of particles, such as water vapour, carbon dioxide and dust in the atmosphere. As these particles absorb and emit back long-wave radiation, the net outgoing radiative heat flux from any object on the earth is equal to the radiation emitted from it towards the sky dome minus that re-emitted from the atmosphere. This difference in radiation rates determines the maximum potential of radiative cooling. Radiance from the clear sky tends to be low in the 8–13 µm spectral region. This high transparency to radiation is known as the infrared 'atmospheric window'.

The envelope of the building, which absorbs heat during the day, emits this heat back to the sky during the night (Figure 2.8). As a horizontal surface is the best radiator to the sky, the roof is the most effective radiator. The radiative potential of a surface is reduced when hot air surrounds it (hot summer nights) because it convects heat to the surface. The convected heat increases with the wind speed and its effect can be decreased by using wind-screens, a screen partially transparent to long-wave radiation.

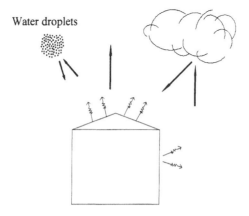

Water droplets

Figure 2.8 Thermal radiation processes

Dependent on the thermal coupling of the radiative surface and the interior of the building, radiative cooling can be classified as:

- **Direct radiative cooling**, where the roof acts as a radiator. The roof is the element of the building that absorbs the biggest part of the solar radiation during summer [27] and has also the best view of the sky dome and thus is an effective radiator. (On a hot summer day the roof of a house could reach 65°C at noon and at this temperature could reradiate almost 750 W m^{-2} towards the sky [28].)

  Use of high-thermal-storage materials can increase the radiative potential of a building, by delaying the transfer to its interior of heat which is radiated back during the night. Traditional architecture in hot, dry and sub-tropical climates can show examples of radiative cooling by using vaulted roofs. The surface of a vault is bigger than its horizontal base (three times bigger for a hemispherical roof) and, thus, there is a bigger storage surface during the day and a bigger radiative surface during the night.

  The '*Roof pond*' or '*Skytherm system*' is the most characteristic contemporary application [29]. The roof is made of structural steel deck plates and

plastic bags, filled with water, are placed on them. Movable insulation panels cover the water bags during the day and expose them to the sky dome
during the night in summer (Figure 2.9).

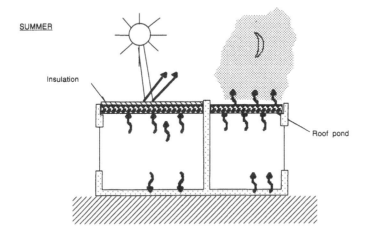

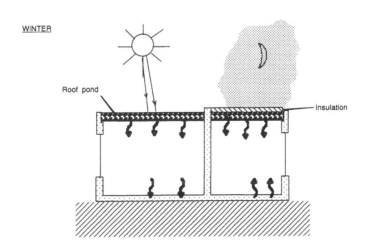

Figure 2.9 The Roof pond system

Similar is the *Movable thermal mass roof*. A pond, filled with water is
situated above an insulated layer during the night and is cooled by radiation.
During the day the water is drained below the insulating layer and absorbs
the solar gains.

Insulation of the roof's structure minimizes the actual radiative cooling potential of the building. Application of movable insulation on top of the roof enhances the radiative potential of the roof, as it is used to protect the roof from solar gains during the day and leaves it exposed during the night.

Efforts have been made to develop selective radiators that enhance the radiative cooling effect [30, 31, 32]. These radiators attempt to exploit the 'atmospheric window', by presenting high emissivity at the long-wave spectrum of the 'atmospheric window' and high reflectivity below and above this range.

The cooling effect of the roof radiative systems is restricted to the space that is in contact with the roof and thus only single storey houses or the top floor of a multi-storey building can benefit.

• **Indirect radiative cooling** is obtained by cooling a fluid (water or air) through radiation to the sky. The coolness is then stored in a specialized storage mass (e.g. rock-bed, water tanks) or in the structural mass of the building. The cool air can also be flushed directly into the building, preferably next to the storage mass (Figure 2.10). A metallic sheet over the roof, with an air gap of 5–10 cm beneath, is a typical radiator.

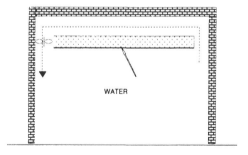

Figure 2.10 Indirect radiative cooling

EVAPORATIVE COOLING

Evaporation is the phase change of water from liquid to vapour. This is accompanied by release of high amounts of heat (sensible heat) from the air that comes in contact with the wet surface or from the surface where evaporation takes place. Depending on fluctuations of the air's moisture content, evaporative cooling is characterized as:

• **Direct evaporative cooling** of the ventilation air. The ventilated air passes over a wet surface (e.g. a pond or fountain) and is cooled by evaporation. The decrease in the dry bulb air temperature (DBT) is accompanied by an increase in the air's moisture content. On the psychometric chart, direct evaporation is characterized by a displacement along a constant wet bulb temperature (WBT) (line *AB* in Figure 2.11).

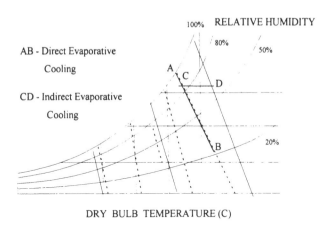

Figure 2.11 Evaporative cooling on the psychometric chart

- **Indirect evaporative cooling**. The evaporation takes place on a surface which is cooled during this process. The air that comes in contact with the cool surface is subsequently cooled without increasing its moisture content. Its representation on the psychometric chart is along a line of constant moisture content (*CD* in Figure 2.11).

The efficiency of the evaporation process depends on the temperature of the air and of the wet surface, the moisture content of the air and the velocity of the air flowing over the wet surface. Evaporative cooling is especially effective in arid and dry climates. Direct evaporative cooling is considered effective in places where the maximum ambient WBT is 22 to 24°C and the maximum DBT is 42 to 44°C. It has been suggested that indirect evaporative cooling can be applied in places where the WBT is 25°C and the maximum DBT is 46°C [24].

Air flow can be induced by natural forces (wind or temperature differences) (natural evaporation) or with simple mechanical systems (hybrid evaporation). The most simple application of natural evaporation is the use of vegetation for evapotranspiration, in fountains, ponds and water streams (Figure 2.12). Traditional architecture can show examples of wet pads made of fibres and placed on windows or wind towers where the incoming air passes through jars with water or through a water spray ('Volume cooler') (Figure 2.13).

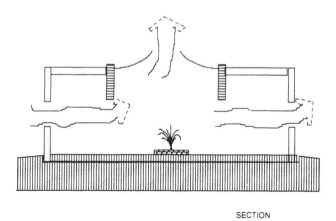

SECTION

PLAN

Figure 2.12 Natural evaporative cooling

Indirect evaporation is used to cool the roof of the building either by keeping a thin layer of water on the roof surface – *Water-film* – or by using a water spray – *Roof spray* – to keep the roof wet. The water-pond technique is also used to cool the roof surface by evaporation. The building is

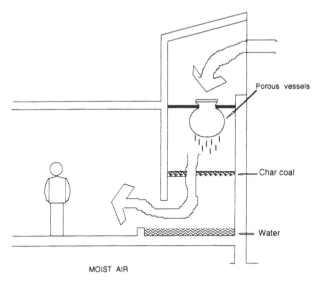

Fig. 2.13 Traditional volume cooler

then cooled, because the radiant temperature of the roof, and of the indoor air coming in contact with the roof, is lowered.

EARTH COOLING

The earth can serve in many climates as a cooling source. Its high thermal capacity keeps the soil temperature, below a certain depth, considerably lower than the ambient air temperature during summer. Seasonal variation of the earth temperature decreases with increase of the depth, moisture content and soil conductivity. It is estimated that a few metres below the surface, the earth temperature remains constant throughout the year [27].

In regions with a temperate climate, the temperature of the soil at a depth of 2.0–3.0 m can be low enough during summer to serve as a cooling source. In hot climates, however, the earth temperature is usually too high to allow the earth to act as a cooling source [24]. Several measures have been proposed to lower the soil temperature, for example shading the earth surface with a layer of gravel or wood chips about 10 cm thick or encouraging evaporation by irrigating the surface of the earth [33].

The cooling potential of the earth can be utilized by (Figure 2.14):

- **Direct earth contact cooling**. The cooling capacity of the earth mass is directly exploited by fully integrating the building within the earth (underground buildings) or by partially earth integrating the building's envelope, for example by covering with earth some of the walls or the roof of

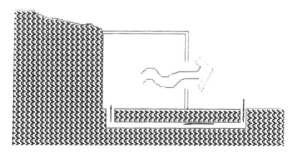

Figure 2.14 Earth cooling

the building. It is most suitable in hot-dry climates with a mild winter, as the direct coupling of the building with the earth, which is necessary for effective cooling, causes high rates of heat loss during winter.

- **Indirect earth cooling**, by precooling the air entering in the building through use of underground pipes. The hot external air or the air circulating through the building (closed-circuit) is cooled by flowing through tubes positioned in the ground. The length and the diameter of the pipes depends on the cooling load of the designed building; they are usually about 10–30 cm diameter and 12–60 m length. When the available earth space is restricted, an array of pipes can be used instead of one big pipe. Circulation of the indoor air through earth pipes can lower the indoor air temperature by 10°C compared to the external ambient one [24].

COOLING POTENTIAL – LIMITATIONS

The operation and efficiency of passive cooling techniques is more dependent on climate than passive heating ones. While the processes for passive solar heating are common throughout the world, passive cooling is based on processes fundamentally linked to climate (air temperature, relevant humidity, velocity and direction of winds). Passive solar heating will always make a positive contribution to the overall thermal performance of a building, whereas improper choice of a cooling technique could create an unpleasant internal environment. In addition, thermal comfort requirements during summer are different for each climate type. As an example, hot dry climates require different cooling strategies to hot humid climates.

The choice of the appropriate cooling technique depends not only on the local climatological conditions, but also on the building type and the occupancy patterns. Additionally, the climatic boundaries for the applicability of various systems differ from region to region [24], in that, as a result of the phenomenon of natural acclimatization, people living in hot regions can tolerate higher temperatures before experiencing distinct discomfort.

Measures for prevention and modulation of heat gains are effective in any climate and type of building. Their necessity and importance in the overall design is associated with the contribution they make to the cooling load. On the other hand heat rejection from the building by natural means depends mainly on [26]:

- sufficient climatological and environmental conditions for providing an appropriate heat sink for rejection of the indoors developed heat;
- appropriate thermal coupling of the building with a heat sink.

The applicability of passive cooling strategies could be limited by :

- **Climate and microclimate**. High night ambient temperature, cloud cover, high humidity and insufficient wind speeds are the main climatic factors affecting the efficiency of the different cooling approaches.
- **Air pollution and noise levels**, especially in the centre of big cities, could discourage the application of some cooling techniques, such as natural ventilation.
- **Site topography and building regulations** in some cases put restrictions in the design of buildings and limit the applicability of several techniques.
- **Lack of national regulations** and guidance for meeting the cooling needs of buildings. There is an emphasis on the heating needs of buildings through regulations on thermal insulation, but the important issue of thermal comfort during the summer period has received less attention.
- **Insufficient information** to both designers and building users on the potential of passive cooling. Also, the number of evaluation tools for the summer thermal performance of buildings is limited.

EPILOGUE

Passive cooling should be considered as part of an overall environmental design strategy that attempts to meet comfort conditions in the interior of buildings and to minimize its dependence on mechanical equipment. Incorporation of passive cooling measures should always be considered in association with the winter thermal needs of the building.

Although climate seems to be a determining factor in the demand for cooling, building type, occupancy patterns, activities and building design are equally important. This makes passive cooling techniques relevant even in northern regions where overglazed offices can experience overheating.

The overall concept behind a strategy for passive cooling can be summarized in the following steps:

1	reduce the solar radiation reaching the building and control the hot ambient air entering its interior;

2	minimize the effect of heat reaching and penetrating the envelope of the building;

3	contribute to the reduction of internal gains, through daylighting and efficient use of appliances;

4	couple the envelope with environmental sinks for heat dissipation.

REFERENCES

1	Agrawal, P.C. (1988). 'A review of passive systems for natural heating and cooling of buildings', *Proceedings of the Conference 'Healthy Buildings 88'*, Stockholm, Sweden, Vol. 2, pp. 585–602.

2	Goulding, J., J. Owen Lewis and T. Steemers (eds), (1993). *Energy in Architecture; The European Passive Solar Handbook*, Publication No EUR 13446, Batsford for the Commission of the European Communities, London.

3	Shaviv, E., (1984). 'The performance of a passive solar house with window sunspace systems', *Energy and Buildings*, Vol. 7, pp. 315–334.

4	*Passive Solar Energy as a Fuel – 1990-2010* (1990). Executive Summary, The Commission of the European Communities, DGXII.

5	Argiriou, A. and M. Santamouris, (1993). 'Passive cooling in hotels', European Seminar on *Advanced Systems in Passive and Active Climatisation*, Barcelona.

6	Akbari, H., J. Huang, P. Martien, L. Rainer, A. Rosenfeld and H. Taha (1988). 'The impact of summer heat islands on cooling energy consumption and global CO_2 concentration', ACEEE Summer Study on 'Energy Efficiency in Buildings', Asilomar, CA.

7	Brown, L.R. (1988). *State of the World, A World Watch Institute Report on Progress toward a Sustainable Society*, Chapter 5, pp.83–100. W.W.Norton & Co, New York.

8	Akbari, H., A. Rosenfeld and H. Taha (1990). 'Summer heat islands, urban trees and white surfaces', ASHRAE Symposium, January.

9	Akbari, H. and A. Rosenfeld (1989). *Urban Trees and White Surface for Saving Energy and Reducing Atmospheric Pollution*, Urban and Community Forestry Act of 1989.

10	*CFCs in the UK Refrigeration and Air-conditioning Industries. Usage and Scope for Substitution* (1992). DOE, HMSO, London.

11	Sherwood Burge, P. (1991).'Sick building syndrome, epidemiological studies and medical aspects', Workshop on *Indoor Air Quality Management*, Lausanne, EUR 137766, May.

12	Goudelas, G., K. Moustris and M. Santamouris (1991). 'Energy audits in public and commercial buildings in Greece', Conference on Renewable Energy Sources for Local Development, Chios (Greece).

13	Hedge, A., E.M. Sterling and T.D. Sterling (1986). 'Building indices based on questionnaire responses', *IAQ 86 Conference, ASHRAE*, Atlanta, GA, pp. 31–43.

14	King, A. (1993). 'The growing market for air conditioning', European Seminar on *Advanced Systems of Passive and Active Climatisation*, Barcelona, June.

15	Santamouris, M. (1990). Workshop on passive cooling, *Proceedings of Passive Cooling Workshop*, Ispra, EUR 13078 EN.

16 Dimoudi, A. and P. Liveris (1988). 'Passive cooling techniques in buildings', 3rd National Conference on *Soft Kinds of Energy*, Thessaloniki, Greece (in Greek).

17 Fairey, P. (1984). *Designing and Installing Radiant Barrier Systems*, Florida Solar Energy Center, DN-7-84.

18 Balcomb, J.D. (1983). *Heat Storage and Distribution inside Passive Solar Buildings*, Report LA 9694 M.S, Los Alamos Laboratory.

19 Givovi, B. (1983). 'A generalized predictive model for direct gain', *Passive Solar Journal*, Vol. 2, pp. 107–115.

20 Shaviv, E. (1988). 'On the optimum design of SHULDING: Devices for windows', PLEA Conference, Porto.

21 Depecher, P., J. Brau and S. Ronssean S. (1990). *L'Inertie Thermique par le Beton, Economie d' Energie et Confort d'Eté*, (INSA de Lyon), Centre d'Information de l'Industrie Cimentiere (CIC).

22 Bahadori, M. (1981). 'Passive and hybrid convective cooling systems', *International Passive and Hybrid Cooling Conference*, Miami Beach, American Section of ISES, pp. 715–727.

23 Dimoudi, A. (1989). 'Evaluation of the use of solar chimnies for promotion of natural ventilation of buildings', MSc Thesis, Cranfield Institute of Technology, England.

24 Givoni, B. (1991). 'Performance and applicability of passive and low-energy cooling systems', *Energy and Buildings*, Vol. 17, pp. 177–199.

25 Givoni, B. (1992). 'Climatic aspects of urban design in tropical regions', *Atmospheric Environment*, Vol. 26B, No 3, pp. 397–402.

26 Santamouris, M (ed.), M. Antinucci, B. Fleury, J. Lopez, E. Maldonado, A. Tombazis and S. Yannas (1990). *Horizontal Study on Passive Cooling*, Commission of the European Communities DGXII, Building 2000 Action.

27 Sodha, M.S., N.K. Bansal, P. K. Bansal, A. Kumar and M.A.S. Malik (1986). *Solar Passive Building; Science and Design*, Pergamon Press, International Series on Building Environmental Engineering, Vol. 2, Oxford and New York.

28 Cook, J. (ed.) (1989). *Passive Cooling*, MIT Press, Cambridge. MA.

29 Hay, H.B. and I.J. Yellot (1969). 'Natural air-conditioning with roof ponds and movable insulation', *ASHRAE Transactions*, Vol. 75, Part I.

30 Head, A.K. (1959). Australian Patent No 239364, and US Patent No 3,043,112.

31 Mitchell, D. (1976). *Selective Radiation Cooling; Another Look*, CSIRO Division of Tribophysics, University of Melbourne, Parkville, Victoria, 3052, Australia.

32 Catalanotti, S., V. Cuomo, G. Piro, D. Ruggi, V. Silvestrini and G. Troise (1975). 'The radiative cooling of selective surfaces', *Solar Energy*, Vol. 7, pp.83–89.

33 Givoni, B. (1987). 'Passive cooling – state of the art', *12th Passive Solar Conference*, Portlant OR, ISES, pp. 11–19.

3

The Mediterranean climate

GEOGRAPHICAL CHARACTERISTICS

The countries surrounding the trapped Mediterranean basin are characterized by special climatological conditions due to the combination of their latitudes and to the influence of the sea on the formation of the area's weather. In addition, a significant role is played by the terrain of the northern part of the basin, which consists of high mountain ranges. Thus, the climate of the European Mediterranean countries introduces special characteristics which are also encountered in certain areas in Australia and in California, USA.

The Mediterranean Sea lies within the temperate zone between 30° and 46°N. Except for a small part of North Africa, east of Tunisia, the rest of the sea is enclosed by mountains over 3000 metres high, lying close to the coast with a coastline that is much indented.

The Mediterranean basin may be regarded as being divided geographically into three parts:

- the eastern part, mainly around Cyprus and Greece, the northern part of which is occupied by the Adriatic Sea and the Aegean Sea;
- the central part between Corsica–Sardinia–Tunisia–Sicily and Greece–Cyrenaica; and
- the western part between the coasts of France–Spain–North Africa and Corsica.

The western basin of the Mediterranean is surrounded by high mountains (Atlas, Pyrenees, Alps). The main gaps in the area through which we can have inflow of air masses from the Atlantic and Northern Europe, are:

- the Strait of Gibraltar and the Alboran Channel, between the mountains of Spain and the Atlas range in North Africa;
- the Garronne–Carcassonne gap, north and west of the Pyrenees which separates them from the Massif Central of France; and
- the narrow Rhone–Sâone gap, which separates the Massif Central of France from the Alps.

All around the western basin the coastal plain is narrow except where the Rhone enters the Gulf of Lyons.

The eastern Mediterranean basin is bordered on the north and east by ranges of mountains of over 1000 metres high. The Adriatic Sea is surrounded on the west by the Apennines, which cross the Italic peninsula, on the north by the Alps and on the east by the Dinaric Alps, which lie along the coastline of Yugoslavia and continue south as the Pindus range of Albania and Greece. The only gaps in the general area are the wide, funnel-shaped Po Valley at the northern end of the Adriatic Sea – separating the Apennines from the Alps – and the Trieste gap between the Alps and the Dinaric Alps.

The Aegean Sea, which lies in the north-eastern part of the Mediterranean, has as a northern boundary the Rhodope range and the Turkish mountains. Between them there is the Dardanelles gap, between 50 and 100 miles wide, which provide an outlet from the Black Sea to the Aegean Sea. To the east, the Mediterranean is surrounded by mountains 1000–2000 metres high, which extend to Lebanon and Syria. To the south the eastern basin is bounded by the mountains of North Africa, which are not very high (< 300 metres) except for the mountains of Cyrenaica, which reach 500 metres above sea level.

As a result of this special topography of the area, the inflow of air into the Mediterranean basin takes place through gaps in the mountain ranges, except over the southern shores east of Tunisia. Strong winds are 'funnelled' through the gaps and influence the Mediterranean areas, thus creating the special conditions that determine the Mediterranean climate.

The most important winds of the Mediterranean, are:

- The north-westerly Mistral, through the Alps–Pyrenees gap
- The north-easterly Bora, in the Adriatic area, which blows through the Trieste gap
- The easterly Levanter and the westerly Vendaval, through the Strait of Gibraltar
- The warm south-easterly to south-westerly wind, which blows from Africa towards the Mediterranean and is best known as the sirocco, ghibli or khamsin.
- The south-easterly winds that blow in the Aegean Sea through the Dardanelles and which, during summertime and under special synoptic conditions, create strong or even stormy local winds mainly during daytime (Etesian winds).

As a result of the topography, the trapped Mediterranean basin and the prevailing synoptic systems in the temperate region, the outstanding features of the weather are mild and wet winters and relatively calm, hot and dry summers. The transitional season of spring is similar to that in England in that the change from the winter to the summer regime takes place with a number of false starts. On the other hand, autumn is relatively short, ending in early November.

In the Mediterranean there are mainly two season categories, the warm season and the cool season. Warm season denotes the months from June to September and cool season the months from October to May, although October and May may be regarded as transitional months.

SOLAR RADIATION

It is widely known that all the observed atmospheric processes are the consequences of the received solar radiation. Thus, the measurement of the solar radiation presents special interest for the environmental researchers.

The prevailing weather conditions in the Mediterranean basin, combined with the latitude, form conditions of uniform distribution of the solar radiation parameters. As a result, the values of the coefficient of variability do not exceed the value of 12%, while in the east Mediterranean region, they are even smaller (4–7%). Thus, despite the lack of a dense network of solar radiation measurements, we can refer with sufficient precision to parameters covering the whole Mediterranean region.

Global radiation

From the Test Reference Years (TRY) of the CEC [1] solar radiation data have been gathered, for 19 on-shore stations in the western and central Mediterranean (Italy and France). In addition, global radiation data has been recorded in Greece for 36 measurement stations using mainly Robitzsch actinographs. From these we have considered data from 11 stations in coastal areas and included this information with the network of the stations mentioned above, as well as data from stations in Croatia, Malta and Cyprus. For all these stations (Table 3.1) we have processed the measurements of temperature, relative humidity, wind and cloudiness (Figure 3.1).

Initially, there appears to be a correlation of the kind:

$$\log y = a + b\varphi \tag{3.1}$$

between the annual values of the global radiation (y) and the latitude (φ). The correlation coefficient that was calculated (–0.76) is statistically significant in the significance level of 0.05. If, using regression analysis, the annual values of the global radiation of the total of all the shore stations in Greece are evaluated, it appears that the differences between the evaluated and the observed annual values are not statistically significant at the significance level of 0.05 in the mean values.

However, evaluation of the annual values of the global radiation, with the help of the above equation, for areas away from the Mediterranean coastal zone, leads to results that differ significantly from existing measurements. The average annual values of the global radiation of the 19 stations of the Mediterranean basin that we examined ranged from 505,080 J cm^{-2} (Thessaloniki, Greece) to 655,590 J cm^{-2} (Trapani, Italy).

Table 3.1 Network of Mediterranean stations examined

Zone A φ< 37° N	Zone B 37°<φ<40° N
1. Gibraltar	10. Alicante,SPAIN
2. Valletta, MALTA	11. Valencia,SPAIN
3. Kithira, GREECE	12. Palma,SPAIN
4. Iraklion,GREECE	13. Palermo, ITALY
5. Milos,GREECE	14. Catania,ITALY
6. Rhodes,GREECE	15. Corfu, GREECE
7. Morphou, CYPRUS	16. Patras,GREECE
8. Paphos,CYPRUS	17. Athens,GREECE
9. Nicosia,CYPRUS	18. Skiros,GREECE
	19. Samos,GREECE

Zone C 40° <φ< 43° N	Zone D φ> 43° N
20. Barcelona, SPAIN	28. Marseilles, FRANCE
21. Pertusato, ITALY	29. Genoa, ITALY
22. Rome, ITALY	30. Leghorn, ITALY
23. Naples, ITALY	31. Ancona, ITALY
24. Taranto, ITALY	32. Venice, ITALY
25. Dubrovnik,CROATIA	33. Trieste, ITALY
26. Thessaloniki,GREECE	34. Fiume, CROATIA
27. Alexandroupolis,GREECE	35. Split, CROATIA

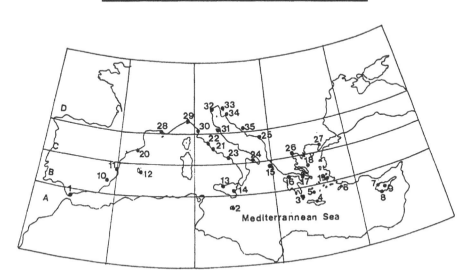

Figure 3.1 Network of the Mediterranean stations

Over the whole Mediterranean zone the annual variation of the global radiation is the same, having a maximum, mainly, in July and a minimum in December. The percentage of the global radiation during the warm period (June–September) covers the 47–50% of the annual values.

We have divided the Mediterranean region into four zones (A, B, C, D) according to latitude [zone A ($\varphi < 37°$ N), zone B ($37°$ N $< \varphi < 40°$ N), zone C ($40°$ N $< \varphi < 43°$ N), zone D ($\varphi > 43°$ N) – see Table 3.1]. Grouping together the measurements, we have calculated for each zone the mean annual variation of the global radiation (Figure 3.2). It is clear from Figure 3.2 that generally the global radiation decreases as the latitude increases. A slight increase, observed during July and August in zone B relative to zone A, can be attributed to the geographical distribution of the stations within zone B.

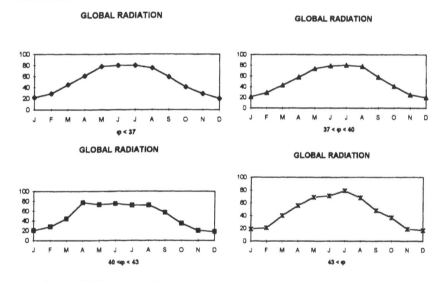

Figure 3.2 Mean annual pattern of the global radiation of zones A, B, C and D

The greatest increase from month to month is observed for the areas with a latitude less than $37°$ N between February and March, while in zones B and C this increase is encountered between April and May. In contrast, the greatest decrease is observed between August–September and September–October.

Diffuse solar radiation

The annual values of the diffuse solar radiation represent 35–44% of the respective global radiation values and some increase of this percentage with latitude is observed. The annual variation of the diffuse global radiation is the same too, having a maximum in July and a minimum in December.

The diffuse solar radiation of the warm period (June–September) ranges from 220,000 to 232,500 J cm^{-2} and represents the 45–49% of the respective annual values. These percentages are of the same order as those referring to the global radiation.

The lack of sufficient data prevents a latitude zonal study of the diffuse solar radiation distribution, but it seems, as expected, that there are other factors which also significantly influence this distribution, so that there is no statistically significant correlation between diffuse solar radiation and latitude.

Direct normal radiation

The annual values of the direct normal radiation range from 0.98 to 1.06 of the respective global radiation values. As far as the annual variation of the direct normal radiation is concerned, it presents a single annual variation with a maximum between June and August and a minimum between December and January.

The total of the direct normal radiation of the warm period (June–September) ranges from 217,000 to 308,000 J cm^{-2} and represents the 41–47% of the annual totals.

The direct normal radiation thus presents statistically significant linear correlation with the latitude (correlation coefficient $r = -0.87$), but the small size of the sample does not permit the definition of a regression line that can efficiently provide correct estimates for the entire Mediterranean zone.

THE AIR TEMPERATURE

The most important meteorological parameter that is relevant to issues of buildings' cooling and of human comfort in external and internal environments, is the air temperature. In the Mediterranean region, owing to the great longitudinal extent of the surrounding countries (30°–45° N latitude) and to the terrain, there are great contrasts in temperature. In addition, temperature is influenced, in the interior, by the site location in relation to the approach of tropical maritime (MT) air masses.

After studying the various expressions of the air temperature in some coastal sites of the European part of the Mediterranean, we make the following observations.

Mean air temperature

The mean annual air temperature for the selected 35 sites in the European region of the Mediterranean Sea (Figure 3.1) is related more to the stations' latitude than to their positions in accordance with the synoptic weather systems appearing in the area.

Thus, the statistically significant linear correlation (correlation coefficient $R = -0.909$) between the latitude and the mean annual air temperature permits the evaluation of the mean annual air temperature ($\langle T \rangle$) in low-altitude sites of the Mediterranean region, with the help of the area's latitude (φ), through the equation

$$\langle T \rangle = 34.23 - 0.437\varphi°. \tag{3.2}$$

Equation (3.2) was used for the estimation of the mean annual air temperature in a number of other sites, including some on the African coast, and the error relative to the real values did not exceed 0.5–1.1°C.

The mean annual temperatures of the network of the selected stations ranged between 13.2° C (Venice) and 19.2° C (Morphou).

In the entire Mediterranean zone there is a single annual variation of the mean air temperature, with a maximum in July or August and a minimum in January or February. The annual thermometric amplitude ranges from 11.7°C (Gibraltar) to 21.7°C (Venice).

The mean values of the warmer months range from 21.7°C in Marseilles, which is under the influence of the north-westerly wind (Mistral), to 28.3°C in Nicosia. Taking the warm season in total (June–September), we find – especially in the northern sites – lower mean monthly temperatures (19.4°C), so that the mean temperature of the warm season ranges from 20.7°C in Marseilles to 26.7°C in Nicosia. In this case, the changes from place to place are related less to the latitude and are influenced by other local factors, such as the altitude. During these months, slight differences in the altitude are obvious in the air temperature changes, given that the vertical lapse rate in the summer is greater.

Furthermore, with the Mediterranean region divided into four zones according to the latitude, so as to eliminate the influence of the local factors, we have calculated the annual variations of the mean monthly air temperatures (Figure 3.3). It seems that there is a definite change of the mean monthly air temperatures in relation to the latitude in all months and the observed differences are less intense during the warm months of the year, which is according to the, previously mentioned, smaller influence of the latitude during the warm season. In addition, greater cooling is observed in all four zones between October and November (4.3–4.8°C) with a delay of one to two months relative to the global radiation. Greater warming is observed between May and June (~3.9°C) with a delay of one month relative to the greater increase of the global radiation for the zone 37°N < φ < 42° N, except for zone D where greater warming is observed between April and May (4.8°C), with a delay of two months relative to the greater increase of the global radiation in this zone.

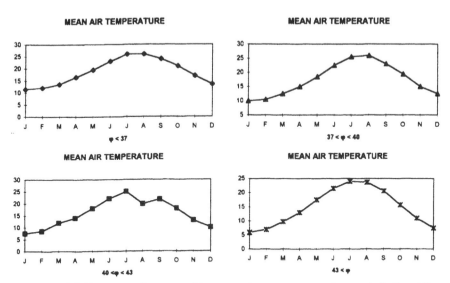

Figure 3.3 Mean annual pattern of the mean air temperatures of zones A, B, C and D

Mean maximum air temperatures

For cooling purposes, the parameters of greater interest are the mean monthly maximum air temperatures, especially those of the warm season (June–September), which are more directly related to the number of hours per day that the air tempera-ture exceeds the comfort level. These are especially influenced by local conditions, such as sea breezes or seasonal winds, e.g. Etesian winds in the Aegean Sea.

The monthly values of the mean maximum air temperatures range in the northern parts of the Mediterranean Sea, from 27.8°C (September) in Venice, where we also observed low mean monthly air temperatures, to 40°C (July and August) in Nicosia. The variation width of the mean maximum temperatures of the warm season is, it seems, greater than that of the mean temperatures, defined by the extreme values of 30–30.3°C (Genoa, Venice) and 38.9°C (Nicosia).

Examining, for the latitude zones, the annual variations of the mean monthly maximum air temperatures (Figure 3.4), we notice in all months a decrease in the values as the latitude increases.This decrease is more intense during the cool season of the year, especially in higher latitudes.

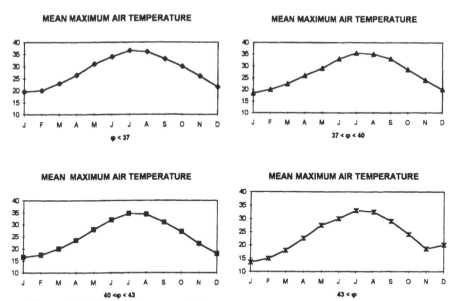

Figure 3.4 Mean annual pattern of the mean maximum air temperatures of zones A, B, C and D

Mean minimum air temperatures

The annual variation of the mean monthly minimum air temperatures in the Mediterranean follows that of the mean monthly temperatures, with minimum values in January or February, ranging from –5.6°C in Alexandroupolis to 6.1°C in Kithira. Generally, lower mean minimum monthly air temperatures are observed in the central Mediterranean and in the northern parts of the east Mediterranean. The same observations are made during the warm season. The mean monthly minimum temperatures for the latter period ranged from 8.3°C (September) again in Alexandroupolis, which is influenced by the north wind current through the gap to the Black Sea as well as by its high latitude, to 18.8°C in Kithira.

The same picture is presented by the zonal annual variation of the mean minimum air temperature (Fig 3.5) as for the mean monthly and the mean maximum air temperatures.

The monthly values of the mean minimum air temperature decrease constantly with increasing of the latitude. However, the differences during the summer months are less intense than those of the mean maximum air temperature. Generally, the difference between mean maximum and mean minimum air temperature during the warm season is greater in the eastern Mediterranean, mainly due to the higher mean maximum temperatures prevailing in this region.

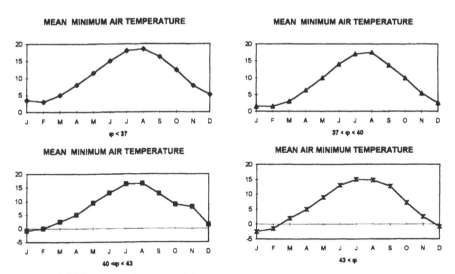

Figure 3.5 Mean annual pattern of the mean minimum air temperatures of zones A, B, C and D

Absolute maximum and minimum air temperatures

Absolute maximum and minimum air temperatures are observed, respectively, in the periods July–August and December–February and are influenced by the latitude as well as by local factors. The absolute maximum temperature of the July–August period ranged from 35.6°C (Barcelona) to 46.7°C (Nicosia), while the absolute minimum of the winter months ranged from –14.4°C (Trieste), to 5.0°C (Malta).

Generally, lower absolute minimum temperatures, during the winter, were observed in the central Mediterranean, while higher absolute maximum temperatures, during the July–August period, were observed in the eastern Mediterranean, mainly in Cyprus.

For a more general covering of the Mediterranean regions, we should note that the absolute minimum air temperatures depend not only on the latitude, but also on the station's continentality factor and on its altitude and orientation.

In Figures 3.6 and 3.7 the observed absolute maximum and minimum air temperatures are reported for each month and each zone. The absolute maximum temperatures present, generally, a decrease as the latitude increases. On the other hand, the absolute minimum air temperatures during the winter months present a certain decrease as the latitude increases (~15.9°C), while, during the May–September period and especially during the summer, lower absolute minimum air temperatures were observed in latitudes below 37°N, particularly in the eastern Mediterranean, where local conditions favour the appearance of low temperatures during summer.

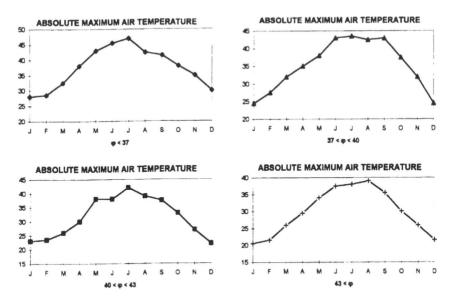

Figure 3.6 Monthly variation of the absolute maximum air temperatures of zones A, B, C and D

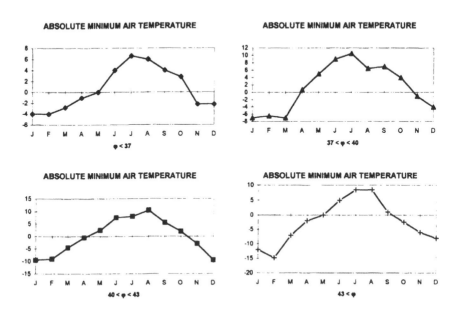

Figure 3.7 Monthly variation of the absolute minimum air temperatures of zones A, B, C and D

RELATIVE HUMIDITY

As has been proved, the conditions of human comfort do not depend only on the prevailing conditions of air temperature. Significant roles are also played by the prevailing atmospheric humidity and by the wind, which contributes to the modification of evaporation. At particular air temperatures, extremely high values of atmospheric humidity, as well as an extremely dry atmosphere, can create discomfort, mainly in the respiratory system.

The Mediterranean Sea is characterized by values of relative humidity, that, in general, do not create conditions of human discomfort. In addition, unlike other meteorological parameters, the regional distribution of the relative humidity is fairly uniform.

The annual variation of the relative humidity does not seem to be as simple as the variation of the air temperature. There mostly seems to be multiple variation, where the maxima and the minima are modulated by the influence of local factors. Generally, we note that, in the western Mediterranean, a minimum is observed during spring and summer months, while in the central and eastern Mediterranean it is observed in July and August. The main maximum in the western Mediterranean is observed in December, while in southern parts the maximum is observed to shift to January or February, when the depression activity in the area is at its maximum.

The mean annual value of the relative humidity ranges from 55% (Nicosia), to 74% (Gibraltar, Venice). The eastern Mediterranean, in general, has slightly lower values of relative humidity than the other regions. The values of the mean relative humidity of the warm season (June–September) are slightly lower than the mean annual ones, ranging from 72% (Gibraltar) to 41% (Nicosia). In the eastern Mediterranean, the values are lower than in the other regions and this difference is much more intense during the warm period.

In studying the annual variation of the relative humidity in the four zones A, B, C and D (Figure 3.8), we note that only in the January–March period is there a slight decrease of the mean monthly values of relative humidity as the latitude increases. In addition, in the July–December period, the relative humidity for zone B is slightly higher than that of zone A. Generally, the differences are slight and do not seem to be related to the latitude.

In the zonal annual pattern of the relative humidity, it is obvious that for zones A, B and C there is a single annual variation, with a maximum in December or January and a minimum in July or August. On the other hand, in high latitudes (zone D) the annual variation of the relative humidity presents a multiple variation related to the area's climate, which is not too similar to the Mediterranean type. Thus, in areas of high altitude, which have a far-from-Mediterranean climate because of the altitude, we often observe an opposite annual variation of relative humidity, with a maximum during summer months and a minimum during winter months.

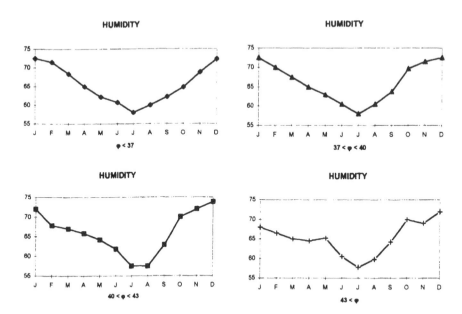

Figure 3.8 Monthly variation of the relative humidity (%) of zones A, B, C and D

CHARACTERISTICS OF THE WIND

The anomalous topography of the northern part of the Mediterranean basin, combined with the African deserts and the indented coasts, results in the development of a multiform wind field in the Mediterranean. This is directly influenced by the gaps between the mountain ranges. The wind is amplified when it is funnelled through the gaps, so that an initial wind field is formed where the basic winds are :

- The north-westerly Mistral, through the Alps–Pyrenees gap
- The north-easterly Bora, in the Adriatic area, which blows through the Trieste gap
- The easterly Levanter and the westerly Vendaval, through the Strait of Gibraltar
- The warm south-easterly to south-westerly wind that blows from Africa towards the Mediterranean and is best known as the sirocco, ghibli or khamsin.

In addition, as a result of the topography and the coastlines, significant anabatic or katabatic winds develop locally, while there is also a significant contribution from sea-breeze phenomena. While general consideration is not possible, the contribution of the local and seasonal (e.g. Etesian winds during summer in the Aegean Sea) winds is apparent.

The wind, with the temperature and the humidity of the air, plays an important role in human comfort in the exterior environment. In addition, proper building design in relation to the prevailing wind directions may contribute to natural cooling of the buildings. Studying the wind over a wide area, even within one particular country, is not possible because of the significant differentiations of this parameter from place to place. Thus, other than the general information regarding the gaps between the mountain ranges, which allow or even amplify the intensity of the traversing wind, it is not possible to study, in general, the annual variation of the wind intensity and direction as they relate to specific topics, such as cooling. In the central and western Mediterranean we record, more often in summer than in winter, low wind speeds, while in the Aegean Sea, the prevalence of the Etesian winds during summer months favours the development of high wind speeds.

The values of the reported wind speed are indicative only of the areas examined. On an annual basis wind speeds are higher in the eastern Mediterranean than in the rest of the region examined. The highest mean annual wind speed (6.6 m s^{-1}) was measured at Skiros (Aegean Sea), but it is known that in the southern part of the Aegean Sea we can measure even higher values. The lowest value (2.2 m s^{-1}) was measured at Palma (west Mediterranean). Similar results were found for the mean wind speeds of the warm season (June–September), where the highest value was measured in the south-east Aegean Sea (9.3 m s^{-1} at Rhodes) and the lowest value (2.8 m s^{-1}) at Valencia.

An effort was made to group the existing data into latitude zones (Figure 3.9), but it is better for comparison between the zones than for studying mean values. It is obvious, throughout the year, that the mean monthly speeds in zone B are lower than those in both zone A and zone C. Such a situation cannot be explained, given that the greater part of the Aegean Sea is located in zone B, where, as is well known, during the year there are higher wind speeds than in other areas. This observation must be attributed to local factors influencing the stations existing within this zone.

Finally, it is reported that between zones D and C there are no significant differences in mean monthly wind speeds.

CLOUDINESS

Theoretically, the percentage of the global radiation received by an area is directly related to the latitude. Thus, the annual values of the global radiation on a horizontal surface indicate that there is a certain increase in the received global radiation as the latitude decreases. It has been calculated that the increase per degree of decreasing latitude is about 0.3%. In spite of this, the actual received global radiation is significantly modified by local geographical and climatic factors. The most important of the climatic factors contributing to this modification, is cloudiness. Thus, for exam-

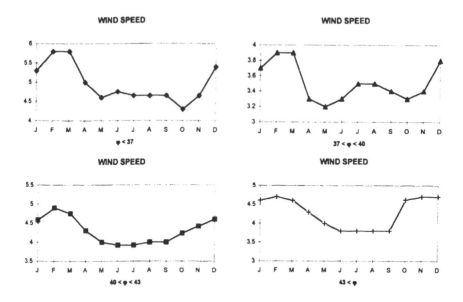

Figure 3.9 Monthly variation of the wind speeds, in m s⁻¹, of zones A, B, C and D

ple in mountainous areas, the development of lower cloud is frequent during the winter, while during the summer the development of convective cloudiness is also frequent, so that the attenuation of global radiation may reach 20%.

As is widely known. the Mediterranean region is characterized by light to medium cloudiness. Towards the northern boundaries of the Mediterranean basin, there is a heavier cloudiness because these areas are not too close to the Mediterranean climate. Thus, it seems that there is a positive correlation between cloudiness and latitude, which has as a result an even greater decrease of the global radiation in higher latitudes.

The main characteristics of the cloudiness in the Mediterranean region are:

- A single annual variation, with a maximum during winter months and a minimum during summer, is observed only in the southern parts of the east Mediterranean.
- The maximum of cloudiness in the west Mediterranean is observed early in spring, while in the Aegean Sea this is observed in January. In the remaining part of the Mediterranean a maximum is observed in December, relating to the maximum of the cyclonic activity that is observed during winter months in this area.
- A minimum of cloudiness in the whole Mediterranean basin is observed in July and August.

The mean annual cloudiness is heavier in the central and western Mediterranean than in the east, but the differences are very small. The maximum mean annual value of cloudiness (5.8/10) was observed in Ancona (similar values were reported at Venice and Trieste and generally in the north Adriatic). The minimum mean annual value was reported at Milos (3.5/10), where it was obvious that the central and southern Aegean Sea is characterized generally by light cloudiness and, as a result, the area has over 3000 hours of sunshine annually.

The cloudiness during the warm period is generally low, especially in the eastern Mediterranean where the minimum is observed in Rhodes (0.9/10). The maximum (4.5/10) is observed in the central Mediterranean (Rijeka) as well as in the western Mediterranean (Barcelona).

From the annual variation of cloudiness within each of the latitude zones (Figure 3.10) it can be seen that in the May–October period there is an increase in cloudiness with the latitude, because of the increase of cloudiness during the early afternoon hours, due to intense convective air movements. During November and December, cloudiness in all zones is almost the same, while during January and February there is a slight decrease, especially in zone D. Finally, during March and April there is no clear variation in the values of cloudiness with latitude.

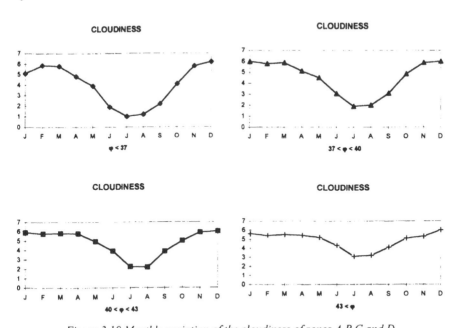

Figure 3.10 Monthly variation of the cloudiness of zones A,B,C and D

SUNSHINE

One safer way of indirect measurment of solar radiation is by evaluation through sunshine, which, unlike cloudiness, can be measured and not just estimated.

Processing the sunshine data from the 19 stations for which measurements of solar radiation exist, a statistically significant correlation between the logarithms of the annual number of sunshine hours and the latitude (correlation coefficient $r = -0.729$) was found.

The estimation of sunshine for the total number of shore stations in Greece, with the help of the regression equation

$$\ln y = 3.816 - 0.0099\varphi, \tag{3.3}$$

where y is the annual sunshine in hours, has indicated that the observed differences between measured and estimated values satisfied the null-hypothesis, that is to say, they are not statistically significant.

On the contrary, for other sites in Europe which are not in the on-shore Mediterranean zone, as for global radiation, it seems that it is not possible to evaluate sunshine through this equation.

The average annual values of sunshine of the stations examined range from 2077 h/year in Venice to 3041 h/year in Rhodes, while on the northern coasts of Crete the values are even higher. Generally, the annual number of sunshine hours in east Mediterranean is about 8% higher than that in the rest of the Mediterranean, which relates to the decreased cloudiness of the east Mediterranean compared to the central and west Mediterranean.

Both a simple annual variation of sunshine in the Mediterranean, with a maximum in July and a minimum in December or January, and a multiple variation, with a first maximum again in July, minimum in December or January and a secondary minimum in February or April, are observed.

The zonal study of the annual variation of sunshine indicates, initially, a flattening of the multiple variation in zones A, B and C but not in zone D, in which there is a secondary minimum in February. In addition, the decrease in the number of sunshine hours as the latitude increases is apparent only in the April–September period. During the remaining period, covering mainly the winter season, the sunshine hours in zone B are more than those in zone A during the January–March period and those in zone D are more than the equivalent ones in zone C during the December–January period (Figure 3.11).

The maximum increase from month to month in zones A, B and C is observed in the April–May period, while in zone D it is in June–July. The decrease rate becomes maximum between September–October and October–November.

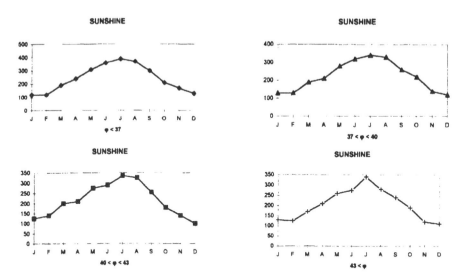

Figure 3.11 Monthly variation of the sunshine hours within zones A, B, C and D

METEOROLOGICAL MEASURES TO REPRESENT
ENVIRONMENTAL STRESS

Meteorologists and physicists have tried for well over four decades to simulate the physiological reactions of the body by analogy with physical reactions. Most designs follow the early experiment of Heberden. First was Hill's katathermometer, used originally in mines. The thermometer body is warmed to body temperature and the time for a specific temperature drop measured. This gives a quantity termed the 'cooling power', which represented the combined effects of ambient temperature, radiation exchange and wind on the thermometer bulb. The effect of the humidity factor on a sweating body can be partially simulated by placing a wet wick around the thermometer bulb.

This system was refined by Dorno and Thilenius in order to obtain continuous records. Their system keeps a metal bulb at a constant temperature by electrical heating. The current needed to maintain this temperature is a measure of the combined cooling effects of the environment. More complex systems, based on the same principle, but simulating the human body by using a dummy, have been designed and primarily used for testing in the laboratory the insulating quality of clothing.

It was an obvious step to try approximating the cooling power (dry or wet), usually given in mcal cm^{-2} s^{-1}, by the various basic meteorological elements. A large number of different formulae was the result. Most of them represent the dry cooling power based on air temperature and wind speed and nearly all of them give cooling power as a linear function of the difference between skin or body temperature and the air temperature, together with a power of the wind speed. In a majority of cases this power is the square root. Cena et al. [2] have performed a most useful service by comparing a large number of these empirical formulae with frigorimeter and katathermometer observations.

The general form of these empirical formulae is

$$H = (a + bv^c)(t_k - t)$$

where H is the cooling power calculated by the various formulae, in which a, b and c are constants; v is wind speed in m s^{-1}, t_k is the body temperature or skin temperature and t is the air temperature.

There are several heat-loss formulae not included in the above mentioned tabulation. One is from Vinje [9]:

$$H = 0.57v^{0.42}(36.5 - t) \times 36 \text{ kcal m}^{-2} \text{ h}^{-1}$$

where v denotes the wind speed in m s^{-1} and t is the air temperature in °C.

This is an alternative to an earlier formula by Siple and Passel [3], which was primarily evolved to establish the wind chill at very low temperatures, essentially for temperatures below freezing. That formula, with skin temperature as a basis, originally gave:

$$H = (\sqrt{100v} + 10.45 - v)(33 - t).$$

The constants were later revised to read:

$$H = (10.9\sqrt{v} + 9 - v)(33 - t).$$

Recent observations in the Antarctic have shown that frostbite ensues with a wind chill index of 1400, according to Vinje's formula, and fingers will freeze when the index exceeds 1500 kcal m^{-2} h^{-1}.

Because thermal quantities mean little to the layman, attempts have been made to express the wind chill at low temperatures as an equivalent temperature by simply transposing the calculated value to an equivalent chilling with zero wind speed.

So far all the formulae given for cooling power refer to dry conditions, i.e. without sweat on the skin. Wet cooling powers have been represented by

$$H' = (0.37 + 0.51v^{0.63})(36.5 - t') \text{ mcal cm}^{-2} \text{ s}^{-1}$$

where t' is the wet-bulb temperature.

Cooling powers have been related to sensation scales. These can be classified according to one system, for dry cooling values as shown in Table 3.2.

Recently, Flach and Morikofer [4] have provided a detailed climatological analysis of frigorimeter measurements at six middle-latitude localities with rather different climatic conditions. In their assessment they indicate that, for the sensation

Table 3.2 Cooling powers and sensation scales

H, mcal cm^{-2} s^{-1}	Sensation
<5	Hot
5-10	Pleasant or Mild
11-15	Cool
16-22	Cold
23-30	Very cold
>30	Extreme cold

of 'cold', wind is the strongest contributing factor, temperature the least effective, with the long- and short-wave radiation fluxes taking an intermediate position.

The two factors already discussed, namely effective temperature and strain index, are distinctly more discriminating at the warm end of the scale than at the cold end. A large number of variants of the effective temperature have been proposed. One of these, which has the virtue of simplicity, is the discomfort index (DI) of E. C. Thom [5, 6]. It is given by

$$DI_f = 0.4(t_d + t_w) + 15 \qquad\qquad\qquad\qquad\qquad\text{(a)}$$

$$DI_c = 0.4(t_d + t_w) + 4.8 \qquad\qquad\qquad\qquad\qquad\text{(b)}$$

where t_d and t_w are the dry- and wet-bulb temperatures respectively given in (a) Fahrenheit and (b) Celsius degrees. Over a considerable range the values of the discomfort index are essentially identical to the effective temperature (ET), between 65 and 88 ETF (18.3–31.1 ETC). Thom was primarily concerned with devising a scheme which would permit ready calculation of air-conditioning needs on the basis of available climatic data. Degree-day sums for DI_f values above 60 ($DI_c = 15.6$) were used to estimate relative monthly cooling requirements.

Other indices based on Thom's pattern, but lacking the relation to effective temperature, include the following:

- An index by Kawamura (7), given by

$$DI_{(K)} = 0.99t + 0.36t_{dp} + 41.5$$

where t and t_{dp} are air and dew-point temperatures, respectively. Kawamura found this index useful in distinguishing regional differences of summer climate in Japan.
- An index by Tennenbaum et al. [8], which is defined as

$$DI_{(T)} = (t_d - t_w)/2$$

where t_d and t_w are dry- and wet-bulb temperatures. According to these authors discomfort starts for $t_d = t_w = 24°C$, i.e. a temperature of 24°C with 100% relative humidity.

- Landsberg [9] defines a biological temperature t_b in which he tries to apply temperature equivalents of wind and humidity. His index is expressed by the following set of equations:

$$t_b = t - \Delta t_v - \Delta t_e$$

where

$$\Delta t_v = \frac{1}{2}v - \frac{t}{10}\left(\frac{v}{10}+1\right)+3\left(\frac{e}{10}-1\right)$$

and

$$\Delta t_e = \frac{1}{T_K}(30-1.5E)(U-30)$$

where t is air temperature in °C, v wind speed in m s^{-1}, T_K temperature in Kelvin, e prevailing vapour pressure in mbar, E maximum vapour pressure at t in mbar and U relative humidity in per cent.

This complex variable is represented as corresponding to sensations of the human skin as the receptor organ.

- Several authors have particularly attempted to represent the sensation of sultriness by vapour pressure alone. Scharlau [9] originally fixed a value of 18.8 mbar (14.1 mm Hg) as the limit of sultriness. This obviously establishes a gradient to the maximum vapour pressure at skin temperature which is about 60 mbar (46 mm Hg). Marinov (1964) [9] discussed this gradient but fixes the comfortable vapour pressure at considerably higher values (23–33 mm Hg). These would perhaps be better interpreted as tolerable rather than comfortable values. Marinov defines a transfer coefficient:

$$W = 2.1\frac{\lambda}{D}C(\text{Re})^m(E_B - e)$$

where λ is the coefficient of thermal conductivity, D body diameter, Re the Reynolds number, E_B the saturated vapour pressure at skin temperature, e the prevailing vapour pressure and C and m are constants (values not given in reference).

- One of the psychological drawbacks of the effective-temperature concept is the fact that the values are lower than the dry-bulb temperature. Several authors have tried to overcome this by establishing a different base. Thom [5] proposed accomplishing this by transforming the ET lines based on

100% relative humidity to a new base of 30% relative humidity. This results in making the effect of the humidity factor additive to the temperature. Thom gave this index the label 'sentient temperature'.

- Another fairly empirical index, first proposed by Lally and Watson [10], labelled 'humiture' and renamed 'humidex' by the Canadian Meteorological Service (Thomas [10]), has the simple form

$$Hi = t + h$$

where $h = e - 10$ with t as air temperature in °F and e vapour pressure in mbar. This has been used in climatological analysis at Toronto, where a five-year interval showed that Hi equalled or exceeded the 100 value for 130 hours.

At the conclusion of this brief look at all the hygrothermal indices for biometeorological purposes it may be noted that the effective temperature (ET) concept is simple and that the other indexes are not clearly superior in use. Many have, of course, a better physical or physiological basis. One might perhaps add that, for outdoor conditions, temperature equivalents of factors not included in ET may provide a significant improvement.

Gregorczuk [7] noted that net effective temperatures even for negative Celsius air temperatures are a meaningful quantity, corresponding to sensations of 'cold'. He evaluated this in terms of the Missenard formula:

$$NET = 37 - \frac{37 - t_a}{0.68 - 0.0014\gamma + \dfrac{1}{1.76 + 1.4v^{0.75}}} - 0.29 t_a \left(1 - \frac{\gamma}{100}\right)$$

where t_a is the air temperature in °C, v is the wind speed in m s^{-1} and λ the relative humidity.

A human comfort diagram based on a construction of Olgyay [9] has a similar coordinate net. It is shown in Figure 3.12, reproduced here from Newman [9]. In this temperature/relative humidity diagram zones of comfort are shown which take some wind and radiation effects qualitatively into account. At the lower temperatures wind would, even in the outlined 'comfort area', undoubtedly cause discomfort, but otherwise this seems to be a simple approach that may be useful for quick assessment of meteorological conditions for outdoor activities.

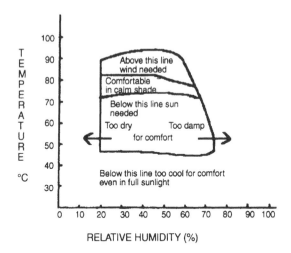

Figure 3.12 A temperature–humidity climatic comfort diagram

COOLING DEGREE DAYS

Apart from the changes of the temperature in the exterior environment, the changes of the air temperature within buildings are of special interest. For the evaluation, though, of the energy needs for maintaining interior environments in conditions of human comfort, the observed changes of the indoor temperatures are essential. For such studies, scientists have, eventually, decided to use a quantity, which may not have physical meaning but is offered for the calculation of the energy needs of a building. This quantity is, in the case where we want to cool a building during the warm period, the cooling degree days (DD), as they are called. The cooling degree days for a month are defined as the sum of the positive deviations of the mean daily temperatures from a base value, the temperature we want to maintain constant in the interior of a building. The base value is defined in terms of our demands, but does not usually exceed 25°C.

Interest in the evaluation of heating and cooling degree days, especially in the European region, is very recent, so that the results are not yet widely known. For this reason, we will refer to the existing studies relating to the cooling degree days of Athens (Greece), which were studied after the heatwaves of the summers of 1987 and 1988, when a rapid increase in the installation of air conditioners was reported.

In Table 3.3 the average number of DD for temperature bases of 25°C and 28°C are given for three stations in the greater Athens area.

Table 3.3 Mean number of cooling degree days for the bases of 25°C and 28°C

	June	July	August	September
BASE 25°C				
N.O.A.	29.9	84.1	78.3	15.2
Hellinikon	19.2	69.1	69.8	11.0
N.Philadelphia	24.1	71.2	64.6	9.4
BASE 28°C				
N.O.A.	4.4	20.8	17.9	1.6
Hellinikon	1.4	13.6	14.2	0.5
N.Philadelphia	2.9	15.4	11.9	0.5

From Figure 3.13 it is obvious that the DD number presents a satisfying linear correlation with the respective values of the mean monthly temperatures, especially during July and August.

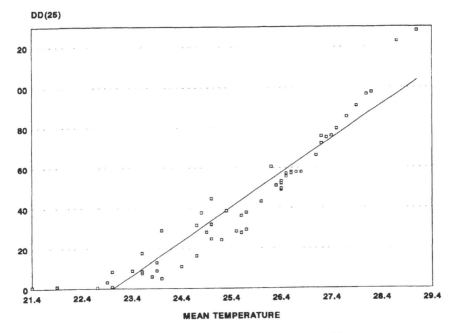

Figure 3.13 Linear correlation between DD and mean monthly temperatures

A more precise approximation of the calculation of the energy needs can be obtained if, instead of the DD, degree hours (DH) are used. This presupposes the existence of hourly measurements of air temperature.

In Athens during the warm period the relative frequency of hours (HTx) having a temperature over a defined base shows an increase in HTx from the first to the third ten-day period of June. An important increase is also observed during the second ten days of July and the first ten days of August. The last twenty days of August, as well as of September, are characterized by a continuous decrease in THx (Figure 3.14).

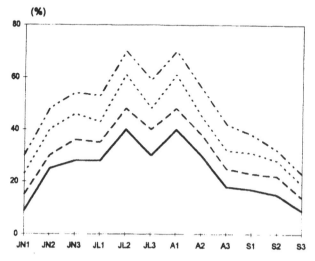

Figure 3.14 Ten-day frequencies of the HTx values. Series 1: T > 25°C (———);
Series 2: T > 26°C (– – –); Series 3: T > 27°C (- - - - -); Series 4: T > 28°C (– - - –)

Night-time temperatures over 25°C were measured during June, July, August and September. The percentages of night-time temperatures over 28°C are nearly zero for June, August and September and only about 5% for July.

As was previously reported for DD, there is a certain linear correlation between the monthly number of hours with temperatures over a certain base and the respective DH values. The average values of DH for the months of the warm period for Athens are presented in Table 3.4.

Table 3.4. Mean number of DH for the 25, 26, 27 and 28°C air-temperature bases

	June	July	August	September
$T > 25°C$	957	1370	1111	387
$T > 26°C$	589	1283	1023	332
$T > 27°C$	523	1204	950	280
$T > 28°C$	372	958	729	181

In order to provide algorithms for the estimation of the cooling degree hours, regression analysis has been used. It was found that the monthly cooling degree hours can be estimated accurately by a multi-linear regression equation, in which DH appears as a function of mean and mean maximum monthly air-temperature values.

If mean air temperature values are not available, a simple linear regression analysis, involving only the use of the monthly mean maximum air temperatures, can be used with significant accuracy for all the summer months. These methods can be applied successfully to any coastal location of the Mediterranean area, but in each case a separately 'local' regression analysis should be applied.

Finally, it is found that the cooling degree days present a strong linear correlation with the corresponding values of the cooling degree hours. Therefore it is possible, using the monthly values of the cooling degree days, to estimate the values of the cooling degree hours.

THOM'S DISCOMFORT INDEX

If we want to focus on issues of human comfort we could refer to some of the proposed equations presented above.

For Greece and especially for Athens, the discomfort index proposed by Thom [11] has been calculated. Table 3.5 shows the mean monthly number of hours with DI > 24°C and DI > 26.7°C, where for values of DI above 24°C 50% of the population feels discomfort and for DI > 26.7°C the discomfort is very strong and dangerous.

Table 3.5 and Figure 3.15 show that the frequencies of hours per day with DI > 24°C and T > 28° C are very high for the warm period. Also high hourly values of DI follow the pattern of the frequencies of hours with T > 28°C.

Table 3.5 Mean number of hours per day with DI>24°C and DI>28°C for Athens

	June	July	August	September
DI>24°C	4.4	8.3	8.0	3.1
DI>28°C	0.4	1.5	1.0	0.1

REFERENCES

1 Commission of the European Communities (1985).*Test Reference Years TRY Weather Data Sets for Computer Simulations of Solar Energy Systems and Energy Consumption in Buildings.* 48pp.
2 Cena, M., M. Gregorczuk and G. Wojeik (1966). 'Proba Wyznaczenia wzoru do obliczania achladzania biometeorologiczenego warunkach klimatycznych Polski' ('An

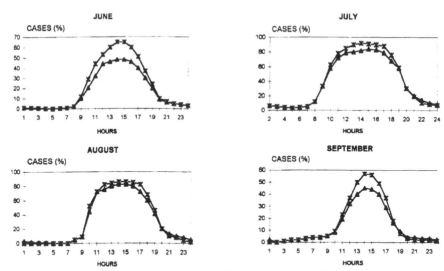

Figure 3.15 Frequency distribution of hours with DI > 24°C and T > 28°C

attempt at formula determinations for computation of biometeorological cooling power in Poland) *Roczniki Nauk Rolniczych*, Vol. 119D, pp. 137–148.

3 Siple, P.A. and C.F. Passel (1945). 'Measurements of dry atmospheric cooling in subfreezing temperatures', *Proceedings of the American Philosophical Society*, Vol. 89, pp. 177–199.

4 Flach, E. and W. Morikofer (1966). *Comprehensive climatology of cooling power measured with the Davos frigorimeter*, Pt I (1962), 71pp., Pt II (1965), 34pp. Pt III (1966) 28pp.

5 Thom, E.C. (1957). 'A new concept for cooling degree days', *Air Conditioning, Heating and Ventilation*, Vol. 54, No. 6, pp. 73–80.

6 Thom, E.C. (1958). 'Cooling degree days' days', *Air Conditioning, Heating and Ventilation*, July, Ref. sec. pp. 65–72.

7 Landsberg, H.E. (1972). 'The assessment of human bioclimate. A limited review of physical parameters', WMO No 331, Tech. Note No 123, World Meteorological Association, Geneva.

8 Tennenbaum J., E. Sohar, R. Adar and T. Gilat (1961). 'The physiological significance of the cumulative discomfort index', *Harefuah Journal of the Medical Assocciation of Israel*, Vol. 60, No. 10, pp. 315–319.

9 Landsberg, H.E. (1977). *World Survey in Climatology*. Vol. 5. *Climates of Northern and Western Europe*. Elsevier, Amsterdam.

10 Thomas, M.K. (1965). 'Humidex at Toronto', Met Branch Canada, CDS No 16-65, 5 pp. (mimeo).

11 Thom, H.C.S. (1957). 'Problems in estimating costs for air conditioning – The climate factor', *2nd Technical Conference of the National Warm Air Heating and Air Conditioning Association Proceedings*, Cleveland, pp. 19–24.

ADDITIONAL READING

1 Air Ministry Meteorological Office (1962). *Weather in the Mediterranean*, Vols I and II. HMSO, London.
2 Katsouyanni, K., A. Pantazopoulou, G. Touloumi, I. Tselepidaki, C. Moustris, D. Asimakopoulos, G. Poulopoulou and D. Trichopoulos (1993). 'Evidence for interaction between air-pollution and high temperature in the causation of excess mortality', *Archives of Environmental Health*, Vol. 48, No. 4, pp. 235–242.
3 Marinov, V. (1964). 'A new method for estimating the thermal conditions and zone of comfort in human organism' (translation of title) *Kurortol i fiziater*, No 2, pp. 22–25.
4 Tselepidaki, I., M. Santamouris, C. Moustris and G. Poulopoulou (1992). 'Analysis of the summer discomfort index in Athens, Greece', *Energy and Buildings* Vol. 18, pp. 51–56.
5 Tselepidaki I. and M. Santamouris (1993). 'Statistical and persistence analysis of high summer ambient temperatures in Athens for cooling purposes', *Energy and Buildings*, Vol. 17, No. 3, pp. 243–251.
6 Tselepidaki I. and M. Santamouris (1993). 'Analysis of summer climatic conditions in Athens, Greece for cooling purposes', *Proceedings of an International Conference of CEC 'Solar Energy in Architecture and Urban Planning*, pp. 122–125.
7 Tselepidaki I., M. Santamouris and D. Melitsiotis (1993). 'Analysis of the summer ambient temperatures for cooling purposes', *Solar Energy*, Vol. 50, No. 3, pp. 197–204.

4

Microclimate

CLIMATE AND MICROCLIMATE

The atmosphere's conditions at any given time are descibed as the *weather*. *Climate* is the average of weather conditions over an extensive period of time, specified over a large region. The principal elements that characterize climate and weather are the air temperature, solar radiation, moisture (humidity and precipitation) and winds [1, 2]. Variations in the amount, intensity and spatial distribution over the earth of the weather elements create a variation of climate types, for example tropical, temperate and arctic climates.

Within a particualr region, deviations in the climate are experienced from place to place within a few kilometres distance, forming a small-scale pattern of climate, called the *microclimate*. Microclimate is affected by the following parameters: topography, soil structure, ground cover and urban forms.

Urban forms can also modify the climate of a city, creating the *urban climate*, and differentiate it from the climate of the surrounding rural areas. The space created by the urban buildings up to their roof level is referred to as the '*urban canopy*' [3] and the volume of air surrounding the city, which is affected by it, is the urban boundary layer, referred to as the '*urban air dome*'. The variation in the height of the buildings and in the wind speed alters the upper boundaries of the urban canopy from one spot to another. The energy needs of buildings and those of human comfort are affected by the climatic conditions prevailing within the urban canopy, which can be different from those in the urban air dome.

From the design point of view, designers should appreciate that the climate of the region is the one that drives microclimate and urban climate. Thus, a systematic analysis of the climate should be the first step in the design of buildings, in order to ensure the building's climatic response. Adjustments in the design should then be considered to reflect local microclimate patterns.

TOPOGRAPHY AS CLIMATE MODIFIER

Climate varies with latitude, since the amount of solar radiation received changes. The unequal distribution of solar radiation over the earth and the fact that continents heat up and cool down more rapidly than oceans, result in a variation of different climate types. Terrain features, such as mountains, valleys and lakes, affect the climatic conditions of a region. Mountains, due to their height impose a different climate than lowlands. In the same area, small differences in terrain can create remarkably large modifications in the microclimate. On a mountain, the climate is different from one side to the other and at different heights as well. Proximity to the sea has also a profound effect on the climate of a region. Deviation in climate from site to site plays an important role in selection of favourable settlement locations. The main climatic elements affected by the topography are described below.

Temperature

Since temperature normally decreases with increasing altitude, 0.65°C for every 100 m in the atmosphere [1], mountains and plateaux are generally characterized by lower temperatures than lowlands. On mountains the temperature drop can be approximately 1°C for each 56 m height rise during summer, and for each 68 m in winter [4]. As cool air is heavier than warm air and heat radiating from the ground during the night causes a cold-air layer to form near the surface of the ground, 'cold islands' are created at the lowest points. This is more profound in valleys where the flow of cool air slides down the valley slopes. Thus late in the evening, the valley floor will be very cold and a plateau will be cold. However, the higher slopes of the hill will be warm, as the low circulation mixes the ground layer with neigh-bouring warm air. As a result, mountain slopes experience smaller temperature changes than the lowlands, while plateaux have higher temperature ranges than mountains (Figure 4.1).

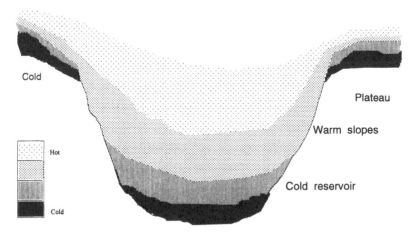

Figure 4.1 Temperature stratification on a mountain slope

Temperature changes over water tend to be more moderate than over land be-cause water, owing to its high capacity, heats up and cools down more slowly than land. Thus, the further one gets from water surfaces, the more extreme are the temperatures.

Radiation

The amount of solar radiation received on the earth varies with latitude. At the same latitude, the amount received on a surface changes with its inclination and orientation. In the Northern hemisphere, south-facing slopes receive more direct sunlight and have a warmer climate than those facing north (Figure 4.2). As an inclined surface receives more solar radiation during winter and less during summer compared to a horizontal one, sites on slopes receive larger amounts of solar radiation during underheated periods and less at overheated periods than a horizontal site. Thus, the same amount of solar radiation received on a southern slope of a mountain will be received on a horizontal level a few weeks later during winter. This results in a shift of the arrival of the seasons at different locations in the same region.

Figure 4.2 Irradiation on south and north hill slopes

Wind

Wind is the result of a pressure difference between two areas, the movement of air being from the area of higher pressure towards the area of lower pressure. Topography influences the wind pattern and the wind speed. Because of friction with the ground, wind speeds are usually less over land than over the sea, and they increase with altitude. As mountains and hills act as obstacles to the movement of air, wind flow is diverted by them in both its horizontal and vertical pattern (Figure 4.3). This results in higher speeds near the top of the mountain or the hill on the windward side and less turbulence on the leeward slope. However, mountains also create local winds that vary from day to night. During the day, the air next to the surface heats up faster than free air at the same height and thus warm air moves up along the slopes. During the night, as mountain surfaces cool down by radiation faster than the free air, cold breezes are formed and slide down the slopes of the mountains (Figure 4.4).

This phenomenon can be intense, especially in narrow valleys, which can experience strong upward winds along the valley floor during the day and down the valley at night.

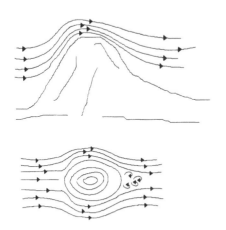

Figure 4.3 Air pattern around a mountain

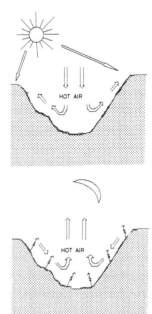

Figure 4.4 Air pattern in a valley throughout the day

Similar but reverse wind patterns between day and night occur near large water surfaces. During the day, the air is hotter over land (low pressure) than over water surface (high pressure) and the resultant pressure difference imposes a breeze, called 'sea breeze', from the sea towards the land (Figure 4.5). Seaside areas benefit from sea breezes, which can lower the temperature up to 8.3°C on a hot day [2]. At night, the air flow reverses but the established breezes are weaker because the temperature difference between land and water is smaller than during the day. Late in the afternoon and early in the morning there is no breeze, as land and water are at approximately the same temperature.

Precipitation

Precipitation (e.g. rain and snow) in an area can be affected by the existence and size of the surrounding hills and mountains. Windward sides of hills experience less precipitation than leeward sides as precipitation on the windward side of a hill is carried over it by the wind as it strikes the slope so that the precipitation falls on the leeward side (Figure 4.6). High mountains on the other hand cause reversed precipitation distributions. Air gets colder as it is forced to ascend on the windward side owing to the volume of the mountain. When the air temperature drops, the relative humidity increases until it reaches saturation point and any further reduction in temperature results in condensation of the moisture in the form of clouds, rain and snow. On the leeward side, the descending air is warmed as it falls down the slope and consequently its relative humidity decreases, allowing the air to suck in moisture rather than release it (Figure 4.7). Thus, the windward side has a higher rate of rain than the leeward and mild winters are experienced at sites positioned at the bottom of the leeward side of high mountains.

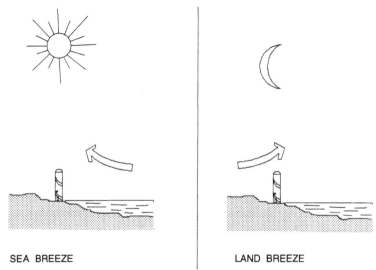

SEA BREEZE   LAND BREEZE

Figure 4.5 Sea and land breezes

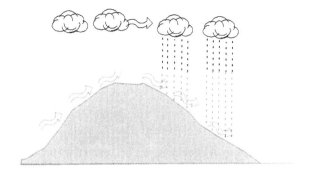

Figure 4.6 Wind and precipitation over a hill

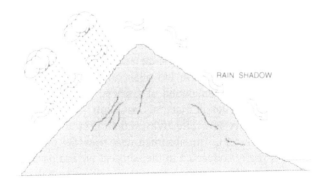

Figure 4.7 Wind and precipitation over a mountain

URBAN FORMS AS CLIMATE MODIFIERS

The climate of cities – *urban climate* – shows considerable differences from that of the surrounding countryside, as big urban areas are warmer than the surrounding suburban and rural areas. This temperature difference between the city and its surroundings is known as the '*Urban Heat Island*' [5] and is responsible for a summer temperature increase of 1 to 4.5°C in cities [6] (Figure 4.8). Urban forms, the structure of the cities and the heat released by vehicles are the main causes of this phenomenon.

The difference in climate between urban areas and countryside is greatly affected by the way the received solar radiation is treated. Countryside is generally characterized by large amounts of green. A large proportion of the solar radiation falling on

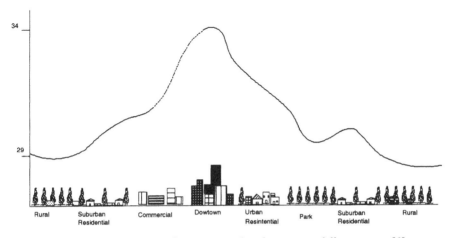

Figure 4.8 Representation of temperature distribution over different areas [6]

plants, which are characterized by high solar absorptivity – about 80% – is used for evapotranspiration, which lowers the temperature of the surrounding air and releases moisture. In addition, solar radiation absorbed by the earth is partly used for evaporation of its moisture, a process that keeps the earth's surface and the air in contact with it at moderate temperatures. On the other hand, cities are characterized by reduced green areas and an accumulation of artificial materials, which are of high absorptive properties. Construction materials have low solar reflectance, 'albedo'; for example, asphalt has very low reflectivity and absorbs almost all the solar radiation falling on it (Figure 4.9). The absorbed heat from the material increases its temperature, which is partly convected to the adjacent air and partly radiated to the surroundings. Landsberg [7] referred to observations where asphalt surfaces reached 51°C for a 37°C air temperature, while Gajago [8] reported that for a maximum ambient temperature of 28°C, asphalt surfaces reached 47°C while dry grass reached only 39°C.

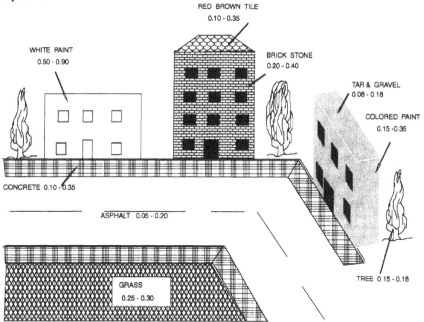

Figure 4.9 Solar reflectance of different materials

The solar radiation absorbed and stored in the structure of buildings is radiated back to the environment after late afternoon. If the buildings are of approximately the same height, the radiant losses from the roofs of the buildings are of about the same magnitude as the losses in an open area. However, if the buildings are of different heights, the higher ones block the sky and reduce the amount of solar reflection and long-wave radiation from the lower buildings. During the night, the vertical surfaces radiate less heat to the sky as a result of their restricted view of the sky dome. The radiated heat from the vertical surfaces of buildings and its repeated reflection from opposite facing surfaces traps part of the heat inside the urban atmosphere, the 'Urban Canopy'. Thus, the denser the urban area with buildings of unequal height, the slower the radiative cooling rate.

In addition, tall buildings in cities, known as urban canyons, block the flowing winds, in this way preventing the removal of the built-up heat inside them. The overall wind speed in cities is about 25% lower than in open areas although in some cases, tall buildings create local turbulence with very high wind speeds [9].

Additionally, air pollution in cities creates a layer that blocks the night heat radiation to the sky dome, contributing to the enhancement of the 'heat island' phenomenon [10]. Air pollution and heat from cars and other mechanical equipment contribute to heat islands during winter. During summer, solar radiation is so intense that it overcomes the exhausted heat. The severity of the summer heat island is mainly caused by solar radiation and the structure of the urban areas [6].

The heat-island phenomenon is more intense during the night, when buildings and ambient air remain hot, resulting in thermal stress in people. The greatest air temperature differences between urban and suburban areas are experienced during evenings, when the air temperature in cities has been observed to be higher by 5°C and sometimes even by 8–10°C [11].

Thus, generally the climate in cities is characterized by [12]:

- high ambient temperatures even during evening hours;
- reduced relative humidity, due to high air temperature and lack of sources of humidity;
- disturbed wind patterns with a reduction of air speeds which contribute to an increase in air pollution;
- reduction of the received direct solar radiation and increase of diffused radiation, owing to pollution particles in the atmosphere.

CLIMATE AND DESIGN REQUIREMENTS

The different climate types impose a variation in the required thermal comfort conditions, as the physical and physiological causes of climatic stress differ from one type of climate to another. This creates a variation in urban design objectives and in the design and structural characteristics of buildings. In addition, the layout and structure of a settlement affect the climate of the area and can even modify it through proper design, thus improving the thermal comfort conditions both outside and inside buildings and even reducing their energy demands for heating and cooling. As the design objectives differ from climate to climate, regional architecture exhibits variations from one region to another to provide a better response of buildings to the existing climate conditions.

In regions with hot summers, control of the ambient temperature, and even lowering it, is desirable. Shade for the pedestrians and the minimum solar exposure of the buildings to the sun should be ensured.

In hot and dry climates, shading is of greater importance than ventilation. The settlements are characterized by a dense layout with the main public spaces embraced by the buildings, in order to provide shading to pedestrians and buildings (Figure 4.10). Often multistorey buildings are built in narrow streets to create the shade desirable for the streets and the buildings. Special features, such as arcades, are also used in wider streets to provide shade. Buildings in many cases are arranged around courtyards, which are used to provide ventilation and shade to adjacent spaces. The walls are of massive construction, such as stone, adobe and brick, to delay the transport of heat into the buildings and to act as a heat sink that cools down during the night and accumulates the heat during the day. Domed roofs are quite common in this type of climate, as they offer a larger area, compared to a flat roof, for heat storage during the day and heat radiation to the sky during the night. Buildings have few and small windows, for reduction of direct solar gains, but which are adequate for internal daylighting. Shutters are used on the windows; these are kept closed in the hours of high solar intensity and opened during evening hours in order to allow night ventilation. External surfaces are painted in light colours, to reduce the absorption of solar radiation. In areas with very low humidity, use of fountains, pools, water streams and plants is desirable to provide evaporative cooling, which can also be combined with cross ventilation of buildings. In this type of climate we also find settlements of underground dwellings (Tunisia, Greece, Turkey, China, etc.) since these types of settlements are kept protected from the solar gains and are using the earth's cooling potential.

In hot and humid climates on the other hand, emphasis is given to cross ventilation as the high humidity of the air creates discomfort for human beings. As air movement over the skin increases the rate of evaporative cooling and offers relief, cross ventilation of streets and buildings is desirable in order to reduce discomfort from excessive humidity. The fabric of the settlements becomes scattered and loose, as the buildings are positioned at a distance apart from one another, in order to channel the winds through the streets and inside the buildings. Although temperatures are not very high, shading is still desirable. Trees along the streets and arcades on the outside of buildings provide shaded walking routes. Buildings have many and big openings, with large overhangs and covered verandas. The ceilings are high to allow space for high windows and stratification of the interior air, with openings in many cases on the ridge of the roof for exhaustion of the hot air. As there is not a big diurnal temperature difference, the structure of the buildings is lightweight.

Figure 4.10 Traditional Mediterranean architecture

Although design strategies during winter – minimization of wind speed, increase of solar exposure – are opposite to those for summer, advantage can be taken of seasonal changes in the sun-path and in the direction of the prevailing winds. The sun moves in a lower orbit during winter and winter winds in many regions usually blow from a different direction than in summer; for example, north winter winds in contrast to south summer breezes.

REFERENCES

1 Trewartha, G.T. and L.H. Horn (1980). *An Introduction to Climate*, 5th ed., McGraw-Hill Book Company, New York (1st ed. 1937).

2 Koeppe, C.E. and G.C. Long (1958). *Weather and Climate'*. McGraw-Hill Book Company, New York.

3 Oke, T.R. (1976). 'The distance between canopy and boundary layer urban heat island', *Atmosphere*, Vol. 14, No 4, pp. 268–277.

4 Olgyay, V. (1963). *Design with Climate*, Princeton University Press, Princeton, NJ.

5 Chandler, T.J. (1976).'Urban climatology and its relevance to urban design', Technical Note No. 149, World Meteorological Organization (WMO), Geneva.

6 Akbari, H., S. Davis, S. Dorsano, J. Huang and S. Winnett (eds) (1992). *Cooling our Communities; A Guidebook on Tree Planting and Light-Coloured Surfacing*. US Environmental Protection Agency, Office of Policy Analysis, Climate Change Division.

7 Landsberg, H. (1947).'Microclimatic research in relation to building construction', *Architectural Forum*, pp. 114–119.

8 Gajago, L. (1973). 'Outdoor microclimate and human comfort', C.I.B. Conference on 'Teaching the Teachers on Building Climatology', Stockholm.

9 Lechner, N. (1991). *Heating, Cooling, Lighting: Design Methods for Architects*. John Wiley & Sons, New York.

10 Fuggle, R.F. and T.R. Oke (1970). 'Infra-red Flux Divergence and the Urban Heat Islands'. In *World Meteorological Organization* (WMO), pp. 70–79. World Meteorological Organization, Geneva.

11 Givoni, B. (1989). 'Urban Design in Different Climates', WMO/TD-No 346. World Meteorological Organization, Geneva.

12 Evmorfopoulou, C. (1988).'Energy savings in buildings with use of plants on the roof and facades', 3rd National Conference on 'Soft Kinds of Energy', Thessaloniki (Greece) (in Greek).

5

Urban design

STAGES OF URBAN DESIGN

Cities are increasing in size very rapidly, and it is estimated that in the future most of the world's population will be living in urban areas. The industrialization, the concentrated activities of the population in the cities and the rapid increase of motor traffic are the main contributors to air pollution and deteriorating environmental and climatic quality. Until now, as Cook [1] emphasizes, the main efforts of bioclimatic planning have been focused on the single building and little attention had been paid to the whole urban area.

Quality of life in urban areas can be improved if *air quality* and *climate quality* can be ensured. Air quality is related to control of the air pollution, resulting from the activities occurring in the urban areas. Climate quality, as Bitan [2] says, means that, by correct usage of different climatological elements and their correct integration at different levels of planning and building design, improvement of local climates could be ensured, for instance in urban built-up areas, and even indoors. Urban areas with poor climatic quality use a lot more energy for air conditioning in summer and for heating in winter, as well as more electricity for lighting, than areas with higher climatic quality.

Design strategies for improvement of thermal comfort conditions during summer should focus on :

- solar protection of buildings and pedestrians;
- reduction of outdoor air temperature;
- wind enhancement, for both pedestrians and the ventilation of buildings;
- improvement of humidity levels.

Design initiatives should aim to:

1   provide protected spaces for outdoor activities which both improve outdoor comfort conditions and promote indoor comfort; and
2   contribute to reduction of the cooling load and to a lesser dependence of buildings on air conditioning.

The design of urban areas should start from a decision taken concerning the *selection of the location* for a new city or expansion of an existing one and the whole procedure could be carried out in the following stages [3]:

- **Urban planning**, where actions are taken on the scale of a whole city or greater;
- **Urban morphology**, where actions are taken for groups of buildings and the spaces between them;
- **Building design**, where individual buildings are treated.

The overall design strategy should be a compromise of measures for both summer and winter thermal requirements.

Site location

The selection of the site is very important for the thermal comfort of a settlement. Correct site selection for a new settlement or for new large housing developments can prevent environmental problems and improve the microclimate of the development. According to the specific climatic conditions of the region and the activities that are planned to be housed in the development, different topographic locations should be selected. Although in cold climates the desirable sites are the protected ones, in overheated climates exposed ones are preferred. This is because exposed locations offer undisturbed linkage with natural cooling resources, such as the sky, air and the ground. Sites affected by natural hazards, such as floods, should be avoided. In addition, the location of a site should be carefully chosen in relation to pollution sources (heavy industry, etc.) and the direction of prevailing winds.

The direction of prevailing winds is a very important parameter in the site selection process, as winds can transfer pollution particles from the sources (industry, city centre) to other sites lying in their path. Especially in hot climates, the effects of air pollution can be very intense if the topography of the area helps temperature inversion, which traps the pollution and, combined with the intense solar radiation, can cause the formation of photo-chemical smog.

As far as cooling needs are concerned, sites with low summer insolation and good ventilation conditions, especially during the evening and the night, should be considered. North- and east-facing slopes get less direct solar radiation than west and south slopes. During summer, a north slope is the coldest and shadiest, while a west slope is the warmest.

The altitude of a site and its distance from the sea affect its air temperature. Air temperature decreases with altitude, at a rate of 0.65°C for every 100 m [4]. Sites with proximity to sea or a lake experience lower air temperature, especially during daytime, and smaller diurnal air temperature differences during summer, than sites on the mainland. In addition, these places benefit from daytime sea breezes and night-time land winds. Sea breezes are associated with high humidity, which can be a problem, especially in humid climates, but in most cases the problem is balanced by the beneficial effect of the increased wind speed on human thermal comfort.

From the ventilation aspect, windward slopes are preferable to leeward slopes. Sites near the crest of a hill, or at high elevations on the windward side near a ridge, are exposed to high air currents. Lower hillside locations benefit from downslope cool air flow and they tend to be cooler than slopes, provided that special arrangements are made to avoid strong wind flow during the winter period [5].

During the night, valley slopes cool down by radiation faster than the free air and cold breezes are formed and slide down the slope of the valley. Thus, valley slopes can experience downslope air current during windless nights, providing in this way more comfortable conditions than on the valley floor [6]. However, the valley floor also benefits from these air currents, especially under hot conditions with no local night winds. If a valley is surrounded by mountains, it is possible to experience poor ventilation conditions, which, under temperature inversion conditions at night, can trap the air pollution produced in the area.

The main points for preferred sites for envelope-dominated buildings, such as residences and small offices are [5, 7]:

- In **hot and dry climates**, the main emphasis is on reduction of heat gains and on potential means of heat rejection from buildings. Positions on lower hillsides are beneficial as they collect cool air. In areas with cold winters, the bottom of a south-facing slope is preferable. In areas with mild winters, east- and north-facing slopes can be used. West slopes should be avoided in any case. Eastern orientation is beneficial because of the large daily ambient temperature range. As shading during the hours of intense insolation (*incoming solar radiation* [8]) is required, a south-east orientation can be considered beneficial in hot-arid climates.
- In **hot and humid climates**, ventilation is desirable and location on the high parts of windward slopes can improve ventilation conditions. West sites should be avoided, as they are associated with high ambient temperatures and exposure to high solar radiation. East slopes are not desirable as during summer east and west slopes collect more solar radiation than any other orientation, owing to their inclined position. Because of their lower exposure to solar radiation, northern and southern oriented slopes are beneficial, but the ventilation aspect should have priority in this climatic type.

The microclimate around a hill is shown in Figure 5.1 and the building sites preferred on the basis of climate are shown in Figure 5.2. Desirable orientations of site locations for different climatic zones were presented by Olgyay [5], who considered the best use of solar radiation throughout the year.

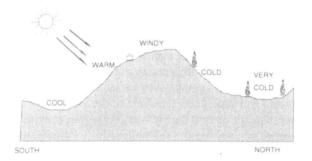

Figure 5.1 Climate pattern around a hill

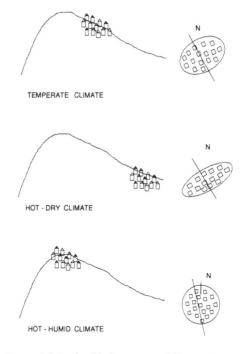

Figure 5.2 Preferable location in different climates

Urban planning

Urban planning is related to actions taken to optimize the energy use at the level of the whole urban area. Planning decisions aim to provide more energy-efficient forms of urban area by properly distributing or altering the mix of land uses. This can be achieved by developing a city in configurations that demand less energy for the thermal needs of buildings, for transportation and for the distribution of fuels.

The different variables affecting the energy requirements of urban spaces rank from the regional to the neighbourhood scale (Table 5.1) [9]. They include the layout of the settlement, location of main traffic routes, distribution of residential and commercial areas, and landscape design. Thus, several scenarios have been considered for the development of more energy-efficient urban regions, as follows (Figure 5.3) [3]:

- Concentration of new developments in the centre of existing cities (nucleated configuration);
- Arrangement of new developments along existing main routes (linear configuration);
- Concentration of new developments along secondary routes (dispersed linear configuration);
- Development of new satellite towns;
- Development of existing villages (dispersed nucleated village configurations).

Although, these scenarios concern savings in fuel and energy, they will not be discussed in detail as they are not directly associated with passive solar design.

Table 5.1 Different variables affecting the energy requirements at different scales of land-use planning [9]

Structural variable	Scale
Settlement pattern (e.g. rank-size, geometrical arrangement, etc.)	Regional
Communication network between settlements	Sub-regional
Size of settlement (area)	Individual
Shape of settlement (circular, linear, etc.)	Settlement
Communication network within settlement (radial, grid, etc.)	Neighbourhood
Interspersion of land uses	Neighbourhood
Degree of centralization of facilities	Neighbourhood

Urban morphology

The climatic zone in which the settlement is located is one of the basic parameters that will guide the overall planning process, along with the architectural character of the area and the functions it is designed to cover.

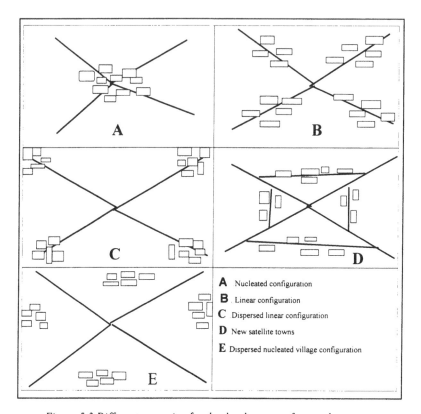

Figure 5.3 Different scenarios for the development of new urban regions

URBAN DENSITY

Urban density is one of the principal factors that affects the microclimate of an area and determines the urban ventilation conditions and urban air temperature. It is not irrelevant that the phenomenon of the 'heat island' is more intense in the city centre, where there are a dense urban plot and concentrated activities. Studies have shown that the phenomenon of the 'heat island' is mainly affected by urban density rather than by the size of the urban area [10] although it can be argued that, the larger the city, the more intense is the summer heat island [11].

In general, the higher the density of buildings in a given area, the poorer its ventilation conditions. On the other hand, a high density is beneficial for reducing the solar exposure of buildings during the summer period. The influence of the urban density on the ventilation conditions depends on the wind direction and the spatial arrangement of the buildings and their height. Givoni [12] proposed a bigger separation between buildings in the north and south directions than east and west, in order to ensure high solar collection during winter and solar protection of the west and east sides, which are not easily shaded, during the summer. Thus, high urban density could be achieved by a row of terrace houses along the east–west axis without preventing ventilation, provided that there are adequate openings in the north- and south-orientated walls.

Hot–arid climates, like the Mediterranean and semi-desert climates, are characterized by high-density settlements with narrow streets, arcades and small enclosed courtyards, to minimize buildings' solar exposure and to provide outdoor shaded areas (Figure 5.4). In *hot and wet climates*, wide streets and open spaces between buildings are found in order to ensure adequate ventilation. The same layout is found in *cold climates* in order to allow maximum solar exposure of buildings and of the open spaces during the winter period.

Figure 5.4 Typical settlement in a hot-dry climate

LAYOUT OF STREETS

The orientation and width of the streets affect the urban ventilation conditions and the solar exposure of buildings. They are of a greater importance in densely built urban areas, such as commercial and high-density residential areas.

An optimum street layout, which provides good ventilation conditions to pedestrians in the streets and to the buildings along the street, is one with wide main avenues orientated at an oblique angle to the prevailing winds [13] (Figure 5.5). In this way, wind can blow through the city. Buildings along such avenues are exposed to differential air pressure between the front and back facades (positive–negative pressure and vice versa), providing in this way potential for natural ventilation. Pedestrians should be protected from solar exposure by the provision of trees planted along the streets and special building features like awnings, overhanging roofs, colonnades, etc. (Figure 5.6).

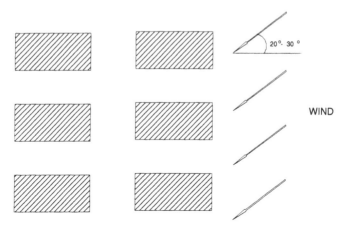

Figure 5.5 Optimum street layout for ventilation

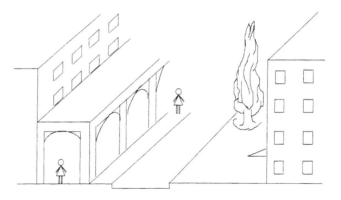

Figure 5.6 Solar protection in outdoor places

When the prevailing winds flow perpendicularly to the streets, they are forced by the facing buildings to flow over them. The air flow in the streets is the result of the friction of the wind on buildings' corners (Figure 5.7) and in this case, the width of streets is of no importance.

In climates where shading is of prime importance, narrow streets ensure the shading of buildings during summer and restricted solar exposure of the buildings along the streets. In southern latitudes [3], effective shading of south-facing buildings can be provided with a street width of around a fifth of the height of the building on the opposite side of the street. For west-facing buildings, the separating distance increases to about 1.5 to 2 times the height of the building that provides the shade.

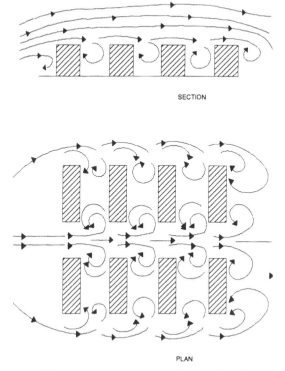

SECTION

PLAN

Figure 5.7 Air flow pattern over and around buildings[14]

HEIGHT OF BUILDINGS

The height of buildings is the most important factor determining the wind pattern over a built-up urban area and this, in association with the distance separating the buildings, characterizes urban ventilation conditions. These parameters also affect the solar exposure of buildings.

Buildings modify the existing wind pattern, as they present a greater roughness to the wind flow than an open area. The air flow around and over the buildings reaches a lower average air speed but has higher turbulence, due to friction on them. Thus, an urban wind pattern is characterized by a lower average wind speed with higher local air speeds and higher turbulence than in an open area. There is a variation of wind speeds inside an urban area which depends on the wind direction and speed, the urban layout and the height of buildings.

As the wind blows against a building, it is diverted from its original direction and zones of positive and negative pressure are created around the building. The pressure on the facades facing the wind (windward sides) exceeds the atmospheric pressure (pressure zone) and the pressure on the leeward sides is reduced (suction zone) (Figure 5.8). This pressure difference drives the air flow around and through the

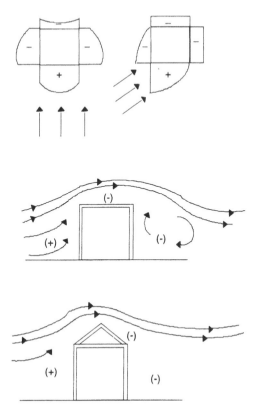

Figure 5.8 Pressure distribution around a building

building, as the air flows from regions of higher pressure to those of lower pressure. For buildings perpendicular to the prevailing winds, the front facade is subjected to positive pressure and all the other sides are under negative pressure. For an oblique wind direction, the two sides facing the wind are subject to positive pressure, with a decreasing magnitude from the windward to the leeward side, while the other two facades are under negative pressure. The leeward side of the roof is always in suction, while the windward side is in suction for small slopes and under pressure for steep ones.

Long rows of high, long buildings of the same height, which are perpendicular to the prevailing wind direction, block the wind in the first row and divert it upwards, creating poor ventilation conditions both along the streets and inside the buildings in the rows behind (Figure 5.9). The wind speed in the spaces between the buildings is gradually decreased after the first row and the result is the same, irrespective of the buildings' height and their separation distance. There is a weak flow in the streets between the rows, which is even weaker in side roads. These flows are caused by the

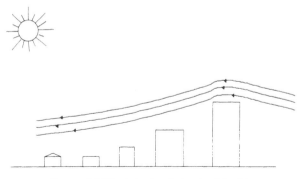

Figure 5.9 Wind shadow behind tall buildings

wind diverted over the roofs and by the turbulence created by the friction effect of the buildings. The average air speed in these places is about 30% of the undisturbed flow at the same height [14].

Individual tall buildings, rising higher than neighbouring ones, disturb the wind pattern of the area and create strong wind currents, especially at pedestrian level (Figure 5.10). The height of such a tall building mainly determines the flow diverted to the sides of the building while its width affects the flow pattern behind it. An increase in the windward surface width results in diversion of a larger air volume around the building, more of it over the roof and a strong upward turbulent flow behind the building. The highest wind speeds are experienced in front of the windward facade and the lowest speeds behind and between tall buildings [14]. The shape of the windward wall also modifies the flow pattern, as a convex wall disperses the flow around the building and a concave wall concentrates the flow along this wall and increases the turbulence. The air flow around buildings has been investigated by several researchers [14–17].

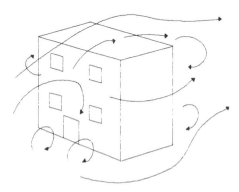

Figure 5.10 Strong wind currents in front of tall buildings

Tall, narrow buildings spread around an urban area increase significantly the air speed in the streets. Consideration should be given to their performance under winter winds, as the high air turbulence developed could cause problems to pedestrians and to the surrounding buildings. Several design solutions have been proposed to overcome the problem, such as breaking the facade with horizontal projections in the form of shading devices [18] or covering the open space between a tall and low building [16,19] (Figure 5.11).

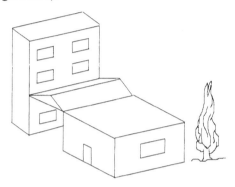

Figure 5.11 Covered spaces in front of tall buildings

A row of buildings of different heights with their facades oblique to the wind direction, improve ventilation conditions both on an urban scale and inside the buildings. If the buildings are of variable height, even a highly dense area may have better ventilation conditions than an area of lower urban density with buildings of the same height.

LANDSCAPING
The microclimate of an urban area can be positively modified with appropriate landscaping techniques by combining use of *vegetation* and *water surfaces*.

Landscaping could be applied to *public places*, in the form of large public parks, small neighbourhood parks, playgrounds, trees along streets [20] or by providing space for landscaping around buildings in the form of *private gardens*.

Vegetation Landscaping through vegetation is a very effective strategy for modifying the microclimate, having a positive impact on the urban climate as well. Apart from the decorative function that vegetation affords, it also modifies the microclimate and the energy use of buildings by lowering the temperature of the air and of surfaces, increasing the relative humidity of the air, functioning as shading devices and channelling the wind flow. In addition, plants can control air pollution, filter dust and reduce the level of nuisance from noise sources (Figure 5.12). Big, green urban areas can positively contribute, under proper maintenance, to the social life of

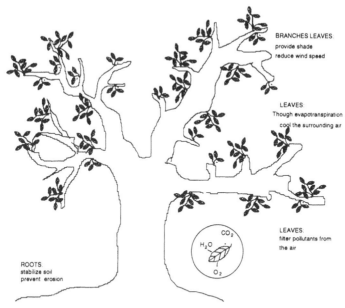

LEAVES, TWIGS
BRANCHES
absorb sound
cause rainfall

BRANCHES LEAVES:
provide shade
reduce wind speed

LEAVES:
Though evapotranspiration
cool the surrounding air

LEAVES:
filter pollutants from
the air

CO_2
H_2O
O_2

ROOTS:
stabilize soil
prevent erosion

Figure 5.12 Cumulative benefits of a tree

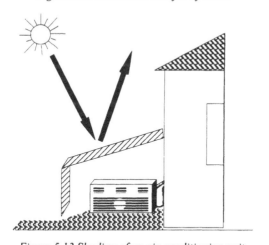

Figure 5.13 Shading of an air-conditioning unit

an area, by housing recreation places, cafés, open cinemas or open theatres. The psychological effect of the green areas on people's moods, their happiness and calm, should not be underestimated either.

Plants offer shade to surfaces and buildings and modulate their temperature. Strategic planting of trees and shrubs next to buildings can reduce summer air-conditioning costs by 15 to 35%, and even up to 50% in specific situations. Shading of an air-conditioning unit with shrubs and vine-covered trellises can reduce annual cooling costs by up to 10% [21] (Figure 5.13).

Vegetation acts as an air-conditioning substitute, by cooling the air through evapotranspiration. This process reduces temperatures and conventional energy use. Although trees absorb most of the solar radiation falling on their leaves – up to 90% – they reradiate back a small portion of it (Figure 5.14). The absorbed solar radiation is mainly consumed for evaporation of the water from the leaves. This process cools down the leaves, and consequently the air in contact with them, and releases vapour, increasing the air humidity. The remaining small portion of the absorbed solar radiation is used for photosynthesis. Leaves with open pores release about 50 to 70% of the amount of vapour that would be released from the same water area, under the same climatic conditions, effectively releasing a quantity of water about five times that of the leaves' weight [22].

Trees can transpire up to 378.54 litres (100 US gallons) of water in one day and the corresponding cooling effect on a hot dry summer day can be equivalent to five

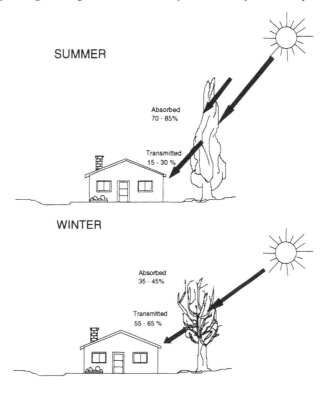

Figure 5.14 Summer and winter performance of a tree

air-conditioning units running for 20 hours [21]. The cooling effect of plants, due to the combined effect of evapotranspiration and the shade they offer, can result, in heavily landscaped areas, in an ambient temperature drop in the immediate surroundings by 5.5 to 8.5°C (10–15°F) in hot dry climates [23]. Increasing the vegetation cover by 10 to 30%, which corresponds to one to three properly placed trees per house, may reduce the required cooling energy by as much as 50% [21]. The comparative benefits of the cooling effect of vegetation as a result of the shade they offer and the lowering of the wind speed (direct effects) and by the evapotranspiration process (indirect effect) are presented by Huang et al. [24] (cited in [21]) (Figure 5.15).

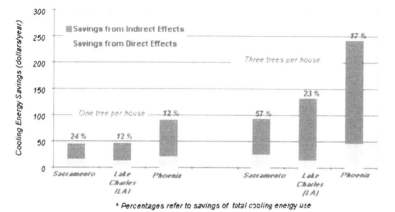

* Percentages refer to savings of total cooling energy use

Figure 5.15 Cooling effect of planting ([24] cited in [21])

The shade provided by a tree is better than any man-made shading device, as plants also modify the properties of the surrounding air. Bowen [25] reports that a shaded area with spread tree canopy, is about 22.2°C (40°F) cooler than an unshaded area owing to evapotranspiration.

During the night, trees block the long-wave radiation from the ground and consequently the night air temperature is warmer and the diurnal temperature range is smaller under trees than in an open field. Vegetation on a hill can modify the wind pattern, for example by decreasing the zone of wind protection on the windward side and increasing the downwind zone of reduced air velocity in the leeward side.

In *hot dry climates*, trees with a dense canopy that give thick shade are desirable as they reduce direct and diffuse irradiance (Figure 5.16). In areas where water is not so scarce, vegetation is a very attractive cooling approach. In *humid climates*, the beneficial cooling effect may be reduced by the undesirable increased relative humidity of the air. The vegetation should be arranged to encourage movement of the air and dispersion of the moist air. Tall plants with light foliage are recommended as they do not block the air and permit less disturbed air movement through the foliage (Figure 5.17).

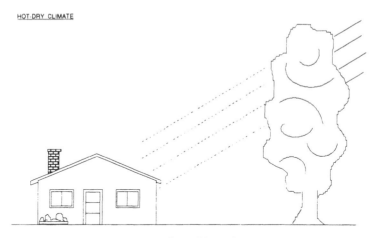

Figure 5.16 Trees with dense foliage for hot dry climates

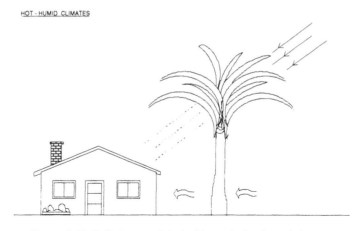

Figure 5.17. Tall plants with light foliage for hot humid climates

The most appropriate plants for good landscape are native plants as they are accustomed to the local climatic conditions. For effective use of plants, consideration should be taken of:

- their specific biological needs such as soil type, watering needs, minimum safe temperature and exposure to sun;
- maintenance costs; and
- availability of water in the area.

Water surfaces   Water surfaces modify the microclimate of the surrounding area by reducing the ambient air temperature either through evaporation (latent heat) or by contact of the hot air with the water surface which is cooler, owing to its high thermal mass. Fountains, ponds, streams and waterfalls or mist sprays may be used as cooling sources in order to lower the air temperature of outdoor spaces and of the air entering a building. The psychological cooling effect offered to people is also important, when added to the cooling result – evaporation of half a litre of water corresponds to about 0.3 kWh. Under average wind speed, dry-bulb and wet-bulb air temperature conditions, the energy released by 1 m^2 of an exposed water surface is about 200 W [26].

As water surfaces increase the air humidity, their existence can be problematic in very humid climates and thus circulation of the air should be promoted. In hot climates, their cooling effect should be maximized through special design features, which prevent diffusion of the air cooled by them and direct this air to inhabited places. Although water surfaces are very beneficial in dry climates, their existence may be limited by scarcity of water. Overheating of the water mass, which encourages algae growth, should be avoided by special features such as shading with trees, bushes, etc. The water should not become stagnant and its circulation is recommended.

PUBLIC SPACES

Air and radiant temperatures behind trees are significantly lower than on hard surfaces like roads, concrete or soil open areas. Provision of trees can minimize the danger of thermal stress and heat stroke in people in cities. Plenty of spaces with trees must be provided along streets, in public parks and in children's playgrounds in order to provide shaded areas. Other functional landscaping elements like trellises, pergolas, alleys and hedgerows can also be used (Figure 5.18). Shade trees and trellises can block sunlight during overheated periods, by reducing the air temperature by as much as 11.5°C (21°F) [27].

Figure 5.18 Landscaping of open spaces for shading

The influence of open green spaces on the ambient conditions extends only a short distance into the surrounding urban area. Thus, it is more effective to have plenty of small green areas spread around an urban area than a few large parks. Their contribution is mainly by creating 'islands of cool' inside cities and by providing areas with a pleasant climate for rest and recreation. Duckworth and Sandbery [28] found that the temperature in the heavily vegetated San Francisco Golden Gate Park was around 8°C (14°F) cooler than nearby less vegetated areas. Development of private gardens around the buildings has the same beneficial effect on human comfort.

Although flowerbeds and lawns contribute to the reduction of the surrounding ambient temperature, their benefits are dependent on the availability of water supplies. They should be preferred in humid climates rather than in dry ones. Montgomery [27] reports that temperatures over grass are about 5.5 to 8.0°C (10–14°F) lower than over exposed soil and that an acre of turf (4,047 m²) on a sunny summer day can transfer more than 13,770 kWh of energy, which is enough to evaporate 25,740 litres (6,800 US gallons) of water.

Green belts are effective both in trapping soil particles and particles generated by motor vehicles and industrial pollution and in purifying the air. Also some nitrogen oxides (NO and NO_x) and airborne ammonia (NH_3) can be absorbed by the foliage, the nitrogen being used by the plants. Some trees, depending on their sensitivity, can also use the sulphur dioxide (SO_2) and the ozone (O_3) – which are products of air pollution – while several other species suffer damage from exposure to high concentrations of these chemicals. It has been estimated that a street planted with trees can reduce airborne dust particles by as much as 7,000 particles/litre of air [29]. It was also found [30] that although industrial pollution is significantly reduced within a green area and in the area behind it, there is little or no reduction outside this vicinity.

The actual effect of vegetation on noise reduction is very limited, as it does not reduce the level of noise reaching the buildings, but it can reduce the reverberation time in the street as a result of sound absorption in the leaves. The actual effect is mostly psychological, by visually hiding the source of noise from the affected person [20]. The use of trees and shrubs is effective for noise reduction if they are planted densely in belts of 6.1 m (20 ft) to 15.2 m (50 ft) width [31].

BUILDING DESIGN

Design approaches at this stage are concentrated on the domain of the site and aim to improve the microclimate around a building and its thermal performance, by reducing heat gains and by promoting natural cooling of the building. Actions are taken in the space surrounding buildings (private gardens, courtyards, atria), in the design (orientation, shape, openings, internal layout, functional elements) and in the construction (material, colour) of each individual building. Successful intervention in the environment surrounding a building can result in minimizing the operating hours of the air conditioning in mechanically ventilated buildings, or reducing the hours of thermal discomfort in naturally ventilated buildings.

Private gardens

Landscaping through vegetation and water surfaces around buildings can reduce the impinging solar radiation and modify the air temperature outdoors and indoors and effectively minimize the cooling load of buildings. It can also extend the living area outdoors under more pleasant thermal conditions. Plants can be effectively used to reduce solar access, to provide shade during summer and to improve ventilation conditions in buildings. They can also improve the quality of daylight entering a building, by softening and diffusing it and by reducing the glare from the bright sky. Trees, bushes, vines and creepers can effectively fulfil these conditions.

Parker [32] has shown that careful landscaping with trees, shrubs and ground cover for maximum cooling, can reduce the air temperature by up to 5.5°C (10°F). As an air conditioner uses 5–7% more energy for each additional degree, a 50% energy reduction can be achieved in buildings that use evaporative coolers or air conditioners and are shaded. Thus, 1 ton of air-conditioning capacity is saved in a shaded building for each 3.5 kW energy consumption.

Choice of trees should be very carefully based on the shape and character of the plant (tree or bush), both during the winter and the summer period and on the shadow shape they provide. Broadleaves and deciduous trees are very useful as they drop their foliage during the autumn and permit solar access during winter. This is especially true of those with dense summer canopy and an almost branchless open winter canopy. Of course, branches of trees, even without leaves, create significant shade, 30 to 60% during winter, which can be as high as 80% for trees that hold their dead leaves like an oak tree [33]. A defoliated Honey Locust was found to block about 60% of the sun [33], while a Silver Maple in full leaf blocks about 75% of the impinging solar radiation [34]. Montgomery [35] has presented a list of selected species for energy conservation (Table 5.2). The principal factors that must be considered in the selection of deciduous plants are [35]:

- mature height and crown;
- growth rate;
- leaf appearance and fall patterns;
- distance of branches from the ground.

Early leaf fall can be encouraged by reduction of late summer pruning, irrigation and fertilization.

The position of plants around the building should be strategically chosen in order to provide shade at the most critical hours of the day. Shading of the windows is the

Table 5.2 Tree species which can be used as windbreaks [35]

Name	Mature height (m)	Growth rate	Site features
Deciduous			
Caragana arborescens – Siberian pea tree	5–11	Moderate	Sunny and well-drained soils
Elaegnus multiflora – Russian olive	6–11	Fast	Full sun, dry area
Populus alba–White poplar	18–21	Fast	Sunny, moist soils
Spiraea – Bridal wreath	5–11	Fast	Sun or shade, moist soils
Syringa vulgaris – Lilac	5–11	Moderate	Sunny/city or seashore
Tamarix parviflora – Salt cedar	5–11	Moderate	Sunny area
Evergreen			
Elaegnus pungen – Silverberry	8–11	Moderate	Sun or shade
Juniperus scopulorum – Rocky mountain juniper	12–17	Moderate	Sunny area
Kalmia latifolia – Mountain laurel	6–11	Moderate	Sun or shade
Pinus nigra – Austrian pine	18–24	Moderate	Sunny area
Pinus thunbergii – Japanese black pine	18–21	Slow	Seashore, moist soils
Thuja occidentalis – Eastern arborvitae	11–18	Moderate	Sunny area

most beneficial. Horizontal overhangs, like pergolas, are preferable at the south side, as the sun is in its high position at this orientation (Figure 5.19). Trees on east-south-east and west-south-west sides give the best performance as the sun is at low altitudes in the morning and late in the afternoon and low sun rays cast long shadows (Figure 5.20). Bushes can also be used to shade east and west windows and additionally act as vertical wingwalls (Figure 5.21). Vertical trellises covered by vines or other creeping plants are very effective on east and west facades. In the case of buildings with solar collectors on the roof, precautions should be taken to control the height of trees positioned at the south side, in order to allow solar access to the collector.

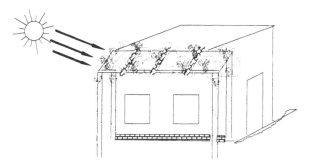

Figure 5.19 Pergolas for shading of south facades

SUMMER

Shade from trees on
E and W sides

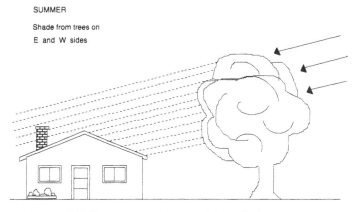

Figure 5.20 Trees at east and west sides for shading in summer

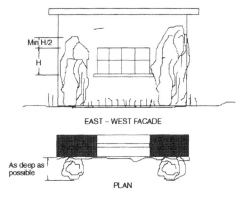

EAST – WEST FACADE

As deep as
possible

PLAN

Figure 5.21 Bushes for shading of east and west windows

Climbing plants on the walls are very effective in reducing the solar heat penetrating through the building's envelope. It is not unrealistic to expect a reduction of up to 50% in the cooling load when an envelope-dominated building is effectively shaded by plants [7]. A 60% reduction in the use of electricity was realized in a Florida school when the walls and windows were shaded by plants [7]. In another study, the temperature inside a mobile home was reduced by 13°C (24°F) when it was well shaded by plants [27]. In tests carried out in South Florida, it was indicated that the shade provided by a large tree reduced the wall temperature by approximately 3.5°C (6.4°F) [36]. When a sparse vine, about 7.5 cm (3 in) thick, covering almost 80% of the wall area, was used on a west wall, an average reduction of 7.5°C (13.8°F) was measured in the surface temperature during the late afternoon compared to an unshaded wall [36].

Trees and bushes can be also positioned in such a way as to improve ventilation conditions inside the building (Figure 5.22) by guiding more of the cooling breezes into the building or even by preventing the wind from spilling around the sides of the building (Figure 5.23). If bushes are used, they should be positioned at a distance

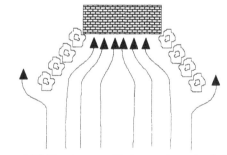

Figure 5.22 Planting to guide breeze through a building

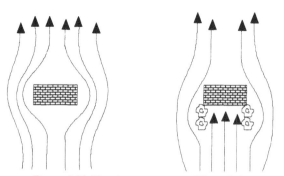

Figure 5.23 Planting to prevent spilling of air

from the building, in order not to deflect the evening breezes above the building (Figure 5.24). However, they can also act as windbreaks to the hot winds if they are located to block the wind reaching the building. Especially in mechanically venti-lated buildings, this arrangement can reduce the cooling load of the buildings, as a dense screen can reduce wind pressure and velocity by 75–85%, which results in lower cooling load due to reduced heat gains by convection and air infiltration from the developed pressure [27]. The vegetative windbreaks on the windward side should be positioned at a distance not more than 1–1.5 times the height of the building. A cluster of canopy trees, shrubs or trees creates a dead air space around it, which decreases the air velocity for about five times its height on the windward side and about 25 times its height on the leeward side [27].

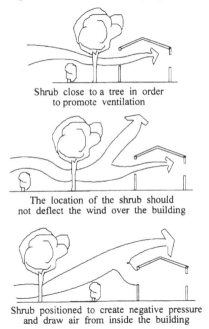

Shrub close to a tree in order
to promote ventilation

The location of the shrub should
not deflect the wind over the building

Shrub positioned to create negative pressure
and draw air from inside the building

Figure 5.24 Planting position can modify ventilation

As the roof is the part of the building that absorbs the highest amount of solar ra-diation during summer, grass planting provides a sharp decrease in the roof's surface temperature and consequently in the air temperature of the space below it. Ohlwein [37](cited in [22]) reported that planting of a normal roof, which presents maximum surface temperature of 80°C, can reduce its temperature to 25° C (Figure 5.25). The development of a roof garden may considerably reduce the roof's temperature, by additionally providing an amenity space and by extending activities at a higher level, where air speed is higher than at the ground. Solutions could range from a simple

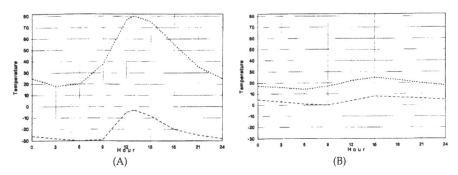

(A)   (B)

Figure 5.25 (A) Normal and (B) planted roof temperature [37] as cited in [22].
Winter – – –; Summer - - - -

horizontal vine-covered trellis, to trees and bushes planted in lightweight soils, made up of perlite or vermiculite, or even creation of fountains, pools, etc. (Figure 5.26). In the Texas area, a house can be up to 11.5°C (21°F) cooler with grape vines on a roof trellis [38].

Grass and other plants covering the ground around a building are also effective in controlling its air temperature and in ensuring that its cooling load will be smaller

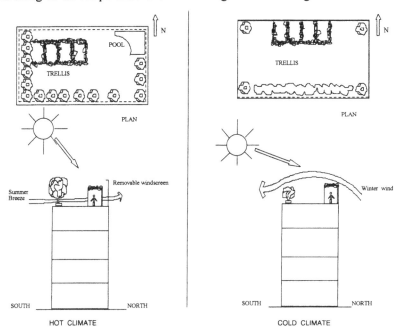

Figure 5.26 Examples of roof arrangements

than a building surrounded by asphalt or concrete surfaces. It was also found that the earth temperature at a depth of 7.6 cm (3 in) under lush grass was about 2.2° C (4° F) cooler than that under soil with about 50% cover of weeds [39].

Water surfaces, in the form of a small fountain, a pond or a stream can be used to cool the air. Evaporative cooling, combined with wind guided to occupied places, can increase human comfort. Distribution of wet surfaces around a garden is particularly effective in hot dry climates. They can also be used in humid climates, but they will be more effective if they are positioned downwind in order to avoid human discomfort from the extra humidity in the air.

Characteristics of buildings

SHAPE

Optimum shape of a building is widely considered to be the one that loses least heat during winter and accepts the least amount of radiation during summer. Volume is approximately related to its thermal capacity – ability to store heat – while exposed surface area is related to the rate at which the building gains or loses heat. Thus, the ratio of volume to exposed surface area is widely used as an indicator of the speed at which the building will heat up during the day and cool down at night. A high volume-to-surface ratio is preferable for a building that it is desired to heat up slowly, as it offers small exposed surface for the control of both heat losses and gains. In hot climates, the area exposed to solar radiation is more important than the total exposed surface area. During summer, in low latitudes, the roof is the surface most exposed to solar radiation, followed by the east and west walls.

Knowles [40] proposed an alternative measure, that takes into account both heat losses and direct solar exposure, by taking the ratio of the surface in solar exposure divided by the overall surface area. The ratio of the summer values of this measure to the winter values reflects the amount of seasonal solar shading inherent in the built forms of the blocks.

A summary of requirements for conditions of optimum heat gain and heat loss for the case of a simple rectangular building form in different climates is presented by Evans (Table 5.3) [41]. In *hot humid climates*, where ventilation is desired, buildings should maximize the area of the exposed surfaces. For *hot arid climates*, buildings with high volume-to-surface ratios are recommended, an example being the compact urban forms seen around the Mediterranean basin.

It is widely believed that the square form is the optimum one, but calculations based on a yearly thermal performance carried out by Olgyay [5] showed that this is not the case for all climates. An elongated form, somewhere along the east–west direction, can perform better. A square form performs better in old massive traditional structures with small openings, whereas the large glazed areas in contemporary buildings lead to big thermal gains in the interior of buildings which eliminate the effect of shape.

Table 5.3 Requirements of building form for different climate types [41]

Climate	Element and requirements	Purpose
Warm humid	Minimize building depth	For ventilation
	Minimize west facing wall	To reduce heat gain
	Maximize south and north walls	To reduce heat gain
	Maximize surface area	For night cooling
	Maximize window wall	For ventilation
Composite	Controlled building depth	For thermal capacity
	Minimize west wall	To reduce heat gain
	Limited south wall	For ventilation and some winter heating
	Medium area of window wall	For controlled ventilation
Hot dry	Minimize south and west walls	To reduce heat gain
	Minimize surface area	To reduce heat gain and loss
	Maximize building depth	To increase thermal capacity
	Minimize window wall	To control ventilation heat gain and light
Mediterranean	Minimize west wall	To reduce heat gain (summer)
	Moderate area of south wall	To allow (winter) heat gain
	Moderate surface area	To control heat gain
	Small to moderate window	To reduce heat gain but allow winter light
Cool temperate	Minimize surface area	To reduce heat loss
	Moderate area of north and west walls	To receive heat gain
	Minimize roof area	To reduce heat loss
	Large window wall	For heat gain and light
Equatorial upland	Maximize north and south walls	To reduce heat gain
	Maximize west-facing walls	To reduce heat gain
	Medium building depth	To increase thermal capacity
	Minimize surface area	To reduce heat loss and gain

Although the ratio of volume to surface [42] provides some indication of the thermal efficiency of the enclosure, it does not take into account either the variations of the thermal characteristics of the building fabric or the effects of solar gains, which can be determinants of the building's thermal performance. It is argued [43] that restriction of the optimum shape could be relaxed by adopting several energy-saving measures, for example extra insulation, allowing in this way architectural freedom in the choice of the building's form. In practice, planning regulations, site conditions, neighbourhood buildings and architectural styles are becoming determinants of a building's form. As high-rise blocks are common in urban areas, characterized by small exposed surface related to occupied area, their summer thermal performance could be improved by giving more importance to the materials used and to the architectural elements such as shading devices, wing walls, courtyards, etc. In addition to thermal characteristics, forms should be explored which enhance wind channelling into the building and permit ventilation throughout the occupied spaces.

INTERNAL LAYOUT
The layout of a building's interior determines the ventilation conditions throughout it. An open plan combined with a proper distribution of openings is preferable for undisturbed ventilation conditions in the interior. As practical needs require separation of spaces, restriction of air-flow paths should be avoided and the positioning of partitions should help channelling of the air motion throughout the occupied space. In general, partitions should be positioned in such a way that creates the larger space on the windward side (Figure 5.27). Design solutions, especially in multi-storey buildings, should explore the possibilities of permitting cross ventilation (Figure 5.28). Distribution of spaces on two floors – maisonette houses – improves ventilation conditions by imposing an additional air movement in the vertical direction as a result of the stack effect (Figure 5.29).

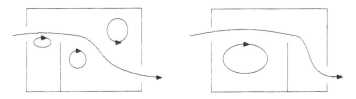

Figure 5.27 Bigger space on the windward side permits better ventilation conditions

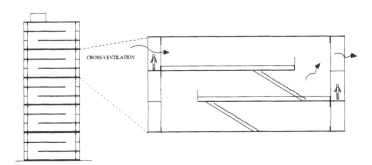

Figure 5.28 Design solutions in multi-storey buildings for improving ventilation conditions

Although high ceilings, which are traditionally met in hot climates, have little effect on the air flow pattern, they allow thermal stratification and decrease the transfer of heat gains through the ceiling. Use of high ceilings to reduce discomfort is based on the assumption that there will be high rates of heat gain through the roof,

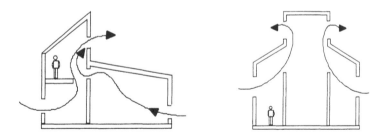

Figure 5.29 Design solutions for stack-effect ventilation

which can be the case with an uninsulated ceiling. The cooling effect of the room height is greater in spaces with very high ceilings where stratification of the air allows the occupants to inhabit the lower cool space. For dwellings that are characterized by moderate ceiling heights, it has been argued [41] that rooms with high ceilings are not significantly more comfortable than rooms with ceiling heights of 2.7m or even 2.5m.

ORIENTATION OF BUILDINGS

The orientation of buildings determines the solar exposure of the building, which should be maximized during winter and offer easy manipulation for shading during summer. The orientation of buildings usually follows the orientation of roads by positioning the main facade facing the main road. As the south side of a building is the one that can be easily shaded during summer and the best one for solar heat collection during winter, buildings on an east–west road are ideal for both summer cooling and winter heating needs (Figure 5.30).

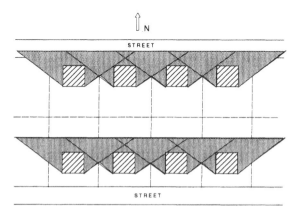

Figure 5.30 East–west orientation of streets is ideal for solar access (winter) and shading (summer) [7]

The thermal performance of buildings on a south–north road could be improved in different ways. Orientating the short facade to the main street (Figure 5.31) or developing a row of houses, detached or terraced, with the long facades facing south–north and permitting access from the central houses to the main street through driveways (Figure 5.32), can offer some solutions to maximizing winter solar exposure and providing shading during summer.

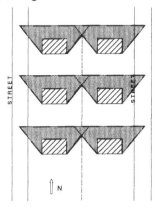

Figure 5.31 Arranging the narrow facade to face the road in south–north orientated streets is advisable [7]

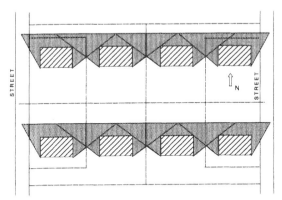

Figure 5.32 Access to the main road through driveways can permit east–west orientation of buildings [7]

In diagonally orientated streets, houses can be positioned diagonally in each lot. (Figure 5.33). This positioning ensures orientation of buildings along the south–north axis, allows unobstructed ventilation in the complex of buildings and also improves privacy inside the houses, as windows of opposite rows of houses do not face each other.

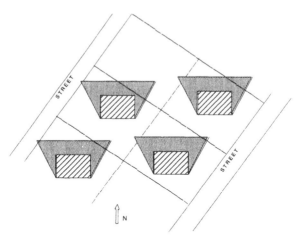

Figure 5.33 South orientation of buildings in diagonally orientated roads [7]

COLOUR OF EXTERNAL SURFACES

The colour of the external surface of a building's envelope influences its thermal performance because it determines the amount of heat absorbed and thus transmitted to its interior. The amount of heat absorbed by a surface is characterized by its absorptivity. Radiation is absorbed selectively, according to the wavelength of the incident radiation (Figure 5.34). The colour of a surface provides a good indication of its solar radiation (short-wave radiation) absorptivity but does not indicate anything about long-wave radiation absorption. In general, the solar absorptivity decreases and the reflectivity increases with lightness of colour. The more solar radiation a surface absorbs, the hotter it gets, while the more it reflects, the colder it stays.

Differences of up to 27°C in the solar temperature on an east facade have been reported between a dark surface of absorptivity 0.9 and a surface with an absorptivity value of 0.2 in a 52° N latitude [44]. The effect of colour on room air temperature depends on the heat resistance and heat capacity of the envelope's materials [45] as well as on the ventilation rate and the direct solar gains [46]. Thus, the effect of surface colour is more intense in lightweight structures, which are characterized by low thermal resistance and low thermal capacity, than in heavyweight structures. Continuous ventilation and solar access to the interior of the building diminishes the effect of colour on both lightweight and heavyweight structures [46].

Orientation effects are reduced with light colours, as differences up to 23.0°C have been reported between different orientations for grey painted walls whereas, for whitewashed walls, deviations were less than 3.0°C [45].

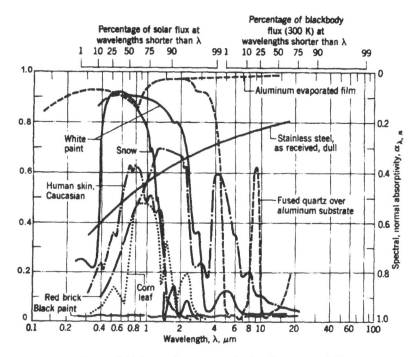

Figure 5.34 Spectral variation of solar absorptivity [47]

Ceramic materials like white cement, gypsum and lime, with the exception of earth-based materials, offer high reflectivity and are suitable for hot climates. Traditional architecture can show examples of reflecting external surfaces, such as the whitewashed buildings and streets around the Mediterranean Sea.

The colour of a building, as well as affecting the building's thermal performance, has a great impact on the urban climate, by modifying the air temperature. Computer simulations of neighbourhoods have shown that changing the colour of the roof, walls and streets can significantly reduce air temperature and cooling energy use. Realistic changes of the colour in hot, sunny climates with many dark colours, could reduce a city's air temperature by as much as 2.8°C (5°F) [21].

The colour of a building's envelope also affects the glare and light in the streets. Special elements on the envelope could reduce the visual discomfort of pedestrians from light-washed buildings. The uniformity of the facades can be reduced with use of overhangs, either short or extended over the entire length of a wall, with other architectural elements on the walls or by trying out different light colours, etc.

REFERENCES

1 Cook, J. (1991). 'Searching for the bioclimatic city', In *Architecture and Urban Space*, eds S. Alvarez et al. Kluwer Academic, Dordrecht, pp. 7–16.

2 Bitan, A. (1992). 'The high climatic quality city of the future', *Atmospheric Environment*, Vol. 26B, No 3, pp. 313–329.

3 Dupagne, A. and L. Matteling (1993).'Passive solar urban design', In *Energy in Architecture; The European Passive Solar Handbook*, J. Goulding, J. Owen Lewis and T. Steemers (eds). Publication No EUR 13446, Batsford for the Commission of the European Communities, London.

4 Trewartha, G.T. and L.H. Horn (1980). *An Introduction to Climate*, 5th ed. McGraw-Hill Book Company, New York (1st ed. 1937).

5 Olgyay, V. (1963). *Design with Climate*, Princeton University Press, Princeton, NJ.

6 Lyons, T.J. (1984). *Climatic Factors in the Siting of New Towns and Specialized Urban Facilities*. World Meteorological Organization, WMO (1986), pp. 473–486.

7 Lechner, N. (1991). *Heating, Cooling, Lighting: Design Methods for Architects*. John Wiley & Sons, New York.

8 Critchfield, H.J. (1960). *General Climatology*. Prentice Hall, Englewood Cliffs, NJ.

9 Owens, S. (1985).'Energy demand: links to land-use and forward planning', *Built Environment*, Vol. 11, No 1, pp. 33–44.

10 Chandler, T.J. (1976).'Urban climatology and its relevance to urban design', Technical Note No 149, World Meteorological Organization (WMO), Geneva.

11 Oke, T.R. (1973). 'City size and the urban heat island', *Atmospheric Environment*, Vol. 7, pp. 769–779.

12 Givoni, B. (1989). 'Urban design in different climates'. WMO/TD-No 346, World Meteorological Organization, Geneva.

13 Givoni, B. (1992). 'Climatic aspects of urban design in tropical regions', *Atmospheric Environment*, Vol. 26B, No 3, pp. 397–402.

14 Givoni, B. and M. Paciuk (1972). 'Effect of high-rise buildings on air flow around them', Building Research Station Technion, Haifa, Israel.

15 Aynsley, R.M. (1976). 'A study of airflow through and around buildings', PhD Thesis, University of New South Wales, School of Building.

16 Arens, E.A. (1982). 'On considering pedestrian winds during building design', In *Wind Tunnel Modelling for Civil Engineering Applications*, ed. E. Reinhord. Cambridge University Press, pp. 3–26.

17 Paciuk M. (1975). 'Urban wind fields - an experimental study on the effects of high rise buildings on air flow around them', MSc Thesis, Technion, Haifa, Israel.

18 Givoni, B. (1989). 'Urban design in different climates', WMO/TD-No 346, World Meteorological Organization, Geneva.

19 Fragoudakis, A. (1985). *Thermal–Moisture Insulation and Wind Protection of Buildings*, University Studio Press, Thesssaloniki, Greece (in Greek).

20 Givoni, B. (1991). 'Impact of planted areas on urban environmental quality: a review', *Atmospheric Environment*, Vol. 25b, No 3, pp.289–291.

21 Akbari, H., S. Davis, S. Dorsano, J. Huang and S. Winnett (eds) (1992). *Cooling our Communities; A Guidebook on Tree Planting and Light-Colored Surfacing*, US Environmental Protection Agency, Office of Policy Analysis, Climate Change Division.

22 Evmorfopoulou, C. (1988). 'Energy savings in buildings with use of plants on the roof and facades', 3rd National Conference on 'Soft Kinds of Energy', Thessaloniki (Greece) (in Greek).

23 Setti, B. (1981). 'Energy conscious landscape design', *In Situ*, Vol. 6, No 1, International Institute of Site Planning.

24 Huang, Y.J., H. Akbari, H.G. Taha and A.H. Rosenfeld.'The potential of vegetation in reducing summer cooling loads in residential buildings', LBL Report 21291, Lawrence Berkeley Laboratory, Berkeley, CA.

25 Bowen, A. (1980). 'Heating and cooling of buildings and sites through landscape planting and design', *Passive Cooling Handbook*, 5th National Passive Conference, AS/ISES, Amheret, MA.

26 Yellot, J.Y. (1983). 'Passive and hybrid cooling research', *Advances in Solar Energy*,

27 Montgomery, D.A., (1981). 'Landscape for passive solar cooling', International Conference on Passive and Hybrid Cooling, Conference, Miami Beach, American Section of ISES.

28 Duckworth, E. and J. Sandberg (1954). 'The effect of horizontal and vertical temperature gradients', *Bulletin of Meteorological Society*.

29 Bernatsky, A. (1978). *Tree Ecology and Preservation*, Elsevier Scientific Publishing, New York.

30 'Green belts and air pollution' (1973). HUD International, (The Netherlands), Information Series 21, 15 March.

31 Bowen, A. (1981).'Sound control for natural energy systems in overheated environments', International Conference on Passive and Hybrid Cooling, Miami Beach, American Section of ISES.

32 Parker, J. (1983). 'The effectiveness of vegetation on residential cooling', *Passive Solar Journal*, Vol. 2, No 2.

33 Zanetto, J. (1978). 'The location and selection of trees for solar neighbourhoods', *Landscape Architectural Magazine*, November.

34 Johnson, B. et al. (1980).'Solar absorption through vegetation: a two season study', Virginia Polytechnic Institute and State University.

35 Montgomery, D.A. (1987). 'Landscaping as a passive solar strategy', *Passive Solar Journal*, Vol. 4, No 1.

36 Parker, J.H. (1981). 'A comparative analysis of the role of various landscape elements in passive cooling in warm humid environments', International Conference on Passive and Hybrid Cooling, Miami Beach, American Section of ISES.

37 Ohlwein, K. (1984). *Dachbegrünung – Ökologisch- und Funktionsgerecht*, p. 119. Bauverlag Gmbh, Wiesbaden und Berlin.

38 Stanford, G. (1980). 'Congressional clearinghouse on the future', Greenhills Research Center, Cedar Hill, TX.

39 Grondzik, W., L. Boyer and T. Johnson (1981). 'Variations in earth covered roof temperature profiles', International Conference on Passive and Hybrid Cooling Conference, Miami Beach, American Section of ISES.

40 Knowles, R.L. (1974). *Energy and Form: an Ecological Approach to Urban Growth*. MIT Press, Cambridge, MA.

41 Evans, M. (1980). *Housing, Climate and Comfort*. The Architectural Press, London.

42 Rickaby, P. (1987). 'An approach to the assessment of the energy efficiency of urban built form', In *Energy and Urban Built Form*, eds D. Hawkes et al. Butterworths, London.

43 Chrissomalidou, N. (1992). 'The form of the building as an energy design parameter', 3rd International Conference on Energy and Building in Mediterranean Area, Thessaloniki (Greece), April, pp. 151–155.

44 Siddiqi, A. (1984). 'Evaluation of the effect of choice of materials on the surface temperatures of buildings in urban built-up areas', 3rd International PLEA Conference, Mexico City, Mexico, 6–11 August.

45 Givoni, B. (1976). *Man, Climate and Architecture*. Elsevier, Amsterdam.

46 Bansal, N.K., S.N. Garg and S. Kothari (1992). 'Effect of exterior surface colour on the thermal performance of buildings', *Building and Environment*, Vol. 27, No 1, pp. 31–37.

47 Incropera, F. P. (1990). *Fundamentals of Heat and Mass Transfer*, 3rd ed. John Wiley & Sons, New York.

6

Thermal comfort

Modern buildings are designed to provide an optimum indoor environment depending on their function (working, leisure etc.). This way occupants can optimize their productivity and fully utilize their human resources. Indoor spaces should provide optimum thermal conditions (temperature, humidity, air movement), visual conditions (proper light levels and pleasant environment), acoustical conditions (low noise levels and disturbances), and air quality (required amounts of fresh air, control of odours and air pollutants). The objective is to achieve all the above conditions in indoor spaces, so that the occupant can be in total comfort (thermal, visual and acoustical), in a 'healthy' environment; this is sensed by the skin (temperature, humidity, air movement and resulting cooling effect), as well as by the eyes (light levels and temperature variations), ears (atmospheric pressure, noise disturbances), and nose (temperature and relative humidity levels in the air and air quality).

Human thermal comfort is defined as the conditions in which a person would prefer neither warmer nor cooler surroundings [1]. It is a rather complex concept, since it depends on various influencing parameters. It is the combination of the various parameters that create the end result of comfort. The dependence of comfort on some of these parameters is stronger and more determinant than for others, as is concluded from the analysis presented in the following sections.

It has become increasingly important, however, to achieve comfort with minimum energy consumption. It is clear that, with the available technology, there are no technical limitations on the heating and air-conditioning industry in providing systems which can result in optimum indoor air conditions. The problem is rather to achieve, maintain and control comfort conditions with a rational use of energy and by optimizing the parameters which influence them.

The following discussion includes a description of the parameters affecting thermal comfort, methods for calculating and controlling it, and some related examples and applications.

THE INFLUENCING PARAMETERS

Building occupants are always in search of thermal comfort, which in turn influences a person's performance (intellectual, manual and perceptual). Recently, a team of researchers at Rensselaer Polytechnic Institute in the USA have completed a scientific study which has shown a direct and measurable link between individual comfort control and job performance [2].

It was found during this investigation that the use of environmentally responsive workstations (ERWs) led to a 2% increase in office productivity. Each ERW allows the employee to control workstation temperature, lighting, background sound and the direction and quantity of constantly filtered fresh air. A panel under each workstation provided radiant heat to warm the lower part of the body. An energy-saving occupancy sensor would automatically shut off the equipment if the employee left the workstation for more than 10 minutes.

Depending on the available means, occupants will attempt several actions to change or control environmental conditions. In order to be most successful in these actions, one must have a thorough quantitative, as well as qualitative, knowledge of the conditions establishing the parameters that influence thermal comfort. This will also enable building designers to provide alternative means to the occupants for controlling their thermal comfort conditions, instead of just lowering the thermostat during summer or increasing it during winter.

The human body is like a complex internal combustion engine. To achieve thermal comfort, the body must balance its heat gains and losses by properly adjusting its functions (i.e. perspiration), while also responding to the prevailing environmental conditions (i.e. temperature and humidity). Under good conditions the human body can function at optimum levels.

There are times, however, when comfort cannot be achieved by the functions of the body itself, owing to the severity of the prevailing conditions. Under such circumstances it is necessary to provide some assistance, by natural, hybrid or mechanical means. It is important, though, for rational use of available energy resources, to exhaust first all means of achieving comfort by natural or hybrid techniques, before having to resort to energy-consuming mechanical systems.

Depending on the function of the building and its various spaces, indoor environment conditions will vary significantly, since occupant needs are different. Clearly, there are significant variations of indoor conditions among houses, offices, factories, shops, hospitals, schools, theatres, restaurants, hotels, athletic halls, museums, computer rooms, etc.

The parameters which influence overall comfort can be grouped into three general categories [1], namely :

- **Physical parameters,** which include the air temperature of the environment (dry bulb), the mean radiant temperature of the indoor wall surfaces, the relative humidity of the air, the relative air velocity of the indoor air, the atmospheric pressure, the colour of the surrounding environment, odours, light intensity, and noise levels;
- **Organic parameters,** which include age, sex and national characteristics of the occupants;
- **External parameters,** which include human activity levels (related to the metabolism), type of clothing, and social conditions.

Among these, the most important parameters influencing thermal comfort are:

- dry bulb temperature
- relative humidity
- air velocity
- barometric pressure
- clothing
- activity.

Thermal comfort can be achieved by many different combinations of the above variables. In all cases, it is the end result that we are interested in achieving, which means that it is the combined effect of these parameters on the human body which is important. The positive or negative affect of one parameter on comfort may be enhanced or counterbalanced by the change of another parameter.

The body's thermal equilibrium is a dynamic balance between heat production (as a result of human metabolic rate) and body heat transfer by convection, conduction, radiation and evaporation to or from the environment, as shown in Figure 6.1. The thermal balance equation between the human body and the environment can be expressed as follows [3]:

$$Q_M - Q_{\text{dif}} - Q_{\text{evap}} - Q_{\text{resp}} = Q_r + Q_c \qquad (6.1)$$

where

Q_M = body's heat production due to metabolic rate (metabolism),

Q_{dif} = vapour diffusion through the skin,

Q_{evap} = sweat evaporation,

Q_{resp} = latent heat from sweating,

Q_r = radiative heat loss from the outer surface of a clothed person to the environment,

Q_c = convective heat transfer to the environment.

Sweating, and the resulting evaporative cooling sensation, is the main mechanism of thermal adjustment for the human body under hot environmental conditions or at a high level of activity. Clothing will directly influence the amount of heat exchange from the body to its environment and, depending on the type of material, will also influence the amount of absorbed solar radiation.

The control of environmental conditions in order to achieve thermal comfort can be performed by:

- passive controls (on the environment, clothing, metabolic rate), and
- active or hybrid controls (on the building).

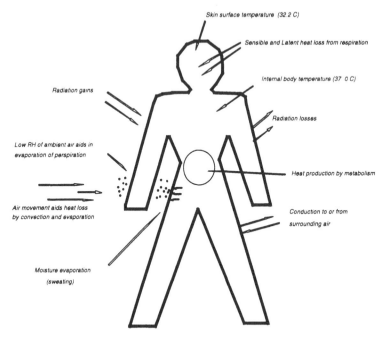

Figure 6.1 Interactions of the human body with the environment

To create thermal comfort and maintain the necessary indoor air-quality levels, in any environment, it is necessary to have air ventilation. Ventilation systems supply the necessary amounts of fresh air, which is controlled in terms of quantity, velocity, quality and thermal conditions.

The quantity of outdoor air that needs to be brought into the space is determined by national standards, depending on the function of the space. The air movement into the space has to be handled with care, since there is a direct influence of air velocity on occupant comfort.

Outdoor air quality will influence indoor conditions, thus one needs to exercise caution when using untreated outdoor air. This is of especially great importance in large metropolitan cities, where outdoor air may be heavily polluted with particulate and gaseous contaminants. Health standards may impose limits on the use of untreated outdoor air, which, as a result, limits the effectiveness of natural ventilation techniques and influences comfort conditions in naturally ventilated buildings. Ventilating a space with polluted outdoor air can significantly affect indoor air quality, which may result in occupant health problems. Alternatively, the use of mechanical ventilation and air-conditioning systems can be used to clean the outdoor air, removing atmospheric dust. However, it is not possible to treat outdoor air for gaseous contaminants. In any event, the use of mechanical ventilation and air conditioning will increase the energy consumption and operational cost of the building. It is also essential that the filtering system is well maintained, to prohibit the growth of micro-organisms, which can even be fatal.

Thermal conditions include indoor temperature and relative humidity. The indoor temperature is defined in terms of air temperature and internal wall surface temperature in a given space (radiant temperature). The relative humidity is the ratio of the mole fraction of water vapour in a given moist air sample to the mole fraction in an air sample that is saturated at the same temperature and pressure. Most air-conditioning systems are, in fact, used primarily to remove the excess water vapour from the air.

In order to understand better the various processes taking place between the human body and the air conditions of the surrounding environment, it is necessary to review some fundamental principles of psychrometrics.

PSYCHROMETRICS

The various thermodynamic properties of moist air can be graphically represented by the so-called psychrometric chart, illustrated in Figure 6.2. Although, the choice of coordinates for a psychrometric chart is arbitrary, a chart with coordinates of temperature and humidity ratio has been the most widely used. ASHRAE [4] has developed such charts for various pressure ranges (sea level pressure, 24.89 in Hg and 22.65 in Hg) and temperature ranges (low, normal and high temperature).

The coordinate system of a psychrometric chart has as the x axis the dry-bulb temperature, while on the y axis the humidity ratio (moisture content of the air, the

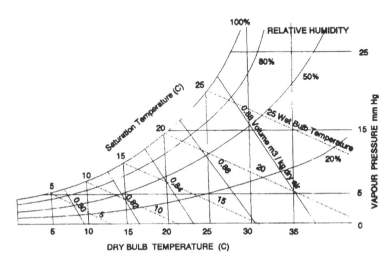

Figure 6.2 An illustration of the psychrometric chart

ratio of the mass of water vapour to the mass of dry air [kg of water/kg of dry air]). The outer left boundary of the chart is a curve representing the saturation line (100% relative humidity). The dry-bulb temperatures are straight lines, not precisely parallel to each other and inclined slightly from the vertical position. The wet-bulb temperature lines are oblique straight lines and differ slightly in direction from the enthalpy lines, which are drawn across the chart precisely parallel to each other. The volume is represented by oblique straight lines, again not precisely parallel to each other.

All the processes which deal with the treatment of moist air (i.e. cooling, dehumidification, mixing of air streams) can be represented on the psychrometric chart. In addition to the use of the chart for defining the various properties of air at a given condition, it can also be used to express the thermal energy content of the air, since 1 kg of water vapour represents 2.26 MJ of latent heat, while it takes 1004 J to increase the temperature of 1 kg of dry air by 1°K.

THERMAL EQUILIBRIUM OF THE HUMAN BODY

Human comfort is not a simple heat balance, but needs to take into account complex psychological processes. The thermal sensation is processed through several mental processes before it leads to an expression of preference or judgement. Of course, the primary parameters are physical, for example environmental conditions, activity, clothing, but there are also other influences, such as the state of acclimatization of the individual, personal expectations and attitudes and behavioural adjustments.

Among the various models and suggestions for the quantitative estimation of thermal comfort, the most widely used is the one suggested by Fanger [3]. This equation has been empirically developed following extensive study and monitoring of human beings under varying conditions, and a comprehensive statistical analysis of their responses.

Fanger's Thermal Equilibrium Equation is given by the following empirical formulation:

$$\frac{M}{A_{DU}}(1-\eta) - 2.6 \times 10^{-3}(256 t_{sk} - 3370 - P_v) - \frac{E_{rsw}}{A_{DU}}$$

$$-1.72 \times 10^{-5} \left(\frac{M}{A_{DU}}\right)(5800 - P_v) - 0.0014 \left(\frac{M}{A_{DU}}\right)(34 - t_a)$$

$$= \frac{(t_{sk} - t_{cl})}{0.155 I_{cl}}$$

$$= 3.96 \times 10^{-8} f_{cl}[(t_{cl} + 273)^4 - (t_{mrt} + 273)^4] + f_{cl} h_c (t_{cl} - t_a) \qquad (6.2)$$

where

M	=	heat generated by metabolism, depending on the activity, kcal h^{-1}
A_{DU}	=	total surface of the human body (average values for an adult ranging between 1.65 to 2.0 m^2)
M/A_{DU}	=	metabolic rate, kcal h m^{-2} (Table 6.1)
h	=	mechanical efficiency (Table 6.1)
t_{sk}	=	skin temperature, °C
P_v	=	vapour pressure, mm Hg
E_{rsw}	=	heat dissipated by sweating, kcal h^{-1}
t_a	=	ambient air temperature, °C
t_{cl}	=	temperature of clothing surface, °C
I_{cl}	=	clothing insulation, minimizes conduction, convection and radiation losses and gains by increasing the surface resistance to heat transfer, clo (Table 6.2)
f_{cl}	=	clothing factor accounting for the relative increase in the clothed body surface over that of the unclothed body (Table 6.2)
t_{mrt}	=	mean radiant temperature, °C (see section below on 'Comfort indices')
h_c	=	convective heat transfer coefficient, W m^{-2}.°C (Table 6.3).

Table 6.1 Data for different human activities [1]

Activity	Metabolic rate (M/A_{DU}) [kcal hr^{-1} m^{-2}]	Mechanical efficiency (η)	Relative velocity in still air [m s^{-1}]
Sleeping	35	0	0
Seated	50	0	0
Standing	60	0	0
Miscellaneous occupations			
Laboratory work			
General work	80	0	0
Setting up apparatus	110	0	0–0.2
Machine work			
Light	100–120	0–0.1	0–0.2
Heavy	200	0–0.1	0–0.2
Heavy work	300	0.2	0.5
Domestic work	100–170	0–0.1	0.1–0.3

Once the equality in Fanger's Thermal Equilibrium Equation is satisfied, then the heat generated by the body is dissipated and there is no increase or decrease of the body's temperature. This is the fundamental requirement for survival.

Table 6.2 Data for different clothing combinations [1]

Clothing combination	I_{cl} [clo]*	f_{cl}
Nude	0	1.0
Shorts	0.1	1.0
Light summer clothing	0.5	1.1
Light working ensemble	0.6	1.1
Typical business suit	1.0	1.15
Typical business suit and cotton coat	1.5	1.15
Light outdoor sportswear	0.9	1.15
Heavy traditional European suit	1.5	1.15–1.2
Heavy wool pile ensemble	3–4	1.3–1.5

* Thermal resistance of clothing is expressed in [clo] units
  (1 clo = 0.155 m^2.C/W)

Table 6.3 Variation in h_c and I_{cl} with activity in still air [1]

Activity [met]*	$h_c^\dagger$ [W m^{-2}.C]	$I_{cl}^\ddagger$ [clo]
Resting (0.85)	3.1	0.83
Sedentary (1.1)	3.3	0.80
Light activity (2.0)	6.0	0.60
Medium activity (3.0)	7.7	0.52

* 1 met = 58.2 W m^{-2} = 50 kcal h^{-1} m^{-2}
† h_c = 5.7 (met-0.85)$^{0.39}$, active in still air [5]. To relate h_c to room air velocity V, while the person is seated with moving air, use h_c=8.3$V^{0.6}$ [6].
‡ I_{cl} = 6.45/(4.7 + h_c)

Thermal comfort can be achieved by adjusting one of the influencing parameters. It is preferable, in order to achieve the desirable end effect, to give priority to the parameters which can be varied with no or low energy requirements.

Clothing is one of the easiest parameters that an individual can adjust in order to achieve thermal comfort in an environment with given conditions. Actions such as removing some garments during summer, or adding some clothing during winter, should be given first priority before attempting to bring the environment (by changing the air temperature for example) to the necessary conditions for thermal comfort [7].

Furthermore, thermal sensation depends on the type of clothing material and on the fit of garments. Research has shown that people are more affected by air humidity than previously assumed and that individuals wearing wool clothing feel the most warmth when entering a humid space [8]. If polyester is worn or people are naked, the effect of changing humidity is of short duration, whereas the thermal effect lasts longer when wool is worn. The increased thermal effect is caused by absorption or desorption in the wool clothing, while polyester is almost unaffected by humidity.

Air movement around the human body can also influence thermal comfort. It determines the convective heat exchange of the body and the evaporative capacity of the air. Convective losses are directly proportional to a power of the air velocity and the temperature difference between the skin and air temperatures. Higher air velocities increase evaporation rates and consequently enhance the cooling sensation and reduce the negative effect of high humidity [9].

During summer, natural ventilation or the use of ceiling fans to enhance and control indoor air movement can shift the thermal comfort area to higher air temperatures. The ASHRAE recommended upper limit of indoor air movement is 0.8 m s^{-1}. Above this value, loose papers may be disturbed. Such air speeds permit one to maintain a space about 2°C warmer, at for example 60% relative humidity, and still maintain optimum comfort. This means that occupants can be in comfort at higher air temperatures. Even in air-conditioned spaces, this will allow us to maintain the thermostat at a higher setting, which means a lower energy consumption of the air-conditioning system, while maintaining comfort conditions.

Humidity is another determinant factor of thermal comfort. It does not affect the thermal load from the environment on the body, but it determines the evaporative capacity of the air. Low relative humidity of ambient air aids the evaporation of perspiration from the human body, which in turn enhances the cooling sensation (a form of evaporative cooling). During an 18-month study sponsored by ASHRAE [8], researchers from the Technical University of Denmark have shown that during transient conditions where people are moving from one space to another with a different humidity, the immediate thermal effect of humidity was felt 2–3 times more strongly than the various standards predicted [5, 6], on the assumption that people are in steady-state conditions. As a result, the thermal effect of humidity is much greater during a change than during a longer stay in a given environment.

The metabolic rate depends on human physiology, sex and type of activity. Although the type of activity is usually specified by the nature of work that humans have to perform for a given task, it is evident from experience that people avoid strenuous work during the hot summer months or at least during the midday hours. Another common practice for the control of metabolic rate is to vary the type of foods consumed during summer. Low-fat, light foods in smaller quantities, and high rates of fluid consumption, are usually recommended in order to keep heat production due to metabolism as low as possible.

COMFORT INDICES

To assess the environmental conditions in terms of comfort, there are three classes of environmental indices [10], namely :

- **Direct indices**
  - dry-bulb temperature,
  - dew-point temperature,
  - wet-bulb temperature,
  - relative humidity, and
  - air movement.
- **Rationally derived indices**
  - mean radiant temperature,
  - operative temperature,
  - heat stress, and
  - thermal stress.
- **Empirical indices**
  - effective temperature,
  - standard effective temperature, and
  - equatorial comfort.

Each one of these indices is explained in the following:

Dry-bulb temperature (t_{db} or t_a). The simplest practical index of cold and warmth under ordinary room conditions is the reading obtained from an ordinary dry-bulb thermometer. For average relative humidity values (40–60%), it can be significant in judging comfort, especially towards lower temperatures.

Dew-point temperature (t_{dp}). A unique measure of humidity in the environment. The temperature at which a given air–water vapour mixture is saturated with water vapour. The saturated vapour pressure at t_{dp} is the ambient vapour pressure.

Wet-bulb temperature (t_{wb}). A temperature, for any state of moist air, at which liquid water may be evaporated into the air to bring it to saturation at exactly the same temperature and pressure. A useful index for severe heat stress, especially when the human body is near its upper limits of temperature regulation by sweating. The upper tolerance limit for t_{wb} occurs at approximately 30°C for both normally clothed and unclothed subjects, with air movement ranging from 0.1 to 0.5 m s^{-1}.

Relative humidity (rh). For any temperature and barometric pressure affecting people's living space, the ratio of the partial pressure of water vapour to the saturation pressure is a measure of rh, as a fraction or as a percentage. Relative humidity has no meaning as an environmental index without knowledge of the accompanying dry-bulb temperature.

When any two of the above four factors are known, then the other two may be predicted by using the Psychrometric Chart. Figure 6.3 shows a detail of a psychrometric chart, concentrating around the comfort region. It includes the dry-bulb temperature (°C) on the x axis and the wet-bulb temperature (°C) on the y axis. Relative humidity lines run from lower left to upper right. The 100% relative humidity line is called the saturation line. The scale of the effective temperatures (discussed in a following section) is marked off along the saturation line.

The assumptions on which this chart is based are that:

- subjects are clothed normally for indoor living,
- subjects are engaged in light activity (office work),
- air velocity is 15 to 25 fpm (0.1 m s^{-1} – almost still air), and
- there are no radiation effects.

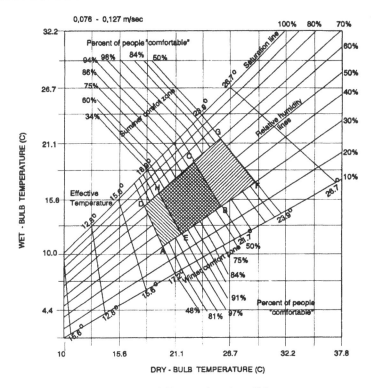

Figure 6.3 The comfort chart [1]

In general, relative humidity values above 70% or below 30% are not considered to be in the optimum comfort zone for the majority, even though some people may prefer one or other humidity extreme. Air conditioners are used to achieve comfort conditions and are primarily intended for mechanically removing the water vapour from the indoor air, rather than lowering the air temperature. The potential effect of humidity on comfort is presented in [11]. An increase in dry-bulb temperature by 0.5°C has the same effect on comfort as a 5.5°C increase in dew point [12]. Lowered humidity levels of indoor air have beneficial effects other than allowing a higher temperature setting, resulting in energy savings, including reduction in mould and/or mildew, reduced sensitivity to odours and a greater feeling of air freshness [11].

The winter comfort zone, shown as area ABCD in Figure 6.3, encompasses the possible combinations of dry bulb temperature and relative humidity which produce winter comfort conditions for most people. The area EFGH shows the summer comfort zone for most individuals.

The reader is cautioned on the use of this chart if the four assumptions listed above are not in effect. The values on this chart will not be reliable, for example, if there are large warm or cold surfaces that will result in radiation heat transfer to or from the people.

Air movement (V). The heat-balance equations are also a function of air movement. The convective heat transfer coefficient is a qualitative measure for the effect of air movement on convective heat exchange.

Figures 6.4 to 6.6 show the combined effects of the various parameters [3]. Comfort lines have been drawn, i.e. curves through various combinations of two variables which create comfort, provided that the remaining parameters are kept constant. In using the comfort charts, one should first estimate the activity level and clothing, taking space use into account. Let us, for example, calculate the comfort temperature in a space where the mean activity corresponds to 1.5 met, relative air velocity is 0.4 m s^{-1}, clothing is 1.0 clo and relative humidity is 50%. Then from the lower chart in Figure 6.5, the air temperature that would satisfy comfort for the given conditions is $t_a = t_{mrt} = 20.8°C$.

Mean radiant temperature (t_{mrt}). The uniform surface temperature of an imaginary black enclosure with which a person (also assumed as a black body) exchanges the same heat by radiation as that in the actual environment.

Let us, for example, consider a warehouse space which must be maintained at an air temperature of 14°C and 50% rh, because of the nature of goods stored. Air velocity is 0.2 m s^{-1}. The comfort of an occupant performing sedentary work and clothed at 1 clo is to be maintained by means of high-intensity infrared heaters placed above their workplace. To calculate the mean radiant temperature necessary for comfort, we use the upper right-hand chart in Figure 6.6, to obtain $t_{mrt} = 38°C$.

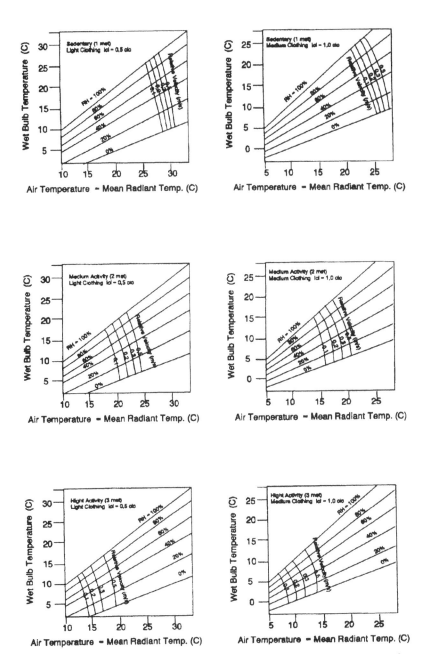

Figure 6.4 Comfort lines as a function of humidity and ambient temperature for people with different clothing and activity levels [3]

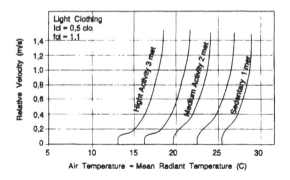

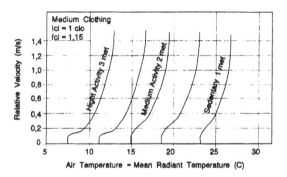

*Figure 6.5 Comfort lines as a function of air velocity and
ambient temperature for people with different clothing and
activity levels [3]*

Operative temperature (t_o). The uniform temperature of an imaginary enclosure
with which a person will exchange the same dry heat by radiation and convection as
that in the actual environment. It can be estimated as follows:

$$t_o = (h_r t_{mrt} + h_c t_a) / (h_r + h_c), \tag{6.3}$$

where, h_r is the radiation transfer coefficient with an average value of 4.7 W m^{-2}.C
in normal environments.

Heat stress (HS). The heat stress index [13] expresses the heat balance equation of
the human body adopted for the given conditions of work and climate. The index is
calculated by :

$$\text{HSI} = (\text{Required evaporation} / \text{Maximum evaporative capacity}) \times 100 \quad (6.4)$$

where the required sweat evaporation can be estimated by the

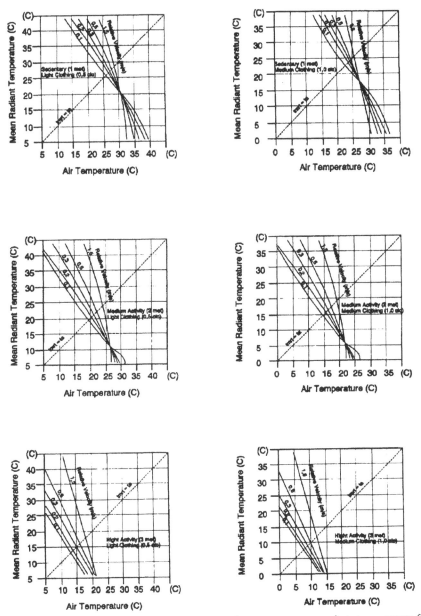

Figure 6.6 Comfort lines as a function of mean radiant temperature and air temperature for people with different clothing and activity levels [3]

total heat stress acting on the human body $=$ metabolism $\pm$ radiation $\pm$ convection

$$= 100 \pm h_r(t_{wb} - t_{sk}) \pm h_c(t_a - t_{sk}) \quad (6.5)$$

and the maximum evaporative capacity $\quad = 24.2V^{0.4}(60 - P_v).$ $\qquad\qquad$ (6.6)

The HSI overestimates the cooling effect of wind and the warming effect of humidity [14].

Thermal stress (TS). This index was developed to cover all the mechanisms of heat loss and heat gain by the human body, at various levels of work and clothing [14]. The general equation for calculating TS is given by:

$$TS = [M - W \pm C]\exp(0.6(E / E_{max} - 0.12)) \qquad\qquad (6.7)$$

where

$$W = 0.2(M - 100)$$

$$C = \alpha V^{0.3}(t_a - 35)$$

$$E = (M - W) \pm C$$

$$E_{max} = pV^{0.3}(42 - P_v).$$

α and p are coefficients depending on the type of clothing ($\alpha = 18.3$, $p = 31.6$ for seminude; $\alpha = 15.1$, $p = 2$ 0.5 for light summer clothing). The thermal stress index has also been modified to include a term for the effect of direct solar radiation on human comfort, for outdoor conditions [14].

Effective temperature (ET or t_{eff}). This index combines the effects of dry- and wet-bulb temperatures and air movement, to yield equivalent sensations of warmth and cold. The effective temperature index was developed for ASHRAE, by Houghten, Yaglou and Miller [15]. The original t_{eff} was developed for subjects wearing 1 clo insulation. However, this scale overemphasizes the effect of humidity in cooler and neutral conditions [16, 17], underemphasizes its effect in warm conditions and does not fully account for air velocity under hot humid conditions [17].

Standard effective temperature (SET). The new effective temperature is defined as the dry-bulb temperature of a uniform enclosure at 50% rh in which humans would have the same net heat exchange by radiation, convection and evaporation, as they would in the varying humidity test environment. Clothing is standardized at 0.6 clo, air movement (still) at 0.2 m s^{-1}, time of exposure 1 h and the chosen activity level sedentary (1 met).

The varying zones of physiological regulation for this standard combination of met–clo–air movement are illustrated in Figure 6.7. The shaded area represents the thermal comfort zone. The human thermal responses corresponding to the various SET temperatures are shown in Table 6.4 [18].

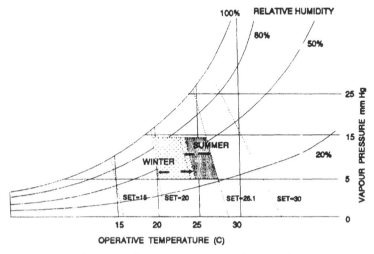

Figure 6.7 Standard effective temperature and the comfort zone [1]

Table 6.4 Human thermal responses to standard effective temperature (SET)

SET [°C]	Temperature sensation	Discomfort	Regulation of body temperature
		Limited tolerance	Failure of free-skin evaporation
40			
	Very Hot	Very uncomfortable	
	Hot	Uncomfortable	
35			
	Warm	Slightly uncomfortable	Increasing sweating
30			
	Slightly warm		
25			
	Neutral	Comfortable	No registered sweating
	Slightly cool		
20			
	Cool	Slightly uncomfortable	
15			Behavioural changes
	Cold		
10	Very cold	Uncomfortable	Shivering begins

Equatorial comfort (EC). The index was based on a thermal comfort study of people in Singapore [19]. It was developed by correlating observed temperature, pressure and air velocity with the temperature of still, saturated air which would produce the same overall sensation. The derived equation for EC is given by:

$$EC = 0.574t_a + 0.2033P_v - 1.81V^{0.5} + 42. \tag{6.8}$$

The index is applicable for conditions with $t_{wb} > 25\ °C$ and when air temperature is equal to the mean radiant temperature.

PREDICTION OF THERMAL COMFORT

Thermal comfort is a complex concept, which is influenced by a number of parameters and is not always perceived the same by all humans. However, several attempts have been made in order to develop empirical correlations for relating comfort perceptions to specific physiological responses, in a more comprehensive way than the previously presented comfort indices.

There are several ways of quantitatively expressing thermal comfort and thermal sensation, the most popular one being the theory of Predicted Mean Vote (PMV), associated with the index of Predicted Percent of Dissatisfied (PPD) people.

The PMV and PPD indices were introduced and empirically derived, by Fanger [3] during the 1970s. However, there have been several field surveys [20] which do not fully agree with the results predicted by the PMV theory. As a result, an adaptive model has been formulated by Humphreys and Nicol [21, 22], in an attempt to account for the fact that people can eventually adapt to the prevailing conditions, while it is in fact sudden or unexpected changes in temperature that are primarily the cause of discomfort [23, 24].

Predicted Mean Vote (PMV)

The PMV index is calculated through a complex mathematical function of human activity, clothing and four environmental parameters. Since 1984 the method has been the basis of the international standard ISO-7730 [9] for assessing thermal comfort in spaces with average temperatures.

PMV relates the imbalance between the actual heat flow from the human body in a given environment and the heat flow required for optimum comfort at the specified activity in terms of the following equation:

$$PMV = \left[0.352 \exp\left(-0.042\left(\frac{M}{A_{DU}}\right)\right) + 0.032\right]$$

$$\times \left[\left(\frac{M}{A_{DU}}\right)(1-\eta) - 0.35\left[43 - 0.061\left(\frac{M}{A_{DU}}\right)(1-\eta) - P_v\right]\right.$$

$$-0.42\left[\left(\frac{M}{A_{DU}}\right)(1-\eta) - 50\right]$$

$$-0.0023\left(\frac{M}{A_{DU}}\right)(44 - P_v) - 0.0014\left(\frac{M}{A_{DU}}\right)(34 - t_a)$$

$$\left. -3.4 \times 10^{-8} f_{cl}[(t_{cl} + 273)^4 - (t_{mrt} + 273)^4 - f_{cl}h_c(t_{cl} - t_a)]\right] \qquad (6.9)$$

where

M/A_{DU} = metabolic rate, kcal h^{-1} m^{-2} (Table 6.1)

η = mechanical efficiency (Table 6.1)

P_v = vapour pressure, mm Hg

t_a = ambient air temperature, °C

t_{cl} = temperature of clothing surface, °C

I_{cl} = clothing insulation, minimizes conduction, convection and radiation losses and gains by increasing the surface resistance to heat transfer, clo (Table 6.2),

f_{cl} = clothing factor accounting for the relative increase in the clothed body surface over that of the unclothed body (from Table 6.2 or the equation below)

t_{mrt} = mean radiant temperature, °C (see previous section)

h_c = convective heat transfer coefficient, W m^{-2} °C^{-1}

$$t_{cl} = 35.7 - 0.032\left(\frac{M}{A_{DU}}\right)(1-\eta)$$

$$-0.181 I_{cl}[3.4 \times 10^{-8} f_{cl}[(t_{cl} + 273)^4 - (t_{mrt} + 273)^4] + f_{cl}h_c(t_{cl} - t_a)]$$

f_{cl} = 1.00 + 1.290I_{cl} for $I_{cl} < 0.078$ [m^2 K W^{-1}]

f_{cl} = 1.05 + 0.645I_{cl} for $I_{cl} > 0.078$ [m^2 K W^{-1}]

h_c = 2.05$(t_{cl} - t_a)^{0.25}$ for 2.05$(t_{cl} - t_a)^{0.25} > 10.4$ $V^{0.5}$

h_c = 10.4 $V^{0.5}$ for 2.05$(t_{cl} - t_a)^{0.25} < 10.4$ $V^{0.5}$

The PMV index can be used to quantify the degree of discomfort, giving the predicted mean vote of a large group of subjects according to the psychological scale listed in Table 6.5. PMV values range between –3 and +3. Negative values indicate an uncomfortable feeling due to a cold sensation, while positive values indicate an uncomfortable feeling due to a hot sensation. Zero is the neutral point, representing comfort.

Table 6.5 Thermal sensation scale for the PMV index

Scale	Sensation
–3	Cold
–2	Cool
–1	Slightly cool
0	Neutral comfort
+1	Slightly warm
+2	Warm
+3	Hot

The PMV index has also been calculated and tabulated as a function of representative values of the various influencing parameters, as listed in Table 6.6. The full expression for PMV, equation (9), can be easily programmed in order to facilitate repetitive calculations. A program listing written in BASIC, for calculating the PMV values or any other variable for balancing Fanger's comfort equation, is given in Appendix A, at the end of this chapter. A manual method is also available in Appendix B at the end of the chapter. Worksheets, along with a worked example, are also provided.

It is also possible to estimate the percentage of people dissatisfied (i.e. voting –3, –2, +2, or +3) with the thermal environment at various temperature conditions. Results of various studies conducted by Fanger [3], led to a relationship between PMV and PPD, as shown in Table 6.7.

The value of PPD is calculated by the following and expressed as a percentage:

$$PPD = 100 - 95\exp[-(0.03353\,PMV^4 + 0.1297\,PMV^2)] \qquad (6.10)$$

where dissatisfied is defined as anybody not voting either –1, +1 or 0. This relationship is shown in Figure 6.8. A PPD of 10% corresponds to the PMV range of –0.5 or +0.5. Even with PMV equal to zero, about 5% of the people remain dissatisfied.

It is very difficult to achieve thermal comfort for all people in a space, since people have different dressing habits, different levels of activities, different metabolic rates and different psychological influences, which also play a role in determining thermal comfort. The objective should be to provide thermal comfort for the majority of occupants in a space. Values ranging between –0.5 < PMV < 0.5 and PPD < 10%, are considered acceptable.

Table 6.6 PMV values for typical conditions

Clothing [clo]	Ambient temperature [C]	Relative velocity [m/s]			
		< 0.1	0.2	0.3	0.4
0.5	23	−1.1	−1.51	−1.78	−1.99
	24	−0.72	−1.11	−1.36	−1.55
	25	−0.34	−0.71	−0.94	−1.11
	26	0.04	−0.31	−0.51	−0.66
	27	0.42	0.09	−0.08	−0.22
	28	0.80	0.49	0.34	0.23
	29	1.17	0.90	0.77	0.68
	30	1.54	1.30	1.20	1.13
0.75	21	−1.11	−1.44	−1.66	−1.82
	22	−0.79	−1.11	−1.31	−1.46
	23	−0.47	−0.78	−0.96	−1.09
	24	−0.15	−0.44	−0.61	−0.73
	25	0.17	−0.11	−0.26	−0.37
	26	0.49	0.23	0.09	0.00
	27	0.81	0.56	0.45	0.36
	28	1.12	0.90	0.80	0.73
1.00	20	−0.85	−1.13	−1.29	−1.41
	21	−0.57	−0.84	−0.99	−1.11
	22	−0.30	−0.55	−0.69	−0.80
	23	−0.02	−0.27	−0.39	−0.49
	24	0.26	0.02	−0.09	−0.18
	25	0.53	0.31	0.21	0.13
	26	0.81	0.60	0.51	0.44
	27	1.08	0.89	0.81	0.75

Table 6.7 Corresponding values of PMV and PPD

PMV	PPD (%)	Percentage of persons predicted to vote:		
		0 (%)	−1, 0 or +1 (%)	−2, −1, 0, +1, or +2 (%)
+2	75	5	25	70
+1	25	27	75	95
0	5	55	95	100
−1	25	27	75	95
−2	75	5	25	70

Although the PMV theory has gained wide acceptance, one needs to clearly understand that the comfort equation is nothing but an attempt to simulate the complex phenomena related to thermal comfort. Fanger has recently stated [25] that 'he has never advocated that one model could describe everything, or even just thermal comfort. The predicted mean vote model just gives you some indication of where

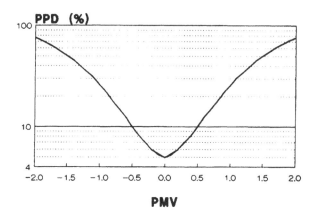

Figure 6.8 The relationship of the PPD with the PMV

you should be'. Clearly, that is the purpose and the limit of most predictive tools and, surely, that is better than nothing.

Satisfaction with indoor conditions is also affected by indoor air quality. To quantify air pollution sources and air pollution perceived by humans, Fanger [26] has introduced two units, the olf and the decipol, respectively.

One olf is the emission rate of air pollutants from a standard person. All other pollution sources are expressed in terms of the equivalent source strength, defined as the number of standard persons (olfs) required to cause the same dissatisfaction as the actual pollution source. The olf is a unit similar to the met unit for metabolic rate or the clo unit for clothing. A correlation, relating the percentage of dissatisfied people (PD) as a function of the ventilation rate per olf during steady state conditions, has been developed as a result of two studies carried out in Denmark and is given by:

$$PD = 395 \exp(-1.83q^{0.25}) \qquad \text{for } q \geq 0.32 \text{ l/s.olf} \qquad (6.11)$$

$$PD = 100\% \qquad \text{for } q < 0.32 \text{ l/s.olf} \qquad (6.12)$$

where PD is the percentage of dissatisfied people and q is the steady-state ventilation rate per olf.

One decipol is the air pollution caused by one standard person (one olf) ventilated by 10 l/s of unpolluted air. In a similar manner, an expression has been developed relating the percentage of dissatisfied people with the perceived air pollution, shown in the following expression as C and expressed in decipol, as follows:

$$PD = 395 \exp(-3.25C^{-0.25}) \qquad \text{for } C \leq 31.3 \text{ decipol} \qquad (6.13)$$

$$PD = 100\% \qquad \text{for } C > 31.3 \text{ decipol} \qquad (6.14)$$

These units provide a rational basis for the identification of pollution sources, for the calculation of ventilation requirements and for the prediction of indoor air quality, which are also closely related to thermal comfort conditions.

Humphreys' adaptive model

Thermal comfort is not an exact concept and human responses with regard to comfort do not occur as a response to an exact temperature. Field studies can provide very useful information in the continuous search for understanding the links between human comfort and other environmental (i.e. temperature, humidity, air velocity) and occupant (i.e. clothing, activity, sex) parameters. A field study of thermal comfort is a survey of the occupants' reports of their subjective warmth, accompanied by objective measurements of the thermal environment within the building [27]. A comparison among the various indices used to assess thermal comfort and actual data from various field studies have not yet indicated an index of universal application. Simple indices such as air temperature appear to be, in some cases, as good as more complex indices.

Using available data from over 30 comfort surveys from around the world, Humphreys has proposed a series of simple correlations for predicting comfort conditions. For free-running buildings, the comfort temperature (T_{co}) can be predicted from the mean monthly outdoor temperature (T_m) in [C], through the following expression

$$T_{co} = 0.53 T_m + 11.9. \tag{6.15}$$

The prediction has a standard error of 1°C and applies to a range of $10 \le T_m \le 34$ C.

For heated or cooled buildings, the comfort temperature is given by a more complex equation,

$$T_{co} = 0.0065\ T_m^2 + 0.32\ T_h + 12.4 \tag{6.16}$$

where T_h is the average daily maximum temperature of the hottest months of the year. The prediction has a standard error of 1.4°C and applies to a range of $-24 \le T_m \le 23$°C and $18 \le T_h \le 30$°C. From the analysis of the available data, Humphreys concluded that the comfort temperature also depends on the country of origin in a way which appears to be unrelated to the climate. Russian and English comfort temperatures were found to be below average, while North American ones were above average.

Figure 6.9 shows the relation between the preferred indoor temperatures obtained from thermal comfort surveys and the mean outdoor air temperature, obtained from meteorological tables, for the seasons at which the surveys were conducted.

Comparing the temperature occurring within a building with the temperatures preferred by the occupants, it was also found that the mean indoor temperatures were higher than the comfort temperatures, by a fixed value. Accordingly, for free-

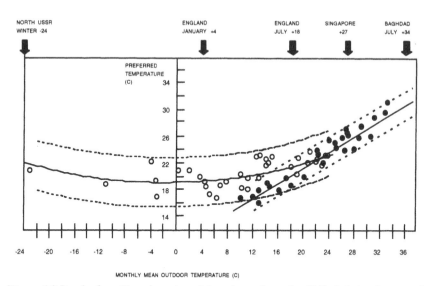

Figure 6.9 Results from Humphreys' work into thermal comfort [26]. Relation between the preferred indoor temperature obtained from thermal comfort surveys and the mean outdoor air temperature, obtained from meteorological tables, for the seasons at which the surveys were conducted. Solid dots represent free running buildings, and circles represent other buildings.

running buildings, mean indoor air temperatures are on average 2.4°C higher than the comfort temperature, while for heated or cooled buildings the corresponding value is 0.6°C higher.

Overall, according to Humphreys, thermal comfort is best ensured by giving as much effective control to occupants as possible. In this way, people can adjust their environment to satisfy their specific requirements, rather than fixing room temperature at some theoretically determined optimum in an attempt to satisfy all occupants. In addition, people appear to be more tolerant of poor conditions if they feel they are in control, or because they can achieve an environment more suitable to their needs [28]. This can be achieved even in naturally ventilated buildings, through assessment and control of shading devices, opening/closing windows and the use of fans. For air-conditioned buildings, space-temperature control or microclimate control at each desk is necessary to satisfy individual needs.

Comfort and behaviour

There is a close relationship between comfort and behavioural activities, through the excitation coefficient, which is directly related to the type of activity.

Some activities are characterized by a higher excitation coefficient. For example, activities which require a higher concentration in comparison with activities which utilize the memory. Small increases of air temperature, while still remaining within the thermal comfort range, cause a decrease of the excitation coefficient. Near the lower limits of comfort, a small decrease of temperature will result in an increase of this coefficient. Any further decrease, though, will result in discomfort. In general, relatively higher temperatures may result in a lower activity level.

Although the previously discussed indices for assessing thermal comfort, and in particular the PMV theory, have been widely used, it appears that there are in some cases significant differences between predicted thermal comfort and the actual comfort sensation which results from monitoring studies under real conditions. These discrepancies can be explained if one considers that there are several behavioural and psychological factors which have not been included in the equations used to predict thermal comfort.

Consequently, the notion of person cooling, as distinct from space cooling, justifies the need for a behavioural comfort model as distinct from a fixed state model [29]. For a complete assessment, one needs both a building model, to predict the effect of the building upon the climatic boundary conditions, and a behavioural model, to predict the response of the occupant.

Current research activities

To investigate further the reported discrepancies between the predictions using Fanger's PMV Theory and the results from Humphreys' work, an extensive field campaign was carried out within the European Research Programme PASCOOL (Passive Cooling in Buildings [30]). This comfort survey was designed by Baker [31] and was carried out in several buildings in southern European countries. The work provided valuable data on the interactions between the occupants and the building. The aim of this survey was to monitor the local thermal environment (indoor conditions, including building data such as thermal mass, ceiling height etc.) with simultaneous behavioural observation and subjective assessment of building occupants. Human activities and behaviour for achieving comfort influence to a great extent the final energy consumption of the building. The results were used to develop a computer simulation tool for relating the reactions of the occupants in indoor spaces to the environment.

Another research activity, which was also carried out within the frame of PAS-COOL, included the work proposed by Berger [32]. The objective of this effort was to relate comfort sensation to the quality of the ambient air (evaporative power of the air, skin wetness, ventilation rates, radiative asymmetries, indoor temperature stratification, indoor surface temperature), psychological aspects (relation to outdoor environment, odours, outside noise, freedom of actions), microclimate (temperature and humidity variations, air velocity), along with other commonly used parameters (level of activity, clothing etc). Correlations resulting from experiments carried out in climatic chambers and from actual monitoring were used to develop software in order to define strategies for improving comfort conditions.

Measurements of thermal comfort

As already discussed, thermal comfort depends on clothing and activity (estimated according to the use of a particular space) and on four physical parameters: air temperature, air velocity, mean radiant temperature and air humidity. The main instruments for measuring these parameters are described in the following:

- **Air temperature** can be measured by various types of thermometers. Placed in a room, the sensor registers a temperature between air temperature and mean radiant temperature. To reduce the radiant error, the sensor should be made as small as possible. This also provides a favourably low time constant. One can also use a shield around the sensor, use a sensor with a low emittance surface or artificially increase the air velocity around the sensor.
- **Air velocities** inside indoor spaces are usually relatively small, ranging from 0 to 0.5 m s^{-1}, unless there is natural or forced ventilation. Since the air velocity fluctuates, the measurement at a given point should be taken over a suitable period, such 3 to 5 minutes, to obtain a reasonable mean value. The most commonly used instrument is the thermal anemometer.
- **Mean radiant temperature** of the enclosure's surfaces can be estimated by measuring the surface temperatures of the surrounding surfaces and determining the angle factors between the person and the N surrounding surfaces. The values are then substituted in the following expression to obtain t_{mrt} :

$$t_{mrt} = t_1 F_{p-1} + t_2 F_{p-2} + \ldots + t_N F_{p-N}. \tag{6.17}$$

However, a commonly used instrument, called Vernon's globe thermometer, can simplify this procedure. The instrument consists of a hollow sphere 152 mm in diameter, coated with flat black paint and having a thermocouple or thermometer bulb at its centre. The temperature assumed by the globe at equilibrium is the result of a balance between heat gained or lost by radiation and convection.
- **Air humidity** can be measured at one location in the space, since the vapour pressure will normally be uniform. Usually a psychrometer will provide sufficient accuracy.
- **Thermal comfort**. When the thermal parameters have been measured, one can calculate their combined effect on humans by using one of the thermal indices previously discussed.

An alternative way is to use an integrated instrument in order to evaluate the combined effect of two or more thermal parameters on thermal comfort. In the early 1980s, a simple instrument was developed at the Technical University of Denmark, which, by direct measurement of the PMV value, gives information on the occupants' expected thermal sensations.

The comfort meter [1] has a heated ellipsoid-shaped sensor that simulates the human body. The sensor then integrates the thermal effect of the air temperature, mean radiant temperature and air velocity in approximately the same way as the human body does. The activity level, clothing and vapour pressure must be specified as input parameters. The electronic instrument calculates the PMV value which is directly indicated on the instrument.

More comprehensive measurements during field tests of individual workstations in working spaces can be performed by a mobile instrumentation cart [33]. This apparatus has been used to collect data over a six month period, in several office buildings, for an ASHRAE research project. The mobile measurement cart includes three systems: a laptop computerized survey system for subjective assessment, a packaged indoor environment measurement system for mid-level physical measurements and a microdatalogger-based measurement system for all additional physical measurements. Temporal indoor conditions were recorded using a fixed-position, microdatalogger-based equipment group, collecting time-series data. A detailed description of the system and the various transducers is provided in [33, 34].

It is evident from the previous discussion that thermal comfort is a complex concept, depending on a number of influencing parameters. However, variations of indoor air temperature and air movement, outside the strictly defined thermal comfort zone, can sometimes be beneficial [16]. They prevent a feeling of monotony and have an invigorating effect. For example, maintaining lower temperatures for some functions and activities (like hard mental work) are preferable in order to maintain alertness. Temperature variations at the perimeters of thermal comfort can also reduce the energy consumption. This of course requires a close control of indoor conditions with an air-conditioning system.

The maximum rate of temperature change should not exceed $0.75°C$ per hour, provided that the maximum deviation from the mean comfort temperature is lower than $2.25°C$ [35]. In another study of evaluating human response to indoor air changes, it was observed that there is no appreciable difference in terms of thermal comfort between an indoor environment at constant temperature and one with a change of 0.5, 1 and $1.5°C$ per hour, provided that the maximum temperature deviation from neutral temperature is no greater than $2.25°C$ [36].

The extent to which comfort zones can be modified, subject to the location and the prevailing climatic conditions, is reviewed in [37]. There are boundaries of applicability of various building design strategies and passive cooling techniques that one must follow to achieve a balance between energy consumption and comfort.

Results from a recently completed survey of comfort in office buildings [38] suggest that occupants reporting discomfort within the extreme thermal sensations were not always simultaneously dissatisfied and that feelings of discomfort were more often associated with a sense of warmth as compared to coolness. It is also possible that the data used as input to predictive models, for clothing and activity levels of the occupants, may account for some of the discrepancy between measured and predicted comfort. For example, the effect of the furniture clo value can cause significant differences between the estimated insulation value based on current comfort standards, which simply assume seasonal clothing worn by occupants, and the actual situation, with additional insulation provided by the chair.

REFERENCES

1 'Physiological principles for comfort and health' (1989). *ASHRAE Fundamentals*, Chapter 8. American Society of Heating Refrigerating and Air Conditioning Engineers, Atlanta.

2 'Study links comfort to job performance' (1992). *ASHRAE Journal*, Vol. 34, July, pp. 11–13.

3 Fanger, P.O. (1970). *Thermal Comfort – Analysis and Applications in Environmental Engineering*, McGraw Hill, New York.

4 'Psychrometrics' (1989). *ASHRAE Fundamentals*, Chapter 6. American Society of Heating Refrigerating and Air Conditioning Engineers, Atlanta.

5 'Thermal environmental conditions for human occupancy' (1981). ASHRAE Standard 55, American Society of Heating Refrigerating and Air Conditioning Engineers, Atlanta.

6 'Moderate thermal environments – Determination of the PMV and PPD indices and the specification of the conditions for thermal comfort' (1984). ISO 7730. International Standards Organization, Geneva.

7 Gagge, A.P., Y. Nishi and R.G. Nevins (1976). 'The role of clothing in meeting FEA energy conservation guidelines', *ASHRAE Transactions*, Vol. 82, Part II, p. 234.

8 Fanger, P.O. (1989). 'Impact of air humidity on thermal comfort during step-changes', *ASHRAE Transactions*, Part II, Paper #3289.

9 Mitchell, D. (1974). *Convective Heat Transfer in Man and Other Animals, Heat Loss from Animals and Man*. Butterworth, London.

10 McQuiston, F.C. and J.D. Parker (1982). *Heating, Ventilating and Air Conditioning*, 2nd ed. John Wiley & Sons, New York.

11 Berglund, L. (1991). 'Comfort benefits for summer air conditioning with ice storage', *ASHRAE Transactions*, Vol. 97, Part 1.

12 Int-Hout, D. (1992). 'Low temperature air, thermal comfort and indoor air quality', *ASHRAE Journal*, Vol. 34, No. 5, pp. 34–39.

13 Belding, H.S. and T.F. Hatch (1955). 'Index for evaluating heat stress in terms of resulting physiological strains', *Heat Pipe and Air Conditioning*, Vol. 27, No. 11, p. 129.

14 Givoni, B. (1976). *Man, Climate and Architecture*, 2nd ed. Applied Science Publishers, London.

15 Houghten, F.C. and C.P. Yaglou (1923). 'Determining lines of equal comfort', *ASHVE Transactions*, Vol. 29, p. 163.

16 Yaglou, C.P. (1947). 'A method for improving the effective temperature index', *ASHVE Transactions*, Vol. 53, p. 307.

17 Givoni, B. (1963). 'Estimation of the effect of climate on man: development of a new thermal index', Research Report to UNESCO, Building Research Station, Technion, Haifa, Israel.

18 Markus, T.A. and E.N. Morris (1980). *Buildings, Climate and Energy*. Pitman, London.

19 Webb, C. (1960). 'Thermal discomfort in an equatorial climate. A monogram for the equatorial comfort index', *Journal of the IHVE*, Vol. 27, p. 10.

20 Humphreys, M.A. (1976). 'Field Studies of thermal comfort compared and applied', *Building Services Engineer*, Vol. 44, pp. 5–27.

21 Humphreys, M.A. and J.F. Nicol (1970). 'An investigation into the thermal comfort of office workers', *Journal of the IHVE*, Vol. 38, pp. 181–189.

22 Nicol, J.F. and M.A. Humphreys (1972). 'Thermal comfort as part of a self-regulating system', *Proceedings of the CIB Commission W45 Symposium: Moderate Comfort and Moderate Heat Stress*. HMSO, London.

23 Humphreys, M.A. (1992). 'Thermal comfort requirements, climate and energy', *Proceedings of the World Renewable Energy Congress*, Reading.

24 Nicol, J.F. (1992). 'Time and thermal comfort', *Proceedings of the World Renewable Energy Congress*, Reading.

25 Bunn, R. (1993). 'Fanger: face to face', *Building Services*, June, pp. 25–28.

26 Fanger, P.O. (1988). 'Introduction of the olf and the decipol units to quantify air pollution perceived by humans indoors and outdoors', *Energy and Buildings*, Vol. 12, pp. 1–6.

27 Humphreys, M.A. (1981). 'The dependence of comfortable temperatures upon indoor and outdoor climates', Chapter 15, in *Bioengineering, Thermal Physiology and Comfort*, K. Cena and S.A. Clark eds. Elsevier, Amsterdam.

28 Oseland, N. (1993). 'UK design rules', *Building Services*, pp. 28–29.

29 Baker, N. (1990). 'Comfort and passive cooling, *Proceedings of the Workshop on Passive Cooling*, Ispra, Italy, 2–4 April, pp. 15–33.

30 Passive Cooling in Buildings – PASCOOL, Research Programme (1992). Commission of the European Communities, Directorate General for Science, Research and Development.

31 Baker, N. and J.F. Nicol (1993). 'Preliminary specification of the comfort monitoring survey', *Proceedings of the 2nd PASCOOL Meeting*, CMF Subgroup, 15–17 February, Glasgow, Scotland.

32 Berger, X. (1993). 'About thermal comfort', *Proceedings of the 3rd PASCOOL Meeting*, CMF Subgroup, Florence, Italy, 21–24 May.

33 Benton, C., F. Bauman and M. Fountain (1990). 'A field measurement system for the study of thermal comfort', *ASHRAE Transactions*, Vol. 96, Part 1, pp. 623–633.

34 Schiller, G., E. Arens, F. Bauman, C. Benton, M. Fountain T. Doherty and K. Craik (1988). 'A field study of thermal environments and comfort in office buildings', Final Report, ASHRAE Research Project RP-462, Berkley Center for Environmental Design Research, University of California.

35 Griffiths, I.D. and D.A. McIntyre (1974). 'Sensitivity to temporal variations in thermal Conditions', *Ergonomics*, Vol. 17, pp. 499–507.

36 Griffiths, I. D. and D.A. McIntyre (1974). 'Changing temperature and comfort', *Building Service Engineering*, Vol. 42, pp. 120–122.

37 Givoni, B. (1992). 'Comfort, climate analysis and building design guidelines', *Energy and Buildings*, Vol. 18, pp. 11–23.

38 Schiller, G. (1990). 'A comparison of measured and predicted comfort in office buildings', *ASHRAE Transactions*, Vol. 96, Part 1, pp. 609–622.

APPENDIX A
CALCULATION OF THERMAL COMFORT CONDITIONS
BASED ON THE PMV THEORY

This Appendix presents a computer program for evaluating indoor thermal comfort conditions based on Fanger's Predicted Mean Vote (PMV) theory.

A computer program for calculating comfort conditions based on the PMV theory

The program listing that follows is written in GWBASIC and can be used to calculate the PMV or any other parameter to balance Fanger's comfort equation.

Once the user has retrieved the program and executed it, the first screen that will appear is the opening page of the program which provides some introductory information. By pressing any key, the program will move to the second screen. This constitutes the input and output workscreen of the program. All input information and results are provided in this screen. The program will provide some initial data by default

There are in total eight parameters included in the calculations, whose names are given on the left hand side of the screen. The numerical data are divided into three columns. The input data are recorded in the first column, under the heading 'INPUT'. The parameter for which the input data are requested each time is highlighted in this column. The result for the corresponding calculated paraeter, appears in the second column under the heading 'RESULT'. The third column, under the heading 'REMARK' is reserved for listing any supplementary information (for example the corresponding value of relative humidity once the air temperature and water vapour pressure are known, or the PPD index that corresponds to the PMV value).

The user must enter the input data for the known parameters, while the parameter to be computed is left blank. At the bottom of the screen, the various F-keys are assigned some specific functions, which control the movement of the cursor. By pressing one of these function keys the user can execute the following actions :

F1 :   Moves the cursor to the previous line (moving up);
F2 :   Moves the cursor to the next line (moving down);
F3 :   Restarts the program;
F4 :   Performs calculations (executes the program);
F5 :   Plots the PMV–PPD graph and identifies the specific conditions;
F6 :   Exits the program (returns to GWBASIC);
F7 :   Restarts the program; and
F8 :   Prints all input and calculated parameters.

To enter a numerical value for a given parameter, simply type the value at the prompt location of 'INPUT LINE : ? _ ' and press the enter key. The value then appears in the INPUT column, on the same line as the corresponding parameter.

Do not enter a value for the parameter to be calculated. Leave it blank by simply pressing the enter key.

Using this program, it is possible to calculate the PMV index by specifying all the other parameters. It is also possible to work in a reverse way. For example, one can fix the desired PMV index and calculate the corresponding value of one of the other parameters, by simply leaving a blank at the corresponding data entry (e.g. air temperature).

```
1 REM ************ Demo-software    COMFORT version 880516 ***********
2 REM
3 REM
10 REM B&K Demonprogram, 1986, Comfortequation -------------------------------
15 COLOR 3,1
16 GOSUB 5000
20 '
30 TA=21 : TR=21 : VAR=.15 : PA=1000 : CLO=1 : MET=1.2 : WME=0 : PMV=0
40 CLS ' clers the screen
50 KEY ON
60 KEY 1,"LAST"+CHR$(13)
70 KEY 2,"NEXT"+CHR$(13)
80 KEY 3,"NEW"+CHR$(13)
90 KEY 4,"CALC."+CHR$(13)
100 KEY 5,"GRAF"+CHR$(13)
110 KEY 6,"BYE"+CHR$(13)
120 KEY 7, "RUN"+CHR$(13)
125 KEY 8, "PRINT"+CHR$(13)
130 COLOR 1,3 ' invers video
150 LOCATE 3,1: PRINT "USER INSTRUCTIONS :"
160 COLOR 3,1 ' normal video
170 LOCATE 4,20: PRINT "Input the values of the known parameters."
180 LOCATE 5,15: PRINT "Leave a blank space for the parameter to be computed."
190 LOCATE 9,38 : PRINT "INPUT    RESULT    REMARK"
200 LOCATE 11,1
210 PRINT "   Air temperature (";CHR$(248);"C)..........."
220 PRINT "   Mean radiant temperature (";CHR$(248);"C).."
230 PRINT "   Relative air velocity (m/s)...."
240 PRINT "   Water vapour presure (Pa)......"
250 PRINT "   Clothing (clo)..(Tcl=       )..."
260 PRINT "   Metabolic rate (met).........."
270 PRINT "   External work (met)..........."
280 PRINT "   PMV..........................."
290 VALG% = 1 : RESNR% = 0 : MODE% = 1
300 FOR NR% = 1 TO 8
310   LOCATE 10+NR%,37
320   GOSUB 2440 ' Display parameter
330 NEXT NR%
340 GOSUB 2620 : GOSUB 2720 ' Display RH and PPD
350 LOCATE 21,4 : PRINT "INPUT LINE : "
360 WHILE S$ <> "BYE"
370   IF MODE% <> 1 THEN GOTO 430
380     COLOR 20,3
390     NR% = VALG%
400     LOCATE 10+NR%,37
410     IF VALG% = RESNR% THEN PRINT "        " ELSE GOSUB 2440 ' Disp. parameter
420     COLOR 3,1
```

```
430    LOCATE 21,17 : INPUT "";S$
440    LOCATE 21,1 : PRINT "   INPUT LINE :                        "
450    IF MODE% <> 1 THEN GOTO 500
460      GOSUB 680 ' Treat input
470      NR% = VALG%
480      LOCATE 10+NR%,37
490      IF VALG% = RESNR% THEN PRINT "        " ELSE GOSUB 2440 ' Disp. parameter
500    IF S$ = "NEXT" THEN VALG% = VALG% + 1
510    IF S$ = "LAST" THEN VALG% = VALG% - 1
515    IF S$="GRAF" THEN GOTO  3000
517    IF S$="PRINT" THEN GOTO 4500
520    IF S$ <> "CALC." THEN GOTO 550
530      MODE% = 2
540      IF RESNR% <> 0 THEN GOSUB 1310 ELSE MODE% = 1
550    IF S$ <> "NEW" THEN GOTO 620
560      ERRNR% = 0
570      FOR I% = 11 TO 18
580        LOCATE I%,46 : PRINT"                          ";
590      NEXT I%
600      MODE% = 1
605 GOTO 10
610      GOSUB 2620 : GOSUB 2720 ' Display RH and PPD
620    IF VALG% > 8 THEN VALG% = 8
630    IF VALG% < 1 THEN VALG% = 1
640 LOCATE 22,4
650 WEND
660 GOTO 2800
670 '
680 REM --- Subroutine for treating input value -----------------------------
690 IF LEN( S$ ) <> 0 THEN GOTO 740
700    NR% = RESNR%
710    RESNR% = VALG%
720    IF NR% > 0 THEN LOCATE 10+NR%,37 : GOSUB 2440 ' Display parameter
730    GOTO 860
740 X = VAL( S$ )
750 IF (X = 0) AND (INSTR(S$,"0") = 0) THEN GOTO 870
760    IF VALG% = RESNR% THEN RESNR% = 0
770    ON VALG% GOTO 780,790,800,810,820,830,840,850
780    TA = X  : GOTO 860
790    TR = X  : GOTO 860
800    VAR = X : GOTO 860
810    PA =  X : GOTO 860
820    CLO = X : GOTO 860
830    MET = X : GOTO 860
840    WME = X : GOTO 860
850    PMV = X
860 GOSUB 2620 : GOSUB 2720 ' Compute and display RH and PPD
870 RETURN
880 '
890 REM --- Subroutine for computing PMV adapted from ISO 7730 ----------------
900 IF TA<-40 OR TA>120 OR TR<-40 OR TR>150 OR VAR<0 OR VAR>10 THEN ERRNR% = 2
910 IF PA<0 OR PA>5000 OR CLO<0 OR CLO>8 OR MET<WME+.7 THEN ERRNR% = 2
920 IF MET>10 OR WME<0 THEN ERRNR% = 2
930 IF ERRNR% <> 0 THEN GOTO 1290
940 ICL = .155 * CLO : M = MET * 58 : W = WME * 58
950 EPS = .00015
960 MW = M - W
970 REM --- Compute the corresponding FCL value
980 IF ICL < .078 THEN FCL = 1 + 1.29 * ICL ELSE FCL = 1.05 + .645 * ICL
990 FCIC = ICL * FCL
1000 P2 = FCIC * 3.96
1010 P3 = FCIC * 100
1020 TRA = TR + 273
```

```
1030 TAA = TA + 273
1040 P1 = FCIC * TAA
1050 P4 = 308.7 - .028 * MW + P2 * (TRA/100)^4
1060 REM --- First guess for surface temperatur
1070 TCLA = TAA + (35.5-TA) / (3.5*(6.45*ICL+.1))
1080 XN = TCLA / 100
1090 XF = XN
1100 HCF = 12.1 * SQR( VAR )
1110 NOI = 0
1120 REM --- Compute surface temperatur of clothing by iterations
1130 XF = (XF+XN) / 2
1140 HCN = 2.38 *  ABS(100*XF-TAA)^.25
1150 IF HCF > HCN THEN HC = HCF ELSE HC = HCN
1160 XN = (P4+P1*HC-P2*XF^4) / (100+P3*HC)
1170 NOI = NOI + 1
1180 IF NOI > 150 THEN ERRNR% = 1 :GOTO 1290
1190 IF ABS(XN-XF) > EPS GOTO 1130
1200 TCL = 100 * XN - 273
1210 REM --- Compute PMV
1220 PM1 = 3.96 * FCL * (XN^4-(TRA/100)^4)
1230 PM2 = FCL * HC * (TCL-TA)
1240 PM3 = .303 * EXP(-.036*M) + .028
1250 IF MW > 58.15 THEN PM4 = .42 * (MW-58.15) ELSE PM4 = 0!
1260 PM5 = 3.05*.001*(5733-6.99*MW-PA)
1270 PM6 = 1.7 * .00001 * M * (5867-PA) + .0014 * M * (34-TA)
1280 PMV = PM3 * (MW-PM5-PM4-PM6-PM1-PM2)
1285 PMV=INT(PMV*100)/100
1290 RETURN
1300 '
1310 REM --- Subroutine for computing the missing parameter -------------------
1320 PMVIND = PMV
1330 ON RESNR% GOTO 1340,1450,1550,1650,1760,1860,1960,2070
1340   REM---Air temperatur
1350     TA=20 : DTA=5
1360     GOSUB 890 : GOSUB 2560 ' Compute and display
1370     WHILE ABS( PMV-PMVIND ) > .01  AND ERRNR% = 0
1380       IF PMV<PMVIND THEN TA=TA+DTA ELSE TA=TA-DTA
1390       OLDGUESS=PMV
1400       GOSUB 890 : GOSUB 2560 ' Compute and display
1410       IF (OLDGUESS-PMVIND) * (PMV-PMVIND) < 0 THEN DTA=DTA/3
1420     WEND
1430     IF ERRNR% = 0 THEN GOSUB 2620 ' Compute and display RH
1440     GOTO 2090
1450   REM---Radiant temperatur
1460     TR=20 : DTR=5
1470     GOSUB 890 : GOSUB 2560 ' Compute and display
1480     WHILE ABS( PMV-PMVIND ) > .01 AND ERRNR% = 0
1490       IF PMV<PMVIND THEN TR=TR+DTR ELSE TR=TR-DTR
1500       OLDGUESS=PMV
1510       GOSUB 890 : GOSUB 2560 ' Compute and display
1520       IF (OLDGUESS-PMVIND) * (PMV-PMVIND) < 0 THEN DTR=DTR/3
1530     WEND
1540     GOTO 2090
1550   REM---Air velocity
1560     VAR=.2001 : DVAR=.2
1570     GOSUB 890 : GOSUB 2560 ' Compute and display
1580     WHILE ABS( PMV-PMVIND ) > .01 AND DVAR>.001 AND ERRNR% = 0
1590       IF PMV<PMVIND AND TA<TCL THEN VAR=VAR-DVAR ELSE VAR=VAR+DVAR
1600       OLDGUESS=PMV
1610       GOSUB 890 : GOSUB 2560 ' Compute and display
1620       IF (OLDGUESS-PMVIND) * (PMV-PMVIND) < 0 THEN DVAR=DVAR/3
1630     WEND
1640     GOTO 2090
```

```
1650    REM---Air humidity
1660     PA=600.1 : DPA=200
1670     GOSUB 890 : GOSUB 2560 ' Compute and display
1680     WHILE ABS( PMV-PMVIND ) > .01 AND ERRNR% = 0
1690      IF PMV<PMVIND THEN PA=PA+DPA ELSE PA=PA-DPA
1700      OLDGUESS=PMV
1710      GOSUB 890 : GOSUB 2560 ' Compute and display
1720      IF (OLDGUESS-PMVIND) * (PMV-PMVIND) < 0 THEN DPA=DPA/3
1730     WEND
1740     IF ERRNR% = 0 THEN GOSUB 2620 ' Compute and display RH
1750     GOTO 2090
1760    REM---Clothing
1770     CLO=1.001 : DCLO=.5
1780     GOSUB 890 : GOSUB 2560 ' Compute and display
1790     WHILE ABS( PMV-PMVIND ) > .01 AND ERRNR% = 0
1800      IF PMV<PMVIND THEN CLO=CLO+DCLO ELSE CLO=CLO-DCLO
1810      OLDGUESS=PMV
1820      GOSUB 890 : GOSUB 2560 ' Compute and display
1830      IF (OLDGUESS-PMVIND) * (PMV-PMVIND) < 0 THEN DCLO=DCLO/3
1840     WEND
1850     GOTO 2090
1860    REM---Metabolic rate
1870     MET=1.201 : DMET=.4
1880     GOSUB 890 : GOSUB 2560 ' Compute and display
1890     WHILE ABS( PMV-PMVIND ) > .01 AND ERRNR% = 0
1900      IF PMV<PMVIND THEN MET=MET+DMET ELSE MET=MET-DMET
1910      OLDGUESS=PMV
1920      GOSUB 890 : GOSUB 2560 ' Compute and display
1930      IF (OLDGUESS-PMVIND) * (PMV-PMVIND) < 0 THEN DMET=DMET/3
1940     WEND
1950     GOTO 2090
1960    REM---External work
1970     W=.001 : DW=.4
1980     GOSUB 890 : GOSUB 2560 ' Compute and display
1990     WHILE ABS( PMV-PMVIND ) > .01 AND ERRNR% = 0
2000      IF PMV<PMVIND THEN W=W-DW ELSE W=W+DW
2010      OLDGUESS=PMV
2020      GOSUB 890 : GOSUB 2560 ' Compute and display
2030      IF (OLDGUESS-PMVIND) * (PMV-PMVIND) < 0 THEN DW=DW/3
2040     WEND
2050     GOTO 2090
2060    REM---PMV value
2070     GOSUB 890 : GOSUB 2560 ' Compute and display
2080     IF ERRNR% = 0 THEN GOSUB 2720 ' Compute and display PPD
2090 GOSUB 2130 ' Error subroutine
2100 IF RESNR% <> 8 THEN PMV = PMVIND
2110 RETURN
2120 '
2130 REM---Error check subroutine--------------------------------------------------
2140 'IF ABS( PMV-PMVIND)>.01 THEN LOCATE 18,46 : PRINT USING "  ##.#";PMV
2150 IF TA<10 THEN LOCATE 11,56 : PRINT "Warning, low value "
2160 IF TA>30 THEN LOCATE 11,56 : PRINT "Warning, high value"
2170 IF TA<-40 THEN LOCATE 11,56 : PRINT "Error, low value    "
2180 IF TA>120 THEN LOCATE 11,56 : PRINT "Error, high value   "
2190 IF TR<10 THEN LOCATE 12,56 : PRINT "Warning, low value "
2200 IF TR>40 THEN LOCATE 12,56 : PRINT "Warning, high value"
2210 IF TR<-40 THEN LOCATE 12,56 : PRINT "Error, low value    "
2220 IF TR>120 THEN LOCATE 12,56 : PRINT "Error, high value   "
2230 IF VAR<0 THEN LOCATE 13,56 : PRINT "Error, low value    "
2240 IF VAR>1 THEN LOCATE 13,56 : PRINT "Warning, high value"
2250 IF VAR>10 THEN LOCATE 13,56 : PRINT "Error, high value   "
2260 IF PA<0 THEN LOCATE 14,56 : PRINT "Error, low value    "
2270 IF PA>2700 THEN LOCATE 14,56 : PRINT "Warning, high value"
```

```
2280 IF PA>PAMAX OR PA>5000 THEN LOCATE 14,56 : PRINT "Error, high value   "
2290 IF CLO<0 THEN LOCATE 15,56 : PRINT "Error, low value      "
2300 IF CLO>2 THEN LOCATE 15,56 : PRINT "Warning, high value"
2310 IF CLO>8 THEN LOCATE 15,56 : PRINT "Error, high value   "
2320 IF MET<.8+WME THEN LOCATE 16,56 : PRINT "Error, low value   "
2330 IF MET>4 THEN LOCATE 16,56 : PRINT "Warning, high value"
2340 IF MET>10 THEN LOCATE 16,56 : PRINT "Error, high value   "
2350 IF WME<0 THEN LOCATE 17,56 : PRINT "Error, low value    "
2360 IF WME>2 THEN LOCATE 17,56 : PRINT "Error, high value   "
2370 IF PMV<-2 THEN LOCATE 18,56 : PRINT "Warning, low value "
2380 IF PMV>2 THEN LOCATE 18,56 : PRINT "Warning, high value"
2390 IF PMV<-3 THEN LOCATE 18,56 : PRINT "Error, low value    "
2400 IF PMV>3 THEN LOCATE 18,56 : PRINT "Error, high value   "
2410 IF ERRNR% = 1 THEN LOCATE 10+RESNR,56 : PRINT "Iteration routine error"
2420 RETURN
2430 '
2440 REM---Display value-----------------------------------------------------
2450 ON NR% GOTO 2460,2470,2480,2490,2500,2510,2520,2530
2460 PRINT USING" ###.#";TA  : GOTO 2540
2470 PRINT USING" ###.#";TR  : GOTO 2540
2480 PRINT USING" ##.##";VAR : GOTO 2540
2490 PRINT USING"#####";PA   : GOTO 2540
2500 PRINT USING" ###.#";CLO : GOTO 2540
2510 PRINT USING" ##.##";MET : GOTO 2540
2520 PRINT USING" ##.##";WME : GOTO 2540
2530 IF ABS(PMV)>1 THEN PRINT USING"  ##.#";PMV ELSE PRINT USING"  #.##";PMV
2540 RETURN
2550 '
2560 REM---Display result----------------------------------------------------
2570 NR% = RESNR%
2580 LOCATE 10+NR%,46
2590 GOSUB 2440 ' Display value
2600 RETURN
2610 '
2620 REM---Compute and display RH----------------------------------------------
2630 IF (MODE% = 2) OR ( (RESNR% <> 1) AND (RESNR% <> 4) ) THEN GOTO 2660
2640   LOCATE 14,56 : PRINT"                    "
2650   GOTO 2700
2660 IF (VALG% <> 1) AND (MODE% = 1) THEN GOTO 2690
2670   TAA = TA + 273.15
2680   PAMAX = EXP(-5800.2206#/TAA+1.3914993#-4.860239E-02*TAA
      +.000041764768#*TAA^2-.000000014452093#*TAA^3+6.5459673#*LOG(TAA))
2690 LOCATE 14,56 : PRINT USING "RH  = ### %";PA*100/PAMAX
2700 RETURN
2710 '
2720 REM---Compute and display PPD----------------------------------------------
2730 IF MODE% = 2 OR RESNR% <> 8 THEN GOTO 2760
2740   LOCATE 18,56 : PRINT "                    "
2750   GOTO 2780
2760 PPD = 100 - 95 * EXP( -.03353*PMV^4 - .2179*PMV^2 )
2770 LOCATE 18,56 : PRINT USING "PPD = ### %";PPD
2775 LOCATE 15,25 : PRINT USING "##.##";TCL
2780 RETURN
2790 '
2800 END
3000 ' --------------------drawing PMV-PPD curve
3001 PMVRES= PMV
3002 KEY OFF
3010 '
3018 SCREEN 1,0
3020 COLOR 1!
3021 CLS
3022 X1=40 : Y1=50
```

```
3024 LINE (40,50)-(40,169)
3025 LINE (40,169)-(286,169)
3026 LINE (286,169)-(286,50)
3027 LINE (286,50)-(40,50)
3030 FOR PMV=-2.95 TO 3 STEP .05
3040 GOSUB 4000
3070 LINE (X1,Y1)-(X2,Y2),2
3080 X1=X2
3090 Y1=Y2
3100 NEXT PMV
3110 LOCATE 7,3 : PRINT"100"
3120 LOCATE 11,4: PRINT"40"
3130 LOCATE 17,4: PRINT"10"
3140 LOCATE 21,5: PRINT"5"
3150 LINE (40,84)-(286,84)
3160 LINE (40,135)-(286,135)
3170 LINE (40,161)-(286,161)
3180 LOCATE 23,5: PRINT "-3"
3190 LOCATE 23,10: PRINT "-2"
3200 LOCATE 23,15: PRINT "-1"
3210 LOCATE 23,21: PRINT "0"
3220 LOCATE 23,26: PRINT "1"
3230 LOCATE 23,31: PRINT "2"
3240 LOCATE 23,36: PRINT "3"
3250 FOR PMV=-2 TO 2 STEP 1
3260 GOSUB 4000
3270 LINE (X2,50)-(X2,169)
3280 NEXT PMV
3290 '--------------------Drawing actual PMV-PPD line
3300 PMV=PMVRES
3310 GOSUB 4000
3320 LINE (40,Y2)-(X2,Y2),2
3330 LINE (X2,Y2)-(X2,169),2
3331 X=X2/8 : Y=Y2/8.333333
3332 IF PMV=0 THEN 3340 ELSE LOCATE 23,X-5 : PRINT "    ";PMV;"  "
3333 LOCATE Y+1,3 : PRINT USING " ##";PPD
3340 LOCATE 1,12 : PRINT "Comfort Equation"
3350 LOCATE 3,1 : PRINT "      Ta    Tr   Va    RH    CLO   MET"
3360 LOCATE 5,6 :PRINT USING" ##.#";TA
3370 LOCATE 5,11:PRINT USING" ##.#";TR
3380 LOCATE 5,16:PRINT VAR
3390 LOCATE 5,21:PRINT USING " ##.#";PA*100/PAMAX
3400 LOCATE 5,26:PRINT USING "  #.#";CLO
3410 LOCATE 5,32:PRINT MET
3420 FOR X=14 TO 40 STEP 13
3430 LINE (43,X)-(283,X)
3440 NEXT X
3450 FOR X=43 TO 300 STEP 40
3460 LINE (X,14)-(X,40)
3470 NEXT X
3473 LOCATE 24,1 :INPUT" ", SSS$
3475 KEY ON
3480 SCREEN 2,0
3490 SCREEN 0,0
3500 GOTO 10
4000 '--------------------Calculating PMV-PPD coordinates
4010 '
4040 PPD = 100 - 95 * EXP( -.03353*PMV^4 - .2179*PMV^2 )
4050 X2=40+((PMV+3)*41)
4060 Y2=220-(LOG(PPD)/LOG(10))*85
4070 RETURN
4500 '
4510 ' ---------------------- Printing parameters
```

```
4520
4524 TA=INT(TA*10)/10
4525 TR=INT(TR*10)/10
4526 TCL=INT(TCL*10)/10
4529 LPRINT TAB(6) "Tair" TAB(13) "Tmrt" TAB(21) "Va" TAB(28) "Pa" TAB(33) "CLO"
     TAB(41) "Tcl "TAB(48) "MET" TAB(54) "PMV" TAB(61) "PPD"
4530 LPRINT TAB(5) TA TAB(12) TR TAB(19) VAR TAB(26) PA TAB(33) CLO TAB(40) TCL
     TAB(47) MET TAB(54) PMV TAB(61) PPD
4540 LPRINT :LPRINT
5000 CLS: PRINT "                      COMMISSION OF THE EUROPEAN COMMUNITIES":
     PRINT: PRINT "                   S A V E     P R O G R A M M E"
5010 LOCATE 5:PRINT "* * * * * * * * * * * * * * * * * * * * * * * * * * * * *
     * * * * * * * * * * * "
5020 FOR I = 1 TO 12
5030 PRINT "*"
5040 LOCATE I+4, 79
5050 PRINT "*"
5060 NEXT
5070 PRINT "* * * * * * * * * * * * * * * * * * * * * * * * * * * * * * * * *
     * * * * * * "
5080 LOCATE 7, 18: PRINT "                    COMFORT                     "
5090 LOCATE 9, 18: PRINT "CALCULATION OF COMFORT CONDITIONS BASED ON FANGER'S"
5100 LOCATE 11, 18:PRINT "PREDICTED MEAN VOTE (PMV) AND PREDICTED PERCENTAGE"
5110 LOCATE 13, 18:PRINT "            DISSATISFIED (PPD) INDICES"
5120 LOCATE 19: PRINT TAB(32) "ATHENS      1993"
5130 LOCATE 23: PRINT TAB(18) "PRINT ANY KEY TO CONTINUE    >>>"
5140 START$=INKEY$
5150 IF START$ = "" THEN GOTO 5140
5160 RETURN
```

Examples

To demonstrate the use of the program, the reader should refer to the following section and try to follow the illustrative examples. It is advisable to try first to apply the
two cases described and verify that there is an agreement in the results, before moving on to further applications.

EXAMPLE 1

Given that the air temperature is 26°C, the mean radiant temperature is 28°C, the
relative indoor air velocity is 0.2 m s^{-1}, the water vapour pressure is 1000 Pa, the
occupants are dressed in light summer clothing with a clothing insulation value of
0.5 clo, a metabolic rate of 1.2 met, and with no external work, calculate the PMV
and PPD indices.

   Enter the given information. Press the F4 key to obtain the following values :

   PMV = 0.26   PPD = 6%

EXAMPLE 2

Given that the desired PMV value is 0.5 (corresponding PPD is 10%) and for a
mean radiant temperature of 28°C, a relative indoor air velocity of 0.2 m s^{-1}, a water
vapour pressure of 1000 Pa, occupants dressed in light summer clothing with a
clothing insulation value of 0.5 clo, a metabolic rate of 1.2 met, and with no external

work, calculate the corresponding indoor air temperature that satisfies these conditions.

Enter the given information. Press the F4 key to obtain the following value :

Air temperature (°C): 28.0

APPENDIX B
A MANUAL METHOD FOR CALCULATING COMFORT
CONDITIONS BASED ON THE PMV THEORY

The following is a manual method for calculating the Predicted Mean Vote (PMV) and Predicted Percentage Dissatisfied (PPD) indices. The reader is first presented with an application, using Fanger's method to calculate the PMV and PPD indices, together with the worksheets supplied. Where necessary, the reader is referred to tables in Chapter 6, in order to obtain the necessary information, in the event that the required data are not available. Additional worksheets are also provided so that they can be used for future applications.

PROJECT :   <u>TYPICAL OFFICE</u>

ANALYST :   <u>C. BALARAS</u>

1. Occupants' activity =   <u>Seated</u>
                                                      (1)

2. For given activity and from Table 6.1,
    determine metabolic rate =   <u>50</u>   [kcal hr^{-1} m^{-2}]
                                                      (2)

3. For given activity and from Table 6.1, determine
    mechanical efficiency =   <u>0</u>
                                                      (3)

4. Determine effective metabolic rate
    = $\underset{\text{Step 2}}{\underline{50}}$ * (1 – $\underset{\text{Step 3}}{\underline{0}}$) =   <u>50</u>   [kcal hr^{-1} m^{-2}]
                                                      (4)

5. Air relative humidity =   <u>50</u>   [%]
                                                      (5)

6. Air temperature =   <u>26</u>   [°C]
                                                      (6)

7. Determine air temperature ratio
    = 273.16 / (273.16 + $\underset{\text{Step 6}}{\underline{26}}$) =   <u>0.9131</u>
                                                      (7)

8. Determine saturated water vapour pressure
    C_1 = 10.79586 C_2 = 5.02808
    C_3 = 1.50474E–04 C_4 = –8.29692
    C_5 = 0.42873E–03 C_6 = 4.76955
    C_7 = 2.219583
    A = C_1 * (1 – $\underset{\text{Step 7}}{\underline{0.9131}}$) =   <u>0.9382</u>
                                                      (8a)

    B = C_2 * log($\underset{\text{Step 7}}{\underline{0.9131}}$) =   <u>–0.1985</u>
                                                      (8b)

    C = C_3 * [1 – 10^{(C_4 /$\underset{\text{Step 7}}{\underline{0.9131}}$) – 1}] =   <u>0.0002</u>
                                                      (8c)

    D = C_5 * 10^{C_6*(1 – $\underset{\text{Step 7}}{\underline{0.9131}}$) - 1} =   <u>0.0001</u>
                                                      (8d)

    P_s = 10^{$\underset{\text{Step 8a}}{\underline{0.9382}}$ +$\underset{\text{Step 8b}}{\underline{–0.1985}}$+$\underset{\text{Step 8c}}{\underline{0.0002}}$+ $\underset{\text{Step 8d}}{\underline{0.0001}}$ – C_7}   <u>0.0331</u>   [atm]
                                                      (8e)

    P_s =$\underset{\text{Step 8e}}{\underline{0.0331}}$* 101.3 10^3 =   <u>3357.6</u>   [Pa]
                                                      (8f)

9. Determine water vapour pressure
    = $\underset{\text{Step 8f}}{\underline{3357.6}}$ * $\underset{\text{Step 5}}{\underline{0.5}}$ * 0.01333 =   <u>22.4</u>   [mm Hg]
                                                      (9)

10. Type of clothing =   <u>Light summer</u>
                                                      (10)

11. For given type of clothing and from Table 6.2
determine clothing insulation =

$$\underline{\hspace{1cm} 0.5 \hspace{1cm}}$$
(11)

12. For given type of clothing and from Table 6.2
determine clothing factor =

$$\underline{\hspace{1cm} 1.1 \hspace{1cm}}$$
(12)

13. Determine air velocity =
For closed windows/still air = 0–0.2 m/s
For open windows or forced ventilation specify wind

$$\underline{\hspace{1cm} 0.1 \hspace{1cm}}$$
(13)

14. Determine mean radiant temperature =
or set it equal to 2°C higher than air temperature

$$\underline{\hspace{1cm} 28 \hspace{1cm}}$$ [°C]
(14)

15. Determine temperature of clothing surface
$$= 12.5 + 0.3350 * (\underset{\text{Step 6}}{26} + \underset{\text{Step 14}}{28}) =$$

$$\underline{\hspace{1cm} 30.6 \hspace{1cm}}$$ [°C]
(15)

16. Determine convective heat transfer coefficient
for air velocity > 0.2 m s^{-1} use $10.4 * (\underset{\text{Step 13}}{\hspace{1cm}})^{0.5} =$

for air velocity ≤ 0.2 m s^{-1} use $2.05 * (\underset{\text{Step 15}}{30.6} - \underset{\text{Step 6}}{26})^{0.25} =$

$$\underline{\hspace{1cm} 3 \hspace{1cm}}$$
(16)

17. Determine intermediate parameters
$$\text{PAR1} = 0.352 * \,^{\wedge}(-0.042 * \underset{\text{Step 2}}{50}) + 0.032$$

$$\underline{\hspace{1cm} 0.0751 \hspace{1cm}}$$
(17a)

$$\text{PAR2} = \underset{\text{Step 4}}{50} \; 0.35 * (43 - 0.061 * \underset{\text{Step 4}}{50} - \underset{\text{Step 9}}{22.4}) =$$

$$\underline{\hspace{1cm} 43.86 \hspace{1cm}}$$
(17b)

$$\text{PAR3} = 0.42 * (\underset{\text{Step 4}}{50} - 50) =$$

$$\underline{\hspace{1cm} 0 \hspace{1cm}}$$
(17c)

$$\text{PAR4} = 0.0023 * \underset{\text{Step 2}}{50} * (44 - \underset{\text{Step 9}}{22.4}) =$$

$$\underline{\hspace{1cm} 2.48 \hspace{1cm}}$$
(17d)

$$\text{PAR5} = 0.0014 * \underset{\text{Step 2}}{50} * (34 - \underset{\text{Step 6}}{26}) =$$

$$\underline{\hspace{1cm} 0.56 \hspace{1cm}}$$
(17e)

$$\text{PAR6} = 3.4 * 10^{-8} * \underset{\text{Step 12}}{1.1} * [(\underset{\text{Step 15}}{30.6} + 273)^{4}$$

$$- (\underset{\text{Step 14}}{28} + 273)^{4}] =$$

$$\underline{\hspace{1cm} 10.7 \hspace{1cm}}$$
(17f)

$$\text{PAR7} = \underset{\text{Step 12}}{1.1} * \underset{\text{Step 16}}{3} * (\underset{\text{Step 15}}{30.6} - \underset{\text{Step 6}}{26}) =$$

$$\underline{\hspace{1cm} 15.18 \hspace{1cm}}$$
(17g)

18. Determine the PMV index =
$$\text{PMV} = \underset{\text{Step 17a}}{0.0751} * [\underset{\text{Step 17b}}{43.86} - \underset{\text{Step 17c}}{0} - \underset{\text{Step 17d}}{2.48} -$$

$$- \underset{\text{Step 17e}}{0.56} - \underset{\text{Step 17f}}{10.7} - \underset{\text{Step 17g}}{15.18}] =$$

$$\underline{\hspace{1cm} 1.1 \hspace{1cm}}$$
(18)

19. Determine the PPD index
$$\text{PPD} = 100 - 95 * e^{\wedge} - \{0.03353 * (\underset{\text{Step 18}}{1.1})^{4}$$

$$+ 0.2179 * (\underset{\text{Step 18}}{1.1})^{2}\} =$$

$$\underline{\hspace{1cm} 30.5 \hspace{1cm}}$$ [%]
(19)

PROJECT : _____

ANALYST : _____

1. Occupants' activity =
    _____
    (1)

2. For given activity and from Table 6.1,
    determine metabolic rate =   _____ [kcal hr^{-1} m^{-2}]
    (2)

3. For given activity and from Table 6.1, determine
    mechanical efficiency =
    _____
    (3)

4. Determine effective metabolic rate
    = _____ * (1 – _____) =   _____ [kcal hr^{-1} m^{-2}]
       Step 2   Step 3   (4)

5. Air relative humidity =   _____ [%]
    (5)

6. Air temperature =   _____ [°C]
    (6)

7. Determine air temperature ratio
    = 273.16 / (273.16 + _____) =   _____
                          Step 6   (7)

8. Determine saturated water vapour pressure
    $C_1 = 10.79586$ $C_2 = 5.02808$
    $C_3 = 1.50474E–04$ $C_4 = –8.29692$
    $C_5 = 0.42873E–03$ $C_6 = 4.76955$
    $C_7 = 2.219583$
    A = C_1 * (1 – _____) =   _____
                    Step 7   (8a)

    B = C_2 * log(_____) =   _____
                    Step 7   (8b)

    C = C_3 * [1 – 10^{(C_4 / _____) – 1}] = _____
                              Step 7   (8c)

    D = C_5 * 10^{C_6*(1 – _____) - 1} =   _____
                           Step 7   (8d)

    P_s = 10^{_____ + _____ + _____ + _____ – C_7} _____ [atm]
              Step 8a   Step 8b   Step 8c   Step 8d   (8e)

    P_s = _____ * 101.3 10^3 =   _____ [Pa]
            Step 8e   (8f)

9. Determine water vapour pressure
    = _____ * _____ * 0.01333 =   _____ [mm Hg]
       Step 8f   Step 5   (9)

10. Type of clothing =   _____
    (10)

11. For given type of clothing and from Table 6.2
    determine clothing insulation =

    _____ (11)

12. For given type of clothing and from Table 6.2
    determine clothing factor =

    _____ (12)

13. Determine air velocity =
    For closed windows/still air = 0–0.2 m/s
    For open windows or forced ventilation specify wind

    _____ (13)

14. Determine mean radiant temperature =   _____ [°C]
    or set it equal to 2°C higher than air temperature   (14)

15. Determine temperature of clothing surface
    $= 12.5 + 0.3350 * ($ _____ $+$ _____ $) =$ _____ [°C]
     Step 6   Step 14   (15)

16. Determine convective heat transfer coefficient
    for air velocity > 0.2 m s^{-1} use $10.4 * ($_____$)^{0.5} =$
     Step 13

    for air velocity ≤ 0.2 m s^{-1} use $2.05 * ($_____ $-$ _____$)^{0.25} =$ _____ (16)
     Step 15 Step 6

17. Determine intermediate parameters
    $PAR1 = 0.352 * {}^{\wedge}(-0.042 *$_____$) + 0.032$
     Step 2   _____ (17a)

    $PAR2 =$_____ $0.35 * (43 - 0.061*$ _____ $-$ _____$) =$ _____ (17b)
     Step 4 Step 4   Step 9

    $PAR3 = 0.42 * ($_____ $- 50) =$ _____ (17c)
     Step 4

    $PAR4 = 0.0023 *$ _____ $* (44 -$ _____$) =$ _____ (17d)
     Step 2 Step 9

    $PAR5 = 0.0014 *$ _____ $* (34 -$ _____$) =$ _____ (17e)
     Step 2 Step 6

    $PAR6 = 3.4 * 10^{-8} *$ _____ $* [($_____ $+ 273)^4$
     Step 12   Step 15

    $- ($_____ $+ 273)^4] =$ _____ (17f)
     Step 14

    $PAR7 =$_____ $*$ _____ $* ($ _____ $-$ _____$) =$ _____ (17g)
     Step 12 Step 16 Step 15 Step 6

18. Determine the PMV index =
    $PMV =$ _____ $* [$ _____ $-$ _____ $-$ _____ $-$
     Step 17a Step 17b Step 17c Step 17d

    $-$ _____ $-$ _____ $-$ _____ $] =$ _____ (18)
    Step 17e Step 17f Step 17g

19. Determine the PPD index
    $PPD = 100 - 95 * e^{\wedge} - \{0.03353*($_____$)^4$
    $\phantom{PPD = 100 - 95 * e^ - {0.03353*(}$ Step 18

    $+ 0.2179*($_____$)^2\} =$ _____ [%]
     Step 18   (19)

7

Cooling load of buildings

Cooling load is the amount of heat that must be removed from a building in order to maintain desirable indoor temperature and humidity conditions. The cooling period, depends on location and therefore may differ significantly from place to place. However, in most areas of the northern hemisphere, it usually extends from May to September.

Indoor conditions are influenced by outside weather conditions, solar radiation and internal heat sources. The load depends on the thermal characteristics of the building's envelope and the difference between the outside and inside conditions.

An accurate survey of the various components which influence the building is a basic requirement for a realistic estimate of cooling loads. The information which is usually required includes:

- **A topographical and architectural drawing of the building**, with the orientation and geographical coordinates of the locality, neighbouring buildings and any other obstructions and local climatological conditions. This information is necessary in order to evaluate the relationship of the building to incident local solar radiation. Architectural details, if available, depending on the design stage of the building, also provide valuable information. Type, number and dimensions of openings determine the amount of solar radiation which enters the space. A description of the building elements is useful for calculating the size of heat gains conducted from outdoors during the day and heat losses during the night. The type of construction can be used for estimating the levels of outdoor air infiltration into the building.
- **A schedule of operation and type of building**, in order to determine operating hours and the types of the various internal heat sources. In particular, the number of occupants, level of activity, occupancy schedule, make it possible to approximate internal heat gains from occupants and fresh outdoor air-ventilation requirements. Type, power and number of items of equipment deployed and the operational schedule, allow the approximation of internal heat sources from equipment. Type, power and number of artificial lighting systems, along with their operating schedule, will give an approximation of the internal heat sources from lights.

The size of the cooling load that is required changes continuously, following the time variation of the various influencing parameters. For practical purposes, it is sufficient to calculate the maximum cooling load, in order to size correctly the cooling equipment or the mean monthly cooling load in order to be able to evaluate the requirements of a building for cooling and the impact of various interventions to reduce the load. Hourly calculations of cooling load values are not common, since they are very time-consuming or require very advanced computational tools. They provide, however, some useful information for research-type applications and detailed sensitivity analysis.

The following discussion focuses on the factors that affect the cooling load of a building and on the calculation of the mean monthly cooling load for residences and small buildings. Alternatively, for calculating the maximum cooling load the reader may refer to [1, 2].

FACTORS AFFECTING THE COOLING LOAD

The space cooling load results from the instantaneous heat gains convected into the indoor air. Heat must be removed from indoor spaces in order to maintain the indoor air temperature at a constant value. This constitutes what is called the *sensible cooling load*. However, indoor comfort conditions are also influenced by air humidity levels, that is, the water content of the air. Air humidity levels must also be maintained within appropriate levels. Excess water vapour must be extracted from the air and this constitutes what is called the *latent cooling load*.

Sensible and latent heat gains are the rates at which heat enters and is generated respectively within a space, at any given moment. Sensible heat gains are manifested by a temperature rise of the air, while latent heat gains occur when moisture is added to the air. Heat penetrates the building from three sources:

- external
- internal
- ventilation.

An analysis of the various processes involved in each of these is presented below.

External loads

The cooling load from outdoors consists of:

- heat from solar radiation conducted through building materials and
- solar radiation entering through transparent openings.

HEAT CONDUCTED THROUGH BUILDING MATERIALS FROM SOLAR RADIATION
Solar radiation striking the outer envelope of buildings (walls and roof), in conjunction with high outdoor air temperatures, causes heat to flow into the interior spaces of buildings. Building surfaces that are exposed to the outdoor environment experi-

ence the temperature variations of ambient air. Depending on the surface orientation and the type of external shading devices, the time of day and the season, exposed building elements also receive variable amounts of solar radiation.

Figure 7.1 illustrates the mechanisms of heat transfer occurring between outdoor and indoor spaces through building elements. A portion of incident solar radiation is either reflected or absorbed by the material, depending on the properties of the outer layer of the element. Absorbed solar radiation increases the outer surface temperature of the material and, as a result, there are temperature differences between the outer surface, the internal side of the element and the environment. Outdoor and indoor boundary conditions influence the amount and rate of heat transfer.

Heat is conducted through the material, or it is convected and radiated to the outdoor environment. For example, heat conduction through a wall depends on the temperature of the outer surface of the wall, the surface temperature of the wall's internal side, how these temperatures vary with time and the energy storage characteristics of the wall construction. Heat is convected to the outdoor air at a rate depending on wind velocity. Heat radiated to the outdoor environment depends on the surface temperature and the surrounding outdoor surfaces and temperatures.

Outside surface temperature depends on the incident solar radiation absorbed by the surface and a heat balance between heat conduction, convection and radiation. Internal side-wall surface temperatures depend on the amount of heat conducted from the outside surface, convected from the surface to the indoor air and radiated from the given surface. Internal furnishing and internal heat sources can also contribute. Therefore, surface temperatures and heat flow through building components do not remain constant.

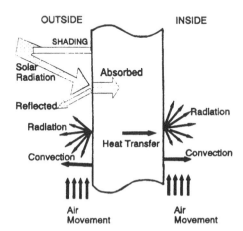

Figure 7.1 Heat transfer mechanisms between outdoor and indoor spaces through an external opaque element (e.g. a wall)

SOLAR RADIATION ENTERING THROUGH TRANSPARENT OPENINGS

Short-wave solar radiation enters indoor spaces through transparent surfaces such as glass windows or doors. The heat transferred through transparent surfaces is a function of the thermal radiation characteristics of a given material. It is less dependent on the thermal storage capacity of such materials owing to the very low mass and absorptance of the materials. Solar radiation that enters into a space is absorbed by the surfaces and contents of the space and does not influence the indoor air temperature until a later time. The radiant energy must first be absorbed by the surfaces that enclose the space, such as walls, floor and internal partitions, and by furniture or other interior objects. As a result, their temperature increases and at a given instant they become warmer than the indoor air. Some heat is then radiated and convected from the surfaces to the indoor air.

The amount of solar heat gain is reduced by using shading devices on the outside or inside of windows, by neighbouring buildings or other obstructions, proper distribution of the openings around the building and selection of appropriate building materials with desirable optical and thermal properties.

Internal loads

Internal heat gains originate from any sources that generate heat inside the building. The internal load or heat generated within a space depends on its type and use. Sometimes, internal gains consist of radiant heat which is partially stored in the material of the building elements, thus reducing the instantaneous cooling load. Generally, internal loads consist of some or all of the following: occupants, artificial lighting and any other mechanical or electrical equipment operating in the space.

- **Occupants**. The human body, through metabolism, generates heat which is released to the environment by radiation, convection and evaporation from the skin, and by convection and evaporation with respiration. The amount of heat generated and released primarily depends on the indoor conditions and on the activity level of the person.
- **Artificial lighting**. Illuminants convert electrical power into light and heat. Some of the heat is radiant and can be partially stored by the building materials, primarily the floor. It is important to note here that, during the daytime operating hours of a building, the use of all lights is infrequent. However, depending on the design and construction of a building, available interior luminance levels from daylight may not be sufficient for certain tasks. As a result, lights may be turned on even during the day. Accordingly, the use of artificial lighting needs to be carefully accounted for by knowing or estimating the operating schedule.
- **Mechanical or electrical equipment**. Appliances and electrical equipment release heat into the space. Normally, not all of the equipment installed in a

building will be in use simultaneously and therefore a usage factor should be applied to the full load heat gain.

Depending on the function of the building, internal gains may represent a significant part of the total cooling load. Large office buildings, with a large number of occupants and high usage of artificial lighting, as well as usually being heavily equipped, exhibit high internal gains. Small buildings and residences, on the other hand, do not usually exhibit significant internal loads.

Ventilation loads

Depending on the type of construction of a building, some outdoor air infiltrates into interior spaces through its various openings (i.e. windows, doors) or cracks. The warm outdoor air enters into the space, as a result of pressure differences induced on external surfaces of the building and outdoor wind velocity.

Depending on their function, indoor spaces also require specific amounts of fresh outdoor air in order to maintain acceptable levels of indoor air quality. Air ventilation imposes an additional cooling and dehumidifying load because the outdoor air has to be treated in terms of its temperature and humidity before it is introduced to indoor spaces.

METHODS FOR CALCULATING THE COOLING LOAD IN BUILDINGS

There are several methods and simulation tools available, which allow for the calculation of the cooling load at different levels of complexity. Usually, for high-accuracy calculations one needs a high-level computational tool, requiring advanced expertise. For easier-to-use approximate methods, one has to sacrifice some accuracy in the calculated values. A comprehensive review of available methods for calculating the cooling load and indoor conditions in buildings is given in Chapter 8. A total of 18 simplified models for estimating the cooling load of a building are presented, along with some brief information on more advanced computational tools.

In general, thermal analysis of buildings can be treated by two methods, namely, the thermal network method and the transfer function method.

The thermal network method breaks down the building into a network of interconnected elements, mainly resistances and capacitances. In other words, it simulates an electrical network representing all the physical processes of heat flow as an electrical current flow. Temperatures are identified as nodes, connecting all building elements which interact with each other, the indoor or the outdoor environment. As the number of nodes increases, these methods become more complex to define and require significant computational time for their solution.

The transfer function method is a simpler approach for the calculation of space cooling loads. The method is based on conduction transfer functions and weighting factors. Both parameters are time series that relate a current variable to past values of itself and other variables, at discreet time intervals, usually one hour. Conduction transfer functions are used to describe the heat flux on the internal side of a building component (wall, roof, partition, ceiling) as a function of previous values of the heat flux and previous values of indoor and outdoor temperatures. Conduction transfer functions may be derived analytically or can be precalculated.

From the available methods, the model of Bida and Kreider [3] has been selected and is recommended as a tool for calculating the monthly average cooling load of buildings. The method provides a satisfactory level of accuracy with a relatively low level of complexity. It is described in more detail in the following section.

CALCULATION OF MONTHLY AVERAGED COOLING LOAD

This simplified design method has been developed by Bida and Kreider [3] and it can be used for calculating the monthly sensible cooling loads of dwellings and small multistorey buildings, with small internal gains. The main source of heat transferred into the building is incident solar radiation. The cooling load component from latent heat is not taken into account. The main characteristics of the method are outlined in the following.

The monthly sensible cooling load for a light building, Q_{CL}, can be calculated as a function of three components:

$$Q_{CL} = Q_{AI} + Q_{S,T} + Q_{S,O} \tag{7.1}$$

where Q_{AI} is the cooling load from ambient and internal gains, $Q_{S,T}$ is the cooling load due to solar gain through transparent elements of the building's envelope and $Q_{S,O}$ is the cooling load due to solar gain through opaque elements of the building's envelope.

A mass effect coefficient (MEC), which is used to account for the thermal mass effect on cooling load, is defined as the ratio of the calculated monthly cooling load ($Q_{CL,a}$) to the monthly cooling load if no thermal mass were present (Q_{CL}, from equation 7.1), as follows:

$$MEC = Q_{CL,a}/Q_{CL} \tag{7.2}$$

The theoretical limits of the MEC values are zero for a building with an infinitely large thermal mass and unity for the case of zero thermal mass. Following a theoretical analysis, two correlations have been proposed for the calculation of the mass effect coefficient.

For periods with large cooling loads

$$MEC = \exp(BAP\ Y^{0.3748}) \tag{7.3}$$

where

BAP = $(6.9477 + 55.7(1 - \Phi))(-0.2089 + 0.2489SR) - 1.0573\Delta T$,

SR = the fraction of the load due to solar gains (SR = 0 if cooling load is due to non-solar gains, SR = 1 if cooling load is due to solar gains through transparent and opaque elements),

DT = building-to-ambient temperature ratio (difference between interior maximum and monthly averaged daily minimum ambient temperatures divided by the difference between monthly averaged daily maximum and minimum temperatures),

Φ = the monthly averaged daily utilizability expression [4], given by:

$$\Phi = \exp[(A + BR_n/R)(X_c + CX_c^2)]$$

where

A = $a_0 + a_1k_t + a_2k_t^2$,

B = $b_0 + b_1k_t + b_2k_t^2$,

C = $c_0 + c_1k_t + c_2k_t^2$,

a, b, c = coefficients from Tables 7.1 and 7.2 as a function of ACR (aperture configuration ratio = (east + west aperture area)/(east + west + south aperture area), for the case of a building with a north–south main axis) [5],

k_t = monthly averaged daily clearness index,

R_n = ratio of the monthly averaged radiation incident on an inclined surface to that on a horizontal surface at noon evaluated at a south-facing surface,

R = ratio of the monthly averaged daily total radiation on the tilted aperture to that on a horizontal surface (evaluated at the actual aperture orientations),

X_c = ratio of the critical ratio intensity (I_{cr}) to the monthly average value of the hourly solar irradiation on a south-facing tilted surface at noon,

I_{cr} = $BLC(T_{i,m} - T_{o,d})/(A_{G,t}\tau\alpha_c)$,

BLC = building load coefficient = $(UA)_b + m_aC_p$, W °C^{-1},

$(UA)_b$ = total thermal conductance of the building envelope, W °C^{-1},

m_a = hourly mass flowrate of outside air into the space, kg/hr,

C_p = specific heat of air, kJ/kg.°K,

$T_{i,m}$ = maximum indoor temperature allowable in the space, °C,

$T_{o,d}$ = monthly average daytime ambient temperature, °C,

$A_{G,t}$ = total glazing area present in the building envelope, m^2,

$\tau\alpha_c$ = monthly-averaged absorptance-transmittance product of all orientations and type of glazing combined.

For periods with intermediate cooling loads

$$MEC = \exp(BAP\ Y^{0.6411}) \tag{7.4}$$

where

BAP = $(5.0454 - 5.3655(1 - \Phi))(0.4847 - 0.3612SR') - 1.1930(2 - \Delta T)$,

SR' = $SR - (Q_{SO}/Q_{CL})$.

Table 7.1 Coefficients for monthly averaged daily utilizability factor, for a building with north–south main axis [5]

Coefs	ACR=0	ACR=0.2	ACR=0.4	ACR=0.6	ACR=0.8	ACR=1.0 EF=WF=0.5	EF or WF=1
a_0	5.177	1.526	2.748	1.814	0.199	−0.81	3.031
a_1	−14.4	−4.13	−11.2	−6.69	−0.69	2.913	−10.9
a_2	10.40	9.377	10.92	6.387	1.179	−1.97	8.588
b_0	−6.62	−3.25	−4.45	−3.98	−2.59	−1.07	−3.24
b_1	14.09	8.501	11.65	8.623	3.391	−1.93	6.329
b_2	−10.1	−9.42	−11.0	−7.75	−3.17	1.325	−5.25
c_0	−1.00	−4.51	−1.37	−2.56	−3.42	−2.42	−2.42
c_1	3.305	5.018	4.533	8.503	11.38	8.108	−3.96
c_2	−1.39	−2.29	−1.38	−4.37	−6.60	−4.03	3.256

EF: east glazing fraction; WF: west glazing fraction

Table 7.2 Coefficients for monthly averaged daily utilizability factor, for a building with a southwest–northeast main axis [5]

Coefs	ACR=0	ACR=0.2	ACR=0.4	ACR=0.6	ACR=0.8	ACR=1.0 EF=WF=0.5	EF or WF=1
a_0	0.636	−0.015	−0.257	−0.855	−0.972	−2.428	−4.501
a_1	−0.953	1.169	1.614	2.854	2.252	5.466	8.686
a_2	0.186	−1.301	−1.331	−1.763	−0.508	−1.839	−4.020
b_0	−1.188	−0.814	−0.842	−0.604	−1.180	−0.089	2.084
b_1	−2.300	−3.427	−3.017	−3.155	−0.286	−2.249	−5.856
b_2	2.109	2.890	2.426	2.130	−0.898	−0.609	−1.721
c_0	−0.320	−0.587	−0.862	−2.537	−2.537	−2.797	−1.663
c_1	1.999	2.893	3.534	8.532	8.532	9.128	5.335
c_2	−0.220	−0.606	−0.650	−4.484	−4.484	−5.040	−3.697

SW: south–west glazing fraction; SE: south–east glazing fraction

Having calculated the value of MEC, one may estimate the monthly cooling load for massive buildings from equation (7.2), since $Q_{CL,a} = Q_{CL}\mathrm{MEC}$.

This method has been validated in [3] against the Solar Energy Research Institute Residential Energy Simulator, SERI-RES [6]. Overall, this method can estimate the annual sensible cooling loads with an acceptable accuracy. For large cooling loads (greater than 10 GJ/year), the maximum error is less than 15%. The calculations of the cooling load can be performed for light and heavy buildings, thus enabling the user to compare easily the effectiveness of thermal mass in reducing the cooling load of the building. Similarly, sensitivity studies of other parameters that are involved in reducing the cooling load of buildings can also be easily performed.

A simplified method, for performing cooling load calculations by hand, is presented in the next section.

A SIMPLIFIED MANUAL METHOD FOR CALCULATING COOLING LOADS

This method was developed by Santamouris [7] as part of a new comprehensive methodology for calculating the thermal performance of buildings. Extracted from this work, the calculation of the cooling load requirements of air conditioned buildings is presented in the following discussion. The method is based on the principles of modified cooling degree hours. A complete description and validation of the method is given in Chapter 14.

The instantaneous cooling load Q_c for an air-conditioned building is given by:

$$Q_c = k\,(T_o - T_i) + Q_s + Q_{in} \tag{7.5}$$

where

k =   building load coefficient, W $°C^{-1}$,
T_o =   ambient outdoor air temperature, °C,
T_i =   desirable indoor air temperature, °C,
Q_s =   solar gains entering into the building space through transparent and opaque elements, W,
Q_{in} =   internal gains, W.

The ambient outdoor temperature should be available from local meteorological data. The indoor air temperature is set at a desirable value, depending on the specific application. The calculation of the remaining parameters can be performed by any available method or the reader may follow the procedures suggested in the following sections.

Summing together solar gains and internal gains, we can define the total amount of heat gains, (Q_T). Substituting back to the above equation, we then obtain that

$$Q_c = k\,(T_o - T_i) + Q_T. \tag{7.6}$$

Rearranging this equation we obtain a linear function of the instantaneous cooling load with outdoor temperature, given by

$$Q_c = k\,(T_o - T_b) \tag{7.7}$$

where

T_b =   balance temperature = $T_i - Q_T/k$.

Integrating the instantaneous value of cooling load, we obtain the monthly cooling load, Q_{cm}, which is given by:

$$Q_{cm} = 3600k\ \text{CDH}(T_{bm}) \tag{7.8}$$

where
CDH(T_{bm}) = cooling degree hours based on the mean monthly value of the balance temperature,

T_{bm} =   mean monthly balance temperature = $T_i - Q_T'/k$,

Q_T' =   mean monthly value of solar and internal gains.

The mean monthly value of solar and internal gains is calculated by summing the mean daily values of solar and internal gains. Information on how to calculate these parameters is provided later in this chapter.

The accuracy of calculating the monthly cooling load of buildings following the above procedure has been verified against the predictions obtained using the widely known computer program TRNSYS [8]. Following a series of simulations for a variety of buildings, the absolute difference between the predicted monthly values ranged between 0 and 25%. On an annual basis, the difference between the predicted values of the cooling load ranged from 0 to 15%, with a mean value close to 6.3%.

Building load coefficient

The building load coefficient is calculated as the product summation of the U-value times the surface area of each external building component, plus the thermal losses of the building from ventilation. Accordingly, for a building with N external components, the building load coefficient is given by:

$$k = \sum_{i=1}^{i=N}(U_i A_i) + (m_a Cn) / 3600 \tag{7.9}$$

where
i =   a number identifying a specific building component,
N =   total number of components that define the building,
U_i =   overall heat transfer coefficient of component i, W m^{-2} °C^{-1},
A_i =   surface area of component i, m^2,
m_a =   mass air flow rate due to ventilation, kg s^{-1},
C =   specific heat of air, kJ kg^{-1} K^{-1},
n =   number of air changes per hour.

Solar gains

Solar radiation contributes to the cooling load of a building with two components. The first component is a result of solar gains through transparent elements, while the second is a result of solar gains through opaque elements. The calculation of each component is described in the following discussion.

The total solar gains through transparent elements, Q_{st}, are calculated as the sum of solar gains from each transparent component of the building envelope. For a building with NT transparent elements, the value of solar gains is calculated by:

$$Q_{st} = \alpha f_1 \sum_{i=1}^{i=NT} (A_i \tau_i H_i SC_i) \tag{7.10}$$

where

α = mean absorptivity of the space,
f_1 = correction factor (Table 7.3),
i = number identifying the specific transparent element,
NT = total number of transparent elements for a given building,
A_i = surface area of transparent element i, m^2,
τ_i = short wave transmissivity of transparent element i,
H_i = mean monthly incident solar radiation on transparent element i, without any shading, W m^{-2},
SC_i = mean monthly shading coefficient of transparent element i.

The correction factor f_1 accounts for the fact that a portion of the heat gains which occur during the period that the building is occupied will be stored by various building components and internal elements and returned to the space during the unoccupied period. This time shift is actually a result of the role of the thermal mass on the cooling load of buildings. The reader should refer to Chapter 8 for a more detailed discussion of this phenomenon. According to [9], this parameter can be estimated as a function of the occupancy patterns and the mass of the building, as shown in Table 7.3.

Table 7.3 Values of correction factor f_1

Occupancy pattern	Type of building	
	Lightweight	Heavy
24 hours	1.0	1.0
7:00–23:00	0.9	0.7
8:00–18:00	0.85	0.55
9:00–15:00	0.8	0.4

The shading coefficient accounts for the effects of shading devices on the amount of solar radiation that finally enters into the space, as a result of a given shading device. It is defined as the ratio of incident solar radiation on the shaded surface to the solar radiation on the same surface without the presence of the shading device (H_i). The shading coefficient depends on the specific shading device and its characteristics. See Chapter 10 for information on how to calculate the shading coefficient.

The instantaneous solar envelope gains, Q_{so}, due to solar radiation absorbed by opaque external building elements, is equal to the heat absorbed by external walls plus the heat absorbed by the roof. For a building envelope with N opaque components (external walls and roof), the value of solar gains is calculated by:

$$Q_{so} = \sum_{i=1}^{i=N} (UA)_i [(T_{eo} - T_{ei}) + \mathrm{df}(T_{eoi} - T_{eo})] \tag{7.11}$$

where

i =   number identifying the specific building component,
N =   total number of external walls and roof of a given building,
$(UA)_i$ =   product of U-value times the surface area of component i, W K^{-1},
T_{eo} =   mean sol-air temperature, °C,
T_{ei} =   mean environmental temperature, °C,
df   =   decrement factor,
T_{eoi} =   sol-air temperature taking into account the component time lag.

The sol-air temperature, T_e, is defined as follows:

$$T_e = \alpha_i H_i / h_o + T_o + h_{or}/h_o(T_{sky} - T_o) \tag{7.12}$$

where

α_i =   solar absorptivity of the exterior surface of component i,
H_i =   rate of incident total solar radiation on component i, W m^{-2},
h_o =   heat transfer coefficient due to long-wave radiation and convection, at the outside surface of component $i = h_{oc} + h_{or}$, W m^{-2} K^{-1},
h_{oc} =   convective heat transfer coefficient = 5.8 + 4.1 V, W m^{-2} K^{-1},
V =   outside wind speed, m s^{-1},
h_{or} =   radiative heat transfer coefficient, average value 4.14 W m^{-2} K^{-1},
T_o =   outdoor ambient air temperature, °C,
T_{sky} =   sky temperature, which is estimated to be 6°C below the outdoor air temperature $(T_o - 6)$, °C.

Internal heat gains

As discussed earlier in this chapter, internal heat gains depend on a number of parameters. For dwellings, internal gains are usually small and their contribution to the cooling load is not significant. An approximate average value will be satisfactory in most cases.

For large buildings, however, internal gains have to be estimated with better accuracy, because they may represent a significant portion of the heat that needs to be removed from the buildings. Recalling from above that internal gains originate from occupants, lights and equipment, the total cooling load from internal heat gains, Q_i, is given by the sum of the three components:

$$Q_i = Q_{io} + Q_{il} + Q_{ie} \tag{7.13}$$

where

Q_{io} = cooling load from occupants, W,
Q_{il} = cooling load from lights, W,
Q_{ie} = cooling load from equipment, W.

The internal cooling load from occupants can be estimated as a function of the number of people in the building (NP) with the following expression:

$$Q_{io} = 65NP. \tag{7.14}$$

The internal cooling load from lights, in watts per square metre of floor area, can be obtained as a function of the indoor illumination level and the type of lighting equipment, from Table 7.4. To calculate the heat generated from the lighting equipment, one should only take into account the number of lamps that are actually used during the specific time that the building is operated.

Table 7.4 Cooling load due to lighting equipment $[W\ m^{-2}]$

Illumination level [Lux]	Type of lighting equipment							
	Type A	Type B	Type C	Type D	Type E	Type F	Type G	Type H
200	25–32	32–45	10–15	7.5	10	10	15–20	17–22
400	50–65	65–90	20–30	15	15–22	20	30–40	32–45
1000			55–75	32–45	42–55	45–65		

Type A: Tungsten, open enamel industrial reflector, 300 W
Type B: Tungsten, general diffuser, 200 W
Type C: Mercury, MFB industrial reflector, 250 W
Type D: White fluorescent lamp, enamel or plastic, 65 W
Type E: White fluorescent lamp with enclosed diffusing fitting, 65 W
Type F: White fluorescent lamp with louvred ceiling fitting, 65 W
Type G: Warm white fluorescent lamp with enclosed diffusing fitting, 65 W
Type H: Warm white fluorescent lamp with louvered ceiling panels, 65 W

The internal heat generation from electrical appliances and equipment depends on the number of units and schedule of operation. Unless the space is heavily equipped, as in the case of computer rooms, the cooling load from equipment can be approximated by an average value. Some typical values for 24-hour operation are given in Table 7.5.

Table 7.5 Average heat gains from electrical equipment

Item	Heat gain [W]	Item	Heat gain [W]
VDU terminal	70	Kettle	20
Intelligent terminal	170	Freezer	70
Photocopier	500	Dishwasher	50
Electric typewriter	15	Washing machine	35
Electronic typewriter	30	Tumble drier	20
Television	25	Humidifier	60
Refrigerator	40		

The mean daily internal heat generation, Q_{in}, which contributes to the cooling load of buildings, can be calculated by the following expression:

$$Q_{in} = f_1 (Q_{io} + Q_{il} + Q_{ie})\qquad(7.15)$$

where f_1 is the correction factor introduced previously which takes into account the time-lag effect due to the thermal inertia of the building.

REFERENCES

1 'Air-conditioning cooling load' (1989). *ASHRAE Handbook*, Chapter 26. American Society of Heating, Refrigerating and Air Conditioning Engineers, Atlanta.

2 McQuiston, F.C. and J.D. Spitler (1992). *Load Calculation Manual*, 2nd ed. American Society of Heating, Refrigerating and Air Conditioning Engineers, Atlanta.

3 Bida, M. and J. Kreider (1987). 'Monthly-averaged cooling load calculations - residential and small commercial buildings', *ASME Journal of Solar Energy Engineering*, Vol. 109, pp. 311–320.

4 Balaras, C.A. (1991). 'Methods for calculating the cooling load of buildings'. Institute for Technological Applications, Hellenic Productivity Centre, Athens.

5 Bida, M. and J. Kreider (1987). 'Monthly-averaged utilizability for multiple aperture solar systems', *ASME Journal of Solar Energy Engineering*, Vol. 109, p. 52.

6 SERI-RES, 'Solar Energy Research Institute residential energy simulator', Version 1.0. The Solar Energy Research Institute, Golden, CO.

7 Santamouris, M. (1993). 'A method to calculate the thermal performance of passively cooled buildings', *Cooling Load of Buildings*, Vol. 5. Department of Applied Physics, University of Athens, Greece.

8 Klein, S. et al. (1988). 'TRNSYS – A transient system simulation program'. University of Wisconsin–Madison, Engineering Experiment Station Report 38-12.

9 Baker, N. (1987). 'Passive and low energy buildings design for tropical island climates'. Report prepared for the Commonwealth Science Council.

8

Heat attenuation

THE ROLE OF THERMAL MASS

Indoor air temperature is primarily influenced by external climatological parameters (solar radiation, outdoor temperature) and highly variable internal loads (human activity, lights, equipment), as well as the building's construction. Under free-floating conditions, a building operates with no assistance from mechanical systems to extract heat from interior spaces. Indoor conditions are determined by a series of complex, but natural, physical phenomena which take place between the outdoor and indoor environment through the building's envelope and inside the building itself. Heat gains and losses between the indoor and outdoor environment occur by conduction, convection and radiation, shown in Figure 8.1. Similar phenomena occur in interior spaces, as a result of internal sources and heat transfer exchanges to or between the building elements. All these phenomena result in a net thermal balance that determines the indoor air conditions.

The temperature variation of indoor air, for a space enclosed by n surfaces, depends on the surface temperature of the surrounding surfaces, the amount of air that is ventilated into the space, internal heat sources and solar-to-air heat flow. This can be expressed by the following thermal balance equation:

$$mc\frac{dT_a}{dt} = \sum_{j=1}^{j=n} Q_{c,j} + Q_{c,v} + Q_{c,i} + Q_{s,r}$$

where

m	=	mass of internal air, kg,
c	=	specific heat of internal air, J kg^{-1} K^{-1},
T_a	=	indoor air temperature, °C,
t	=	time, s,
n	=	number of surfaces,
$Q_{c,j}$	=	convected heat flow rate by each surface j, W,
$Q_{c,v}$	=	heat flow rate exchanged by ventilation, W,
$Q_{c,i}$	=	convected heat flow rate from internal sources, W,
$Q_{s,r}$	=	solar to air heat flow rate, W.

Heat conduction through building components, in the x direction, for elements with constant thermal properties and no internal heat generation, is described by the following partial differential equation:

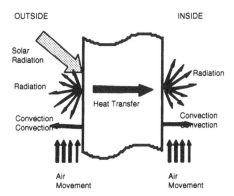

Figure 8.1 Heat transfer mechanism through building materials

$$\frac{d}{dx}\left(\lambda \frac{dT}{dx}\right) = \rho c\left(\frac{dT}{dt}\right)$$

Longitudinal　　Thermal
conduction　　　inertia

where

λ　　= 　thermal conductivity of the material, W m^{-1} K^{-1},

T　　= 　temperature depending on the location and time,

ρ　　= 　density of the material, kg m^{-3},

c　　= 　specific heat of the material, J/kg^{-1} K^{-1}.

The energy balance at any point inside the element depends on two effects; the net heat transfer by longitudinal conduction and the retarding effect of thermal inertia. The product (ρc) represents the thermal inertia per unit volume, or the specific heat capacity of the material. In the event that heat is conducted other than in the x direction, then similar terms are added to the left-hand side of the above equation for the other directions.

When the variation of temperature along a bar is small enough so that the thermal conductivity can be treated as a constant, independent of temperature, the one-dimensional conduction equation, can be simplified as follows :

$$\frac{d^2 T}{dx^2} = \frac{1}{\alpha}\frac{dT}{dt}$$

where

α　　= 　thermal diffusivity of the material = $\lambda/\rho c$, m^2 s^{-1}.

Figure 8.2 illustrates the temperature distribution and heat transfer in a building element. Initially the building element is at thermal equilibrium. As time progresses, the outside surface temperature of the element increases and, owing to the temperature difference within the element, heat is transferred from the outside to inside. The temperature of the element increases, with higher temperatures closer to the outside surface. At noon, the outside temperature reaches a maximum value and from there on decreases, as the incident solar radiation also decreases. As a result of the material's thermal inertia, interior locations inside the building element retain higher temperatures. As time progresses, the heat wave is transferred to the right, towards lower temperatures. A similar cycle follows the next day.

Indoor conditions are directly linked to prevailing outdoor conditions. This results in temperature swings, with peaks occurring during the most unfavourable periods of the day, around noon and early afternoon hours. During these periods of the day, high outdoor temperatures eliminate the use of simple alternative passive cooling techniques, like natural ventilation. Consequently, in order to maintain thermal comfort during these hours, it is necessary to remove all excess heat from the space immediately upon entry, with an energy-consuming mechanical system. This will require an oversized system capable of handling the short-period peaks of the cooling load.

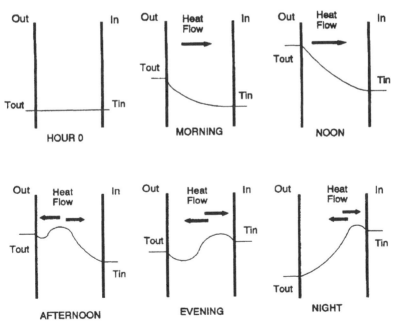

*Figure 8.2 Diurnal temperature distribution and heat transfer mechanism
of a building element*

These problems are more severe in passive solar buildings and buildings with large openings. The direct solar gains through the windows of the building, the heat conducted through the building's non-opaque elements and the internal heat gains are useful for passive heating purposes in winter, but they are undesirable during the summer months. Of course, the role of solar radiation in providing the necessary natural daylighting must also be kept in mind.

To reduce indoor air temperature and cooling load peaks and to transfer the load to later in the day, it is possible to store the heat in the structural materials of the external cell and the interior surfaces of the building. The storage material is the construction mass of the building itself, which is referred to as thermal mass. It is typically contained in walls, partitions, ceilings and floors of the building, constructed of material with high heat capacity, such as poured concrete, bricks and tiles.

The thermal mass of the building has a positive effect on the indoor conditions during the summer and winter periods. The energy available from the high solar gains during the day is stored and then is slowly released into the indoor environment at a later time. In winter, the stored heat is transferred back into the room in late afternoon and evening hours, when it is mostly needed, satisfying part of the heating load and avoiding overheating and discomfort conditions during the high solar radiation periods of the day.

In summer, heat is stored in the thermal mass, thus reducing the cooling load peaks. A reduced portion of the load will need to be removed from the interior space, while the remaining portion of the external and internal heat gains is contained within the thermal storage materials. The stored heat is progressively released to the interior of the building at a later time. The cooling load in any case will remain the same. There is, however, a time shift of the peak load, a reduction of the maximum load value and a time lag of the heat release from the material to the indoor air.

The heat flow for light and massive buildings is illustrated in Figure 8.3, which exhibits the moderating influence of thermal mass under ideal conditions [1]. When the curves are above the x axis the building is losing heat while, when the curves are below the x axis, the building is gaining heat. A massive building may require no additional cooling or heating, under certain conditions. For example, during fall and spring, light buildings may require cooling for part of the day and heating for another part of the day. Massive buildings do not exhibit this variability and, depending on the outdoor conditions, they may provide satisfactorily comfortable indoor conditions.

Another positive effect of thermal mass is the time delay of heat flow. For example, heat transfer from the outside surface of a building material, which is at a higher temperature, towards the lower-temperature inside surface, is first used to raise the temperature of the material. Only after the wall has substantially warmed up can heat exit from the other side. The time it will take for the heat to flow through the

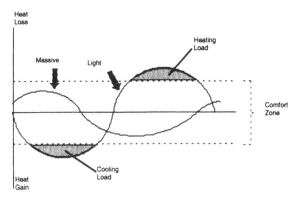

*Figure 8.3 Daily building heating and cooling loads for
buildings of massive and light construction*

material itself depends on its thermal resistance. This parameter describes the opposition of materials and air spaces to heat transfer. For example, under steady-state
conditions, 2.5 cm of wood has the same thermal resistance as 30.5 cm of concrete,
mainly because of the air spaces created by the cells in the wood [2]. However, the
delay in heat conduction is very short for 2.5 cm of wood because of its low heat
capacity. Concrete, on the other hand, exhibits a much longer time delay because of
its high heat capacity. The time lag for some common building materials, 30.5 cm
thick, is ten hours for common brick, six hours for face brick, eight hours for
heavyweight concrete, and 20 hours for wood because of its moisture content [2].

The daily variation in cooling load for light and heavy buildings is illustrated in
Figure 8.4. The cooling load values are only shown as relative data for comparative

DAILY COOLING LOAD VARIATION

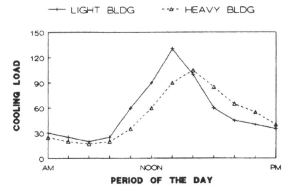

*Figure 8.4 Time variation of cooling load for light and
heavy building construction*

purposes. During the early morning and midday hours, the load for the heavy building is lower than for the light building, as a result of the external loads. For the massive building the peak load is lower than for the light construction and occurs at a later time. During the afternoon and evening hours the relative cooling load values are higher for the heavy building, as the stored heat is progressively returned to the indoor spaces. The areas under the two curves are the same, meaning that the cooling load remains the same. The positive end result is a reduction of the peak cooling load value and a time shift of the whole load distribution curve to the right, towards late afternoon and evening hours.

This time lag is actually desirable, since at the time when the heat is returned to the interior, the indoor temperature is lower than the desirable. In the event that the building is unoccupied during the evening hours, as in the case of most office buildings, it is possible to relax the restrictions on indoor thermal comfort. In any event, outdoor conditions are more favourable for passive cooling techniques, natural, hybrid or indirect ventilation [3], which can be used to remove portions of the load.

HEAT TRANSFER MECHANISMS

During the day, external wall temperature increases, as the balance of incident solar radiation and losses by convection and radiation result in a net heat flux. As time progresses, the absorbed heat is dissipated inside the wall, as shown in Figure 8.1. Temperature increases at different locations within the wall depend on the material's thermal properties and on the boundary conditions on either side of the wall. Eventually, temperature and heat excitations move through the wall structure, towards the inside surface.

During the night, a reverse process takes place. Outside boundary conditions change, as outdoor temperatures decrease and there is no incident solar radiation. The temperature of the wall decreases with time, at a rate depending on the location, the thermal properties of the materials and the boundary conditions. The temperature closer to the outside decreases first, followed by the internal wall temperatures. The boundary conditions, convective and/or radiative gains or losses, determine the rate of temperature changes. For example, high outdoor wind velocity enhances convective losses. Outside wall surface temperature drops, followed by a temperature drop of interior wall locations closer to the outside. Heat transfer from the higher-temperature locations occurs, in an attempt to reach equilibrium. Similarly, on the inside wall surface, increased air movement can increase heat losses and the dissipation of the stored heat.

Forced and natural convection are the two primary heat transfer mechanisms for heat dissipation from building material surfaces to the indoor air. In buildings, hot surfaces result from the incident solar radiation on the outer envelope of the building, from solar illumination and from the direct solar gains and heating on the interior surfaces. Heat is first conducted from the hot surface to an interior thin air layer adjacent to it. The warm air then becomes lighter and begins to rise. Air velocity and temperature changes are limited to a thin region next to the surface, the so-called boundary layer. At the outset, the boundary layer is laminar, but at some distance from the leading edge, depending upon fluid properties and the thermal gradient, transition to turbulent flow may occur.

Whenever the air motion is due entirely to the action of gravitational forces, the heat transfer mechanism is known as 'natural or free' convection. The heat transfer rate may be enhanced by introducing a forced flow over the surface, resulting in what is known as 'forced convection'. The heat transfer depends on the temperature difference between the surface and the environment, the heat transfer coefficient and the overall geometry.

Free convection heat transfer data are usually correlated in terms of two or three dimensionless parameters, namely the Nusselt, Prandtl and Rayleigh numbers. Sometimes the Grashof number is used instead, which is the ratio of the Rayleigh to the Prandtl number. Each parameter is defined as follows:

Prandtl	$Pr = v/\alpha$	Ratio of momentum to thermal diffusion,
Grashof	$Gr = g\Delta TL^3/v^2$	Ratio of buoyancy to viscous forces,
Rayleigh	$Ra = GrPr = g\beta\Delta TL^3/v\alpha$	Defines the temperature boundary condition on a particular surface,
Rayleigh	$Ra^* = g\beta L^4 q/v\alpha\lambda$	Defines the heat flux boundary condition on a particular surface,
Nusselt	$Nu = hL/\lambda$	Dimensionless heat transfer coefficient,

where,

g = gravitational acceleration, m s^{-2},
β = coefficient of thermal expansion (=$1/T$ for ideal gases), K^{-1},
ΔT = characteristic temperature difference, K,
v = kinematic viscosity, m^2 s^{-1},
q = surface heat flux, W,
L = characteristic length, m,
h = average convective heat transfer coefficient, W m^{-2} K^{-1}.

In natural convection the flow may be laminar or turbulent depending on the distance from the leading edge, the fluid properties, the geometry of the system and the temperature difference between the surface and the fluid. Transition generally occurs in the range $10^7 \leq Ra \leq 10^9$. In building applications, Rayleigh numbers are of the order of 1×10^{10} [4].

The Prandtl number defines the impact of fluid properties: air has a Prandtl number of 0.7 and water has a Prandtl number of about 7. Prandtl number changes over the range $0.7 \leq Pr \leq 7$ affect average heat transfer results by only 10 to 15% when the flow is laminar [5].

In general, the approach in handling natural convection problems is first to determine an applicable expression of the Nusselt number for the given configuration. Such expressions have been developed by numerical analysis, while the majority of the available correlations have resulted from experimental studies. Using measured data (temperature and velocity variation), a correlation for the Nusselt number can be derived. Such correlations apply under the specific conditions of each experiment, so one must be careful before applying them under different conditions. The Nusselt number is, by definition, a function of the heat transfer coefficient. Accordingly, Nusselt number correlations can then be used to calculate the heat transfer coefficient, which is applicable for the specific experimental configuration and within the given limits (usually given as ranges of Re and Ra numbers).

The majority of experiments that have been carried out so far use vertical or horizontal flat-plate geometries and temperature difference conditions that are not applicable to buildings' applications. Recently, however, in an effort to understand better the nature of convective heat transfer processes in buildings, experiments have also been based on real-size configurations and representative thermal conditions.

A recent literature review [6] has revealed several relationships, which have been developed for calculating the natural convection heat transfer coefficient. The various correlations are presented in Tables 8.A1–8.A6, in Appendix A at the end of this chapter.

A comparative statistical analysis of these correlations is given in Appendix B. Most of the existing correlations used to determine the natural convection heat transfer coefficient have been extracted from experiments in controlled environments. Small vertical and horizontal heated test plates have been studied under laboratory conditions, which differ from the conditions in real buildings. However, there is a good agreement between these correlations, as shown in Tables 8.B1–8.B3 in Appendix B at the end of this chapter. All correlations have been modified for air with Pr=0.72 and the results are given for representative Gr values (ranging between 10,000 and 1E+12). Correlations derived from experiments in small and simplified test enclosures and laboratory test cells are also included in the tables. As expected [1], the values of the Nusselt number obtained using these correlations are smaller. This may be attributed to the fact that, in this case, the flow is confined, which affects its characteristics.

PARAMETERS INFLUENCING THERMAL MASS EFFECTIVENESS

Thermal mass in buildings can be used to avoid dealing with instantaneous high cooling loads, to reduce energy consumption by up to 20% in commercial buildings

[7] and to attenuate indoor temperature swings, caused by rapid changes in the ambient conditions during the day. According to [8], the effectiveness of thermal storage is acceptable where the diurnal variation of ambient temperatures exceeds 10°K. Under such circumstances, i.e. ample night ventilation and closing the building during the day, as a result of the thermal mass effect it is possible to achieve a temperature difference that is greater than 10°K in the negative direction between the mean indoor and outdoor temperatures and an indoor temperature swing of only 2.5°K.

Thermal mass also has a positive effect on occupant comfort. High-mass buildings prohibit high interior air and wall temperature variations and sustain a more steady overall thermal environment. This increases comfort, particularly during mild seasons (spring and fall), during large air-temperature changes (high solar gain), and in areas with large day–night temperature swings [1]. Traditional architecture, especially in southern European countries, has successfully demonstrated these positive effects.

The rate of heat transfer through building materials and the effectiveness of thermal mass is determined by a number of parameters and conditions. To achieve the best possible results, one needs to follow some general guidelines and take appropriate actions that fall within the overall procedure of energy-efficient building. It is important, though, also to understand the relationship of these parameters to the performance of thermal mass in order to achieve the best possible results. Optimization of thermal-mass levels depends on building-material properties, building orientation for the location and distribution, thermal insulation, ventilation, climatic conditions and use of an auxiliary cooling system, and occupancy patterns. The most important features are analysed in the following discussion.

Material thermal properties and performance

The temperature distribution within the structural materials varies with time, boundary conditions and thermal properties of the wall material. The phenomena which take place during day and night periods differ significantly, depending on whether the mass material is being charged (temperature increase) or discharged (temperature decrease), respectively.

The thermophysical properties of the heat-storing material can influence the performance of the system. For the wall material to store heat effectively, it must exhibit a proper density (ρ), high thermal capacity (C) and a high thermal conductivity value (λ), so that heat may penetrate through all the wall material during the specific time of heat charging and discharging. As a result, a wall with a low value of ($\rho C \lambda$) will have a low heat storage capacity, even though it may be quite thick.

Consequently, the very properties that make a wall perform well as an interior thermal storage element make it perform poorly as an exterior envelope insulating element. In a similar manner, the depth that the diurnal heat wave reaches within the storage material depends on thermal diffusivity. This is the controlling transport property for transient heat transfer and is equal to $\lambda/\rho c_p$, where c_p is the specific heat of the material.

Materials with higher thermal diffusivity values can be more effective for cyclic heat storage at greater depth than materials with lower values. Beyond a certain material thickness, the heat flow into the indoor air does not take place during the night hours, but is delayed until the following daytime hours. This is undesirable during the cooling season, since the inflow of heat during the early hours of the day will result in unpleasant comfort conditions and fail to satisfy the early-day cooling loads. Similarly, during the heating season, the efficiency of the heat absorption and storage by the structural mass is reduced. However, it may prove beneficial by providing additional heat during successive cloudy days.

Thermal mass location and distribution

Placing the thermal mass in different building locations can result in distinctively different behaviours. Accordingly, it is important to make the proper decisions. One may distinguish two cases, based on whether the heat storage material receives energy by solar radiation (direct) or by infrared radiation and room air convection (indirect). Direct heat gains are experienced by the outer cell of the building, which is exposed to solar radiation, and by the interior room surfaces, which absorb the incident solar radiation as it enters through the building's openings. Indirect heat gains are experienced by opaque elements inside a space from the energy which is transferred from direct-gain surfaces. Direct locations are much more effective for heat storage mass than indirect locations.

According to [2], thermal mass must be properly distributed, depending on the orientation of the given surface and the desirable time lag. A surface with north orientation has little need for time lag, since it only exhibits small heat gains. For east orientations it is desirable to have either a very long time lag, greater than 14 hours so that heat transfer is delayed until the late evening hours, or a very short one. The latter case is recommended for east orientations since high-time-lag mass is economically unfeasible. For south orientations, an eight-hour time lag is sufficient to delay the heat from midday until the evening hours. For west orientations, an eight-hour time lag is again sufficient, since there are only a few hours until sunset. The roof, which is exposed to solar radiation during most hours of the day, would require a very long time lag. However, because it is very expensive to construct heavy roofs, the use of additional insulation is usually recommended instead.

Thermal mass and insulation

Thermal insulation has long been mandatory in practically all building codes around the world, to reduce heat penetration from the outside of exterior surfaces, through the building materials, into the interior space. In general, both thermal mass and insulation are important in the thermal performance of a building [4].

For buildings where heating is the major factor, insulation will be the predominant effective envelope factor. In climates where cooling is of primary concern, thermal mass can reduce energy consumption, provided the building is unused in the evening hours and the stored heat can be dissipated during this idle period. However, common insulation material (like expanded polystyrene) will degrade the performance of a thermal storage wall [5], because it reduces its effectiveness in the portion that is positioned inside the wall insulation. Overall, because thermal mass stores and releases heat, it interacts with the building operation more than just by the simple addition of insulation [9]. This makes analysing thermal mass performance and the overall thermal performance of the building a more difficult problem to treat.

The role of ventilation

The role of the thermal mass, during the cooling period, is also extended into the night period. Once the stored heat is dissipated into the indoor environment, the process is reversed. During the night, when the outdoor temperatures are lower than the indoor temperatures, it is possible to cool the structural mass of the building by night ventilation, known as convective cooling. Ceiling fans can also be used to raise the convective heat transfer coefficient [10], which facilitates the process of rejecting heat from massive walls at night. It is also recommended that the night ventilation rate be increased to 1.5 m^3 min^{-1} m^{-2} of floor area [10]. The objective is to maximize the building thermal transmittance during the night-time and minimize it during the daytime.

The cold-storage capability of building material can benefit from night-time ventilation, although this may introduce some problems. Leaving the windows open during the night, to allow for nocturnal ventilation, introduces safety- and privacy-related concerns. Also, high wind speeds may cause indoor problems since the inflow of the air through the openings cannot be manually controlled and adjusted. Appropriate measures must be taken in order to minimize these disadvantages of night ventilation, for example using safety window screens, ground floor windows that do not open and specially designed windows with top openings. Alternatively, mechanical ventilation systems can be operated during the night, to introduce cool outdoor air into the space.

Instead of natural night ventilation or when the climatic conditions do not permit it, for example in areas with high humidity levels, one can still precool a building during off-peak hours, using an air-conditioning system. This, in fact, results in considerable savings to the user and to the electric utility [11–14]. Electricity rates at night are usually much lower, since the demand drops considerably. On the other hand, the utility company also benefits, since eventually there will be a smaller demand during the daytime, which in turn reduces the need for additional generating capacity and maintains a more uniform load on the power generating facility.

Savings are achieved by subcooling the mass immediately before the peak utility rate period, thereby shifting a portion of the cooling load off-peak [13]. The method formulates an objective function based on the energy consumption of a simple building model. The effects of thermal mass on HVAC systems are quantified in [14]. A simplified model for including the effects of structural thermal mass in an above-ceiling unducted return air plenum on the air temperature returning to the HVAC equipment from the zone was also proposed. The results show that thermal mass effects in building zones can be significant and that the mass in the return plenum can effectively temper building air temperatures.

In any event, at the beginning of the following day, the cooled mass is utilized as a heat sink. The cooling load will be partly covered, passively, by the mass of the building, which is at a lower temperature, provided that the building is well insulated and is not ventilated during the daytime. This means that it is possible to reduce the energy consumption of the mechanical cooling system, since it will reduce its operation time. At a certain point, the temperature of the thermal mass will rise again and the same cycle will repeat.

Occupancy patterns

A designer should keep in mind that the occupant will be the final determinant factor on the extend of the utilizability of any building system, including thermal mass. Clearly, by changing the use of internal spaces and surfaces, one can drastically reduce the effectiveness of thermal storage. This implies that one has to consider carefully the final use of the space when making calculations of the cooling load and incorporating the possible savings from thermal-mass effects.

CALCULATIONS OF THERMAL MASS EFFECTIVENESS

The prediction of thermal performance of buildings can be treated by two methods, namely the thermal network method and the transfer function method [15]. The thermal network method considers the building as a network of interconnected elements. Each element is assigned an electrical analogue. Heat conduction paths through walls, roof, and floor are represented by resistances and capacitances. For windows, these paths are represented by resistances, implying that windows do not have significant thermal mass.

To construct a thermal network, one starts with a simple sketch of the object to be modelled. Then one identifies and locates a point (node) on the network for which it is desirable to calculate detailed information, or the points which affect the temperature of the nodes about which detailed information is required. Each node is then connected to all other nodes with which it has direct contact through a thermal resistance between each node. Each node also has a capacitance due to its finite mass, which requires a finite quantity of energy to change its temperature.

Figure 8.5 shows, in terms of an electrical analogue, the factors involved in the thermal performance of a building element [16]. The resistance of the material to heat transfer is represented by electrical resistances and the thermal capacity of the wall is represented by electrical capacitors. When the calculations are performed for steady-state conditions, the effect of thermal capacity is ignored.

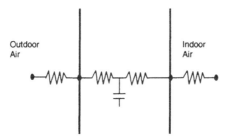

Figure 8.5 Electrical analogue of the thermal performance of a building element

Multidimensional effects, wall thickness and physical properties of building materials can be taken into account. However, if the number of nodes increases, the electrical analogue method requires significant effort to construct the analogue and solve for the unknown temperatures. For this reason, the simplified transfer function methods can be used with acceptable accuracy for calculating the heating and cooling loads of buildings.

The heat transfer through a composite surface can be calculated by examining how the energy is transferred through the surface when it is subjected to a continuous, piecewise linear curve or equivalently a series of triangular pulses of energy. Transfer functions relate the output of a linear, time-invariant system to a time series of current and past inputs, and past outputs. The heat response of a given wall construction is equivalent to the summation of the individual responses for each excitation of the materials composing the wall. In general, transfer function methods are more efficient for solving long-time transient heat-transfer problems than are other classical techniques, such as Euler or Crank–Nicolson.

There are a number of methods available for calculating transfer functions. Mitalas [17–19] has presented a method for calculating transfer functions for one-dimensional heat transfer through multilayered slabs by solving the conduction equation with Laplace and z-transform theory. The transfer coefficients, for a variety of constructions, have been developed by the American Society of Heating Refrigerating and Air Conditioning Engineers (ASHRAE) following a simplified procedure and are presented in the ASHRAE *Handbook of Fundamentals* [20].

According to the ASHRAE method, the effect of thermal mass on the cooling load is taken into account by various factors, called cooling load temperature differences (CLTD) and cooling load factors (CLF). These factors were generated for each component of the space-cooling load using the methodology and basic equations of the transfer function method (TFM). Following the TFM procedure, extensive computer simulations were first carried out and then, based on these results, the representative multipliers (CLTD and CLF values) were extracted by dividing component results by U-factors. Values of the space-cooling load components are calculated directly, through use of the tabulated CLTDs and CLFs factors, which include the effect of time lag due to thermal storage.

In particular, the transfer function method (TFM) was used to compute the one-dimensional transient heat flow through various sunlit roofs and walls. The heat gain was converted to cooling load by using the room transfer functions for three types of rooms having light, medium and heavy thermal characteristics, although the variations with room construction were small. The results were generalized by dividing the cooling load by the U-factor for each roof or wall. The results are in units of total equivalent CLTD. It is assumed that the heat flow through a similar roof or wall is obtained by using the tabulated values of CLTD in the following expression,

$$q = UA \text{ CLTD}$$

where U is the overall heat transfer coefficient of a given element (i.e. roof, wall) in W m^{-2} °C^{-1}, with an effective surface area A, expressed in square metres. A similar approach is followed using the CLF coefficients for all the other components of the cooling load (from windows, internal gains etc). Although the transfer function method has been validated by several experimental efforts with satisfactory results, it does exhibit some limitations which are outlined in the following discussion.

The tabulated transfer functions are calculated for a constant room temperature. Consequently, they are not valid for free-floating buildings. Another limitation of the method is that the functions are categorized in terms of light, medium or heavy building constructions. As a result, the user of this method cannot define with high accuracy the precise amount of mass of a given building. Passive buildings with high thermal mass cannot be treated accurately, since the given mass may be greater than the mass of the heavy building constructions defined by ASHRAE. To overcome this problem, it is possible to calculate the transfer functions for a given construction according to the Mitalas method [17–19].

In calculating the thermal-mass effectiveness, the primary objective is to calculate the optimum thermal mass of a building. Then, it is critical to distribute the thermal mass in such a way that its temperature fluctuations and heat return rate into the indoor environment are in accordance with the previously discussed principles. Thus, the correct selection for locating the thermal mass is a fundamental topic that needs to be addressed. That is, one needs to determine the most effective way to add mass. For example, it is possible to add more thickness, more surface, or both, in the external shell or the internal spaces of the building.

A review of various studies of heat attenuation, including thermal mass, in the region of southern Europe, is presented in [21]. The primary need for additional research in this field is identified in the area of information dissemination, through simulation work, development of simplified design tools and compilation of guidelines. Similar observations are made in [22], which presents several methods for simulating the performance of buildings, taking into account the thermal mass. The optimum amount of thermal mass, for winter and summer conditions in the Mediterranean climate, has been investigated and results presented in [23-25].

There are several design and computational tools available for passive and hybrid cooling purposes, which take into account the impact of thermal mass on the building's thermal performance. In a recently completed study [26], 128 programs have been identified which have some capabilities to calculate the cooling load of a building. Among them, 54 programs can also calculate the variation of the indoor air temperature, 45 programs can also take into account the impact of mass and shading, while only 23 programs can also simulate natural ventilation strategies.

A total of 19 models for estimating the cooling load of a building, taking into account the thermal mass of the building, have been reviewed and classified in [27]. Overall, one may classify the available methods into two categories. The simplified design tools and the advanced programs.

Simplified design tools [16, 20, 28–42] are necessary for designers during the first stages of the design process in order to evaluate new techniques and systems and their performance under specific conditions. These tools are easy to use and do not require much input, but they provide accurate information. They may require the use of a computer or, in some cases, the calculations can also be performed by hand.

Advanced programs, such as for example those described in [43–45], are more complex tools. They require a great amount of input information, but provide in return the capability for treating complicated phenomena with high accuracy. Advanced programs are suitable for simulating the behaviour of buildings and evaluating the performance of the various systems with great accuracy. They are more likely to be used by people with advanced knowledge of these subjects, for scientific and research activities.

The most popular methods are tabulated in Table 8.1, which includes some relevant information for each one. For each method, one can identify the inputs and outputs, the parameter used to account for the thermal mass effect, the restrictions and whether the method has been used to develop a software package. The list of the advanced programs is a representative one and by no means complete. There are several other software packages available on the market, which have been developed by national research organizations and institutions, universities and private firms. The ones listed represent the most popular software used in Europe (ESP) and the United States (TRNSYS).

Table 8.1 Simplified design tools and advanced programs for calculating the cooling load in buildings

Model	Inputs	Outputs	Parameter for thermal mass effects	Restrictions	Soft-ware
Ashrae [20]	BG,BC,IL, EL,AT/M, VE/S&L, IN/S&L	Hourly and peak	CLTD values for wall construction	Simplified treatment	No
Athienitis [28]	BG,BC, AT/M,AT/H, SR/M or SR/H,IL	Hourly or monthly for optimum thermostat set point	Elements in a thermal network		Yes
Bida & Kreider [29]	BG,BC,IL, EL,VE/S, SR/M,AT/M, CL,SS	Monthly loads	Mass Effect Coefficient	Monozone, dwellings, small multistorey buildings, small IL, no latent heat loads, no wind effects	Yes
BRE [16]	BG,BC, EL,IL	Peak summer temperatures	Admittance factor	Office buildings with one external wall, no wind effects	No
Catani [30]	BG,BC	Performance of massive walls	M-factor	Manual method for calculating the thermal mass performance	No
Givoni [31] & Balcomb [5]	BG,BC	Long-term performance of massive walls	Effective heat capacity	Method for calculating thermal mass performance	Yes
Givoni & Hoffman [32]	BG,BC,VE, SR/M,AT/M	Internal air temperature	Total thermal time constant	All external input functions are taken as equal, no wind effects	Yes
Kusuda [33]	BG,BC,EL, IL,VE,TA/H, SR/H Previous temperature history	Transient sensible cooling load	Conduction transfer functions		No
LANL [34]	BG,BC, SR/M,DD	Monthly cooling load. Daily average indoor temperature	Solar saving fraction	Restricted to certain type of buildings, no VE, IL, wind effects	Yes
van der Maas & Roulet [35]	BG,BC,IL, AT,VE/S	Indoor air temperature. Cooling rate	Thermal effusivity	Monozone, stack effect driven flows, short periods, homogeneous walls, no radiation, no convective effects, constant heat flow rates, no wind effects	Yes
Mathews & Richard [36]	BG,BC,IL, EL,VE, SR/M,AT/M	Hourly indoor air temperature and sensible loads	Effective heat storage	Single zone, no temperature stratification, interior surfaces at same temperature, no wind effects	Yes

Table 8.1 (continued)

Model	Inputs	Outputs	Parameter for thermal mass effects	Restrictions	Soft-ware
PASSPORT [37]	BG,BC,IL, EL,VE, ADT,SR/M, AT/M,WV, CDD	Indoor air temperature, monthly cooling load, distribution of heat losses	Effective thermal capacity	Number of zones, elements	Yes
PASSPORT Plus [38]	BG,BC,EL, IL,IN/S/L, VE/S,AT/H, SR/H,WV/H, WD/H (Meteoro-logical Library files)	Hourly indoor air and surface temperatures. Cooling/heating load. Air flows. Comfort	Transfer functions (Next release to include finite differences)	One-dimensional heat conduction (Next release to include 2-D conduction). Number of zones, elements	Yes
PMDG [34]	BG, BC, SR, DD	Monthly cooling load, daily average cooling load	Heat capacity	No VE, IL, wind effects	Yes
Seem [39]	Previous temperature and flux history, transfer function coefficients, cooling loads from AT,IL,EL	Cooling load	Comprehensive transfer functions	Requires calculations for input information	No
Shaviv [40]	BG,BC, SR/H,AT/H	Maximum indoor air		Isothermal internal mass	No
SPIEL [41]	BG,BC,SS, AT/M,IN	Mean monthly hourly indoor air, cooling air	Capacitance	Convergence of calculated values, no direct account of SR levels, no IN, no wind effects	Yes
Steady State [42]	BG,BC, AT/M,IN	Average daily space cooling	Envelope heat transfer coefficient	Steady state conditions, no IL,EL,VE	No

Advanced programs	Inputs	Outputs	Parameter for thermal mass effects	Restrictions	Soft-ware
ESP [43]	BG,BC, indoor-outdoor conditions, IL,VE,WV,IN	Hourly indoor temperature and cooling load	Solves all heat and mass flow paths and flow path interactions	Detailed inputs and analysis, UNIX environment	Yes
TRNSYS [45]	BG,BC,IL,EL, SR,AT,IN	Sensible cooling load	Wall transfer functions	One dimensional heat transfer through building materials	Yes

Table 8.1 (continued)

Nomenclature

ADT:	Monthly ambient daytime temperature
BC:	Detailed information on the building construction (U-values, thermal properties)
BG:	Detailed information on the building geometry and orientation
AT/M AT/H:	Ambient temperature monthly data and hourly data
IN/S IN/L:	Infiltration loads from sensible heat, infiltration loads from latent heat
IN:	Total infiltration loads (sensible and latent)
SR/M SR/H:	Solar radiation monthly data, solar radiation hourly data
CDD:	Monthly cooling degree days
DD:	Monthly degree days
SS:	Mean daily sunshine hours
CL:	Monthly clearness index
VE/S VE/L:	Ventilation loads from sensible heat, ventilation loads from latent heat
DD:	Monthly degree days
WV/M WV/H:	Wind velocity monthly data, wind velocity hourly data
EL:	External loads (solar radiation, temperature)
IL:	Internal loads (human activity, equipment, lights)
WD/M WD/H:	Wind direction monthly data, wind direction hourly data

For the different simplified tools and advanced programs, thermal-mass effects on the cooling load and overall behaviour of buildings are accounted for by a number of parameters. These parameters and their physical meaning are summarized in Table 8.2.

CASE STUDIES

Results from numerical simulations

The performance of massive walls is affected by the type of building, its operating schedule, internal loads and mass, external shading and the climatic conditions. A series of simulations of three building types (a warehouse, a low-rise office building and a retail building) were carried out for four climate zones [9], using the simplified design method adapted by ASHRAE [20]. The buildings were modelled for the climatic conditions of south and north USA (Atlanta, Phoenix, New York City and Minneapolis). According to the results reported in [9], the wall thermal mass is less effective in warehouses and buildings with high internal mass levels. It is more effective in office and retail buildings.

A series of simulations, performed for a building with a fixed orientation, concrete walls with a U value of 0.682 W m^{-2} °C^{-1}, and a fixed overall heat transfer coefficient, has provided some useful information [23]. The study was performed for a Mediterranean climate, with climatic data coming from Haifa, Israel. Accordingly, the effect of the internal thermal mass in reducing the cooling load of the building is largest when the building is in use day and night. The optimum thickness of the concrete layer of an internal partition should be 10 cm, while the recommended thickness of an external wall is 15 cm. The addition of internal partitions dramati-

Table 8.2 Parameters for describing thermal mass effects

Parameter	Physical meaning	Ref.
Admittance factor	Represents the extent to which heat enters the surface of materials in a 24-hour cycle of temperature variation	16
Capacitance	Accounts for the ability of the external and internal materials to store heat	40
Comprehensive transfer functions	Describes heat flows in building elements, combining individual wall transfer functions for an enclosure	39
Conduction transfer functions	Expresses the decay of temperature through the material	33
Cooling load temperature difference	Includes the effect of time lag in the propagation of heat through the material, due to thermal storage	20
Diurnal heat capacity	Measures the effectiveness of the material for heat storage during a continuous 24-hour cycle	5
Effective heat capacity	Accounts for the effects of the building materials' thermal properties and design factors on the long-term energy performance	34
Effective heat storage	Accounts for the effects of thermal capacity and thermal resistance of the building elements, and the exterior resistance, for exterior and interior building elements	36
Effective thermal capacity	Accounts for the effective mass of the building elements	46
Envelope heat transfer coefficient	Includes the effects of thermal transmittance of the material along with heat transfer rate due to infiltration	42
Heat capacity	Introduces the effect of heat storage for different building types in the correlation coefficients used in calculating the solar saving fraction	34
Mass effect coefficient	Accounts for the temperature fluctuation allowable in the space, the amount of heat gain due to ambient conditions, and the degree of exposure of the space to ambient conditions	29
M-factor	Corrects steady state U-values for the building materials	30
Solar saving fraction	Correlation coefficients as a function of the heat capacity for specific building types	34
Thermal capacity	Determines the heat flow in unit time by conduction through unit thickness of a unit area material, across a unit temperature gradient, defined as the product of density by specific heat	20
Thermal effusivity	Accounts for the response of a surface temperature to a change of the heat flow density at the surface	35
Total thermal time constant	The heat stored in a whole enclosure per unit of heat transmitted to or from the outside through the elements surrounding the enclosure and by ventilation	32

cally improves the energy consumption of the building during the summer. Thermal mass must be added in such a way as to provide adequate surface area in order to facilitate the exchange of energy with the indoor environment.

The effect of mass on cooling load and on insulation requirements for buildings in different climates has also been investigated in [47]. The cooling loads are calculated for three types of buildings; low-rise residential buildings having internal mass and low internal heat gains; non-residential buildings having, during short occupied periods, high internal heat gains and no internal mass; small models to show wall mass effects with and without the influence of internal heat sources and solar radiation through windows, with no internal mass. Accordingly, results from the first and third case studies indicate that adding insulation to walls can increase the cooling loads, while the mass effects decrease the cooling loads. Mass is most effective in reducing cooling loads when internal heat gains are present. Results from the second case study indicate that the mass is effective in reducing cooling loads at all locations for non-residential buildings. The cooling requirements of the buildings studied in this series averaged 22% less for heavier constructions.

The role of thermal mass effects in Greek buildings has been investigated in a recently completed study [48]. The thermal behaviour of two types of buildings (monozone and two-zone) was calculated for varying amounts of the buildings' thermal mass. The simulations were carried out by using the ESP program [43], an official computational tool for the European Commission, developed by the Energy Simulation Research Unit at the University of Strathclyde in Scotland.

The calculations were performed for Athens, while in total 13 different types of building materials were used. The geometry of the buildings remained the same, but the building materials changed, thus varying the effective amount of the building's thermal mass. The thermal conductivity of the materials ranged between 0.19 and 1.63 W m^{-1} K^{-1}, the specific heat between 796 and 1014 J kg^{-1} °C^{-1} and the density between 600 and 2300 kg m^{-3}.

The cooling load was first estimated for a monozone building (5 by 5 by 3 m), with a single window on the south side of the building. The results of the annual cooling load per square metre of floor area as a function of the effective thermal mass, are shown in Figure 8.6. The reduction of the cooling load is significant up to a value of thermal mass approaching 300 kJ m^{-2} K^{-1}. An additional increase beyond this value does not appear to have any significant impact on the cooling load.

For the two-zone building (15 by 8 by 3 m), the additional effects of interior thermal mass were also investigated. The building included additional windows, in all sides, as shown in Figure 8.7, although the construction materials remained the same. The results for the annual cooling load per square metre of floor area, as a function of the effective thermal mass, are also shown in Figure 8.7.

The reduction of the cooling load is again significant up to a value of thermal mass approaching 300 kJ m^{-2} K^{-1}. The absolute values of the cooling are relatively higher than the monozone building, since there are additional openings which increase the direct solar gains and air infiltration. The thermal mass effectiveness, however, remains the same.

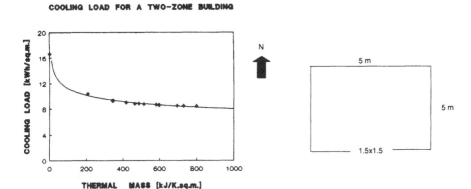

Figure 8.6 Calculated annual cooling load for different levels of thermal mass in monozone building located in Athens

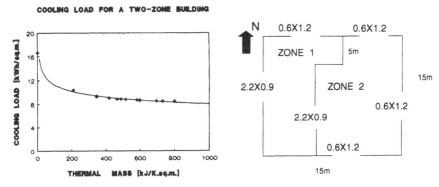

Figure 8.7 Calculated annual cooling load for different levels of thermal mass in a two-zone building located in Athens

Results from experimental investigations

A theoretical analysis of the data collected during an extensive monitoring campaign of an office building in northern USA, using variable levels of thermal mass in similar buildings, has been reported in [7]. First, a small building model was defined for a two-storey small office, with masonry wall construction and steel roof truss system and reinforced concrete deck system. A series of alternative thermal mass systems was defined for application to the building model. Winter and summer conditions were examined. The various cases were examined with a building simulation program. Accordingly, an increase of the thermal mass from 21 to 201 kg/m^2 of floor area, in closed and in ventilated buildings, can reduce the peak indoor temperature by approximately 1 and 2°C, respectively. Important reductions in peak heating and cooling loads can be achieved for high-thermal-mass designs. Energy savings can vary between 18 and 20%, compared to the base case.

The effect of building thermal storage on the peak cooling load has been evaluated during an experiment on a large office building in Jacksonville, Florida [49]. The building's exterior consisted of a glass curtain wall with structural steel supports, while the interior construction included a carpeted concrete floor, gypsum walls, interior partitions and a concrete block core. The overall building thermal capacitance was approximately 23 Btu ft^{-2} °F^{-1} (130.6 W m^{-2} K^{-1}). The objective of this study was to precool the building at night and during the weekend to reduce daytime cooling loads. Diurnal heat capacity calculations were used in analysing the experimental results. The results showed an 18% reduction in cooling energy supplied during the daytime.

A high-mass test house with outside dimensions 6.5 m by 5.3 m by 4.7 m, was monitored in an environmental chamber under a night-time cooling operation [50]. The inside floor area was 34.4 m^2 and ceiling height 4.06 m. The U-value of the envelope components was 0.284 W m^{-2} K^{-1}. The cooling load was measured for simple cyclic outdoor climatic conditions. The simulated outdoor conditions were for a desert-type climate (low temperatures and high humidity at night, and hot and dry conditions during the day). As a result, night-time cooling was proved to be much more efficient than 24-hour cooling. However, it appears that the large latent cooling load, resulting from the release of moisture from the structural concrete during the daytime warm-up period, cannot be ignored under these climatic conditions.

Six small single-room detached buildings, constructed over an insulated slab-on-grade floor, were used to investigate the effect of wall mass on summer space cooling [42]. The buildings were located in the mid-eastern USA at the outdoor facilities of the National Bureau of Standards in Gaithersburg, MD. The wall U-values varied between 0.415 and 0.55 W m^{-2} K^{-1}. The buildings were operated at a fixed indoor temperature setting of 18°C or, to simulate building performance in hotter climates, 24°C with night ventilation cooling. It was concluded that thermal mass has an unimportant effect on the performance of a building when exposed to sufficiently hot weather conditions. For an indoor temperature set at 24 °C, heavyweight buildings were observed to consume less cooling energy than comparable lightweight buildings having equivalent thermal resistance in their walls. The thermal mass was more effective when it was positioned inside the wall insulation. Night-time ventilation reduced the cooling load by 27 to 36%, depending on the wall construction of the building. Larger savings occurred in heavyweight rather than lightweight buildings. Again, the wall mass was more effective when it was positioned inside the wall insulation.

A number of test cells have been used to illustrate experimentally the thermal performance and the role of thermal mass on various passive systems [51]. Specific recommendations on how best to use thermal-mass materials in residential and small commercial buildings, for southwestern US locations, have been presented in [1], as a result of an extensive experimental thermal-mass study funded by the US

Department of Energy. The test site included eight test cells sized 6 by 6 by 2.4 m. Different constructions (wall materials and wall thicknesses) were used, along with a series of experiments carried out on cells with and without openings (windows, doors), heavily insulated ceilings, and very low air infiltration rates, so that the behaviour of the walls could be isolated and their performance better evaluated. Dense and usually more conductive materials are better.

A 5–10 cm wall thickness is usable for heat absorption, storage and release on a daily basis. Thicker walls, up to 25 cm, may provide longer periods of thermal storage. For heating purposes, the more thermal mass and surface area, the better. For direct-gain heating, radiative coupling is more effective than convective coupling and will reduce overheating. External insulation should be used, since thermal contact of the mass with the interior space is necessary. Massive structures are less susceptible to any air infiltration that is present, because the heat is stored in the mass, not the air. Ventilation during the cooling season should be performed by natural means during the night or when the outdoor conditions are favourable. Natural ventilation should be avoided when outdoor temperatures or humidity are high. Thermostat set-back (during winter) or set-forward (during summer) is less effective in massive structures, because they take longer to heat up or cool down. The effect of thermal mass on night-temperature set-back savings is also discussed in [52]. Numerical simulations have also shown that the annual reductions in sensible heating and cooling loads, due to the thermal mass levels of the walls (floors and foundations were assumed to be massless), can be as high as 40% in mild climates.

The French Centre for Information of the Cement Industry has published two booklets [53, 54] on results for the use of mass in dwellings, and its role on energy conservation and comfort. The booklets include specific guidelines for the use of thermal mass under the different climatic conditions in France.

REFERENCES

1 Robertson, D. (1986). 'The performance of adobe and other thermal mass materials in residential buildings', *Passive Solar Journal*, pp. 387–417.
2 Lechner, N. (1991). *Heating, Cooling, Lighting*. John Wiley & Sons, New York.
3 Kammerud, R., E. Ceballos, B. Curtis, W. Place and B. Andersson (1984). 'Ventilation cooling of residential buildings', *ASHRAE Transactions*, Vol. 90, No. 1B, pp. 226–252.
4 Hopkins, V., G. Gross and D. Ellifritt (1979). 'Comparing the thermal performances of buildings of high and low masses', *ASHRAE Transactions*, Vol. 85, No. 1, pp. 885–902.
5 Balcomb, J. and R. Jones (1988). *Workbook for Workshop on Advanced Passive Solar Design*. Balcomb Solar Associates, Bled, Yugoslavia.
6 Dascalaki, E. (1992). 'Natural convection heat transfer from vertical and horizontal surfaces. Correlations and proposed experimental procedures within PASCOOL', *Proceedings of the Model Development Subgroup Meeting, PASCOOL* Programme, 23–27 November, Lyon, France.

7 Brown, M. (1990). 'Optimization of thermal mass in commercial building applications', *ASME Journal of Solar Energy Engineering*, Vol. 112, pp. 273–279.

8 Szokolay, S. (1985). 'Passive and low energy design for thermal and visual comfort. Passive and Low Energy Ecotechniques', *3rd International PLEA Conference, Mexico City*, 6-11 August 1984, eds A. Bowen and S. Yannas, pp. 11–28. Pergamon Press, Oxford.

9 Ober, D. and D. Wortman (1991). 'Effect of building type on wall thermal mass performance', *ASHRAE Journal*, Vol. 33, No. 11, pp. 18–21.

10 Baer, S. (1983). 'Raising the open *U* value by passive means', *Proceedings of the 8th National Passive Solar Conference*, pp. 839–842, Glorieta, NM.

11 Braun, J. (1990). 'Reducing energy costs and peak electrical demand through optimal control of building thermal storage', *ASHRAE Transactions*, Vol. 96, No. 2, pp. 876–888.

12 MacCracken, C. (1991). 'Off-peak air conditioning: a major energy saver', *ASHRAE Journal*, Vol. 33, No. 12, pp. 12–23.

13 Snyder, M. and T. Newell (1990). 'Cooling cost minimization using building mass for thermal storage', *ASHRAE Transactions*, Vol. 96, No. 2, pp. 830–838.

14 Brandemuehl, M., M. Lepore and J. Kreider (1990). 'Modelling and testing the interaction of conditioned air with building thermal mass', *ASHRAE Transactions*, 96, No. 2, pp. 871–875.

15 Haghighat, F. and M. Chandrashekar (1987). 'System-theoretic models for building thermal analysis', *ASME Journal of Solar Energy Engineering*, Vol. 109, pp. 79–88.

16 Burberry, P. (1983). *Practical Thermal Design in Buildings*, Batsford Academic and Educational Ltd, London.

17 Stephenson, D. and G. Mitalas (1967). 'Cooling load calculation by thermal response factor method', *ASHRAE Transactions*, Vol. 73, No. 2, p. III.1.1.

18 Mitalas, G. (1972). 'Transfer function method of calculating cooling loads, heat extraction rate and space temperature', *ASHRAE Journal*, Vol. 14, No. 12, p. 52.

19 Mitalas, G. and J. Arseneault (1971). 'Fortran IV program to calculate Z-transfer functions for the calculation of transient heat transfer through walls and roofs', *Proceedings of the Conference on the Use of Computers for Environmental Engineering Related to Buildings*. NBS Building Science Series No. 39. (October), Gaithersburg, MD.

20 *Handbook of Fundamentals*, (1981). American Society of Heating Refrigerating and Air Conditioning Engineers, Atlanta.

21 Antinucci, M. (1990). 'Heat Attenuation as a means to limit cooling load and improve comfort', *Workshop on Passive Cooling*, eds E. Aranovitch, E. Fernandes and T. Steemers, pp. 111–119. Ispra, Italy.

22 Paparsenos, G. (1990). 'Heat attenuation and the thermal mass effect', *Workshop on Passive Cooling*, eds E. Aranovitch, E. Fernandes and T. Steemers, pp. 111–119. Ispra, Italy.

23 Shaviv, E. (1988). 'On the determination of the optimum thermal mass in the Mediterranean climate', *Energy and Buildings for Temperate Climates*, 6th International PLEA Conference, Porto, Portugal, eds E.de Oliveira Fernandes and S. Yannas, pp. 385–390. Pergamon Press, Oxford.

24 Guglielmini, G., M. Misale, E. Nannei and G. Tanda (1988). 'The relative merits of envelope and interior structures on the weekly thermal inertia of buildings', *Energy and Buildings for Temperate Climates*, 6th International PLEA Conference, Porto, eds E.de Oliveira Fernandes and S. Yannas, pp. 391–396. Pergamon Press, Oxford.

25 Lefebvre, G. and I. Sommereux (1988). 'Analysis of building inertia using a modal description', *Energy and Buildings for Temperate Climates*, 6th International PLEA Conference, Porto, Portugal, eds E.de Oliveira Fernandes and S. Yannas, pp. 397–407. Pergamon Press, Oxford.

26 Antinucci, M., B. Fleury, J. Lopez d'Asiain, E. Maldonado, M. Santamouris, A. Tombazis and S. Yannas (1990). *Horizontal Study on Passive Cooling*, ed. M. Santamouris. Building 2000 Research Programme, EEC, DG 12.

27 Balaras, C.A. (1992). 'The role of thermal mass on the cooling load of buildings. An overview of computational methods and proposed experimental work within PASCOOL', *Proceedings of the Model Development Subgroup Meeting*, PASCOOL Programme, Lyon, France, 23–27 November.

28 Athienitis, A. (1988). 'A predictive control algorithm for massive buildings', *ASHRAE Transactions*, Vol. 94, No. 2, pp. 1050–1067.

29 Bida, M. and J. Kreider (1987). 'Monthly averaged cooling load calculations – residential and small commercial buildings', *ASME Journal of Solar Energy Engineering*, Vol. 109, pp. 311–320.

30 Catani, M. (1978). 'Insulation and the *M* factor', *ASHRAE Journal*, June, pp. 50–55.

31 Givoni, B. (1987). 'The effect of heat capacity in direct gain buildings', *Passive Solar Journal*, Vol. 4, No. 1, pp. 25–40.

32 Givoni, B. (1981). *Man, Climate and Architecture*, 2nd ed. Applied Science Publishers, London.

33 Kusuda, T. (1985). 'Heat transfer in buildings', *Handbook of Heat Transfer Applications*, eds W. Rohsenow, J. Hartnett and E. Ganic, Ch. 9. McGraw Hill, New York.

34 Givoni, B. (1987). 'A generalized predictive model for direct gain', *Passive Solar Journal*, Vol. 4, No. 1, pp. 5–24.

35 Maas, J. van der and C. Roulet (1991). 'Night time ventilation by stack effect', *Ventilation and Infiltration, ASHRAE Technical Data Bulletin*, Vol. 7, No.1, NY-91-5-3, pp. 32–40. ASHRAE, Atlanta.

36 Mathews, E. and P. Richards (1989). 'A tool for predicting hourly air temperatures and sensible energy load in buildings at sketch design stage', *Energy and Buildings*, Vol. 14, pp. 61–80.

37 *PASSPORT Technical Manual*, version 1.4.C (1992). ed. M. Santamouris. Commission of the European Communities, DG for Science Research and Development.

38 *PASSPORT Plus User's Manual* (1995). eds C.A. Balaras and S. Alvarez. PASCOOL Project, Commission of the European Communities, Joule II - Programme, JOU2-CT92-0013.

39 Seem, J., S. Klein, W. Beckman and J. Mitchell (1998). 'Comprehensive room transfer functions for efficient calculation of the transient heat transfer processes in buildings', 1987 *ASME/AICHE National Heat Transfer Conference, Technical Session on Heat Transfer in Buildings*, pp. 35–45, Pittsburgh, PA, August.

40 Shaviv, E. and G. Shaviv (1978). 'Designing buildings for minimal energy consumption', *Computer Aided Design Journal*, Vol. 10, No. 4, pp. 239–247.

41 Green, C. (1985). *The Simulation Tool SPIEL*, Ecotech, UK.

42 Burch, D., S. Malcolm and K. Davis (1984). 'The effect of wall mass on the summer space cooling of six test buildings', *ASHRAE Transactions*, Vol. 90, No. 2B, pp. 5–21.

43 Clarke, J. (1985). *Energy Simulation in Building Design*, Adam Hilger Ltd, Bristol.

44 Clarke, J. (1989). 'Building energy simulation: state-of-the-art', *Solar and Wind Technology*, Vol. 6, pp. 345–355.

45 Klein, S. et al. (1988). 'TRNSYS – A transient system simulation program'. University of Wisconsin-Madison, Engineering Experiment Station Report 38-12.

46 Mathews, E., P. Rousseau, P. Richards and C. Lombard (1991). 'A procedure to estimate the effective heat storage capability of a building', *Building Environment*, Vol. 26, p. 179.

47 Goodwin, S. and M. Catani (1979). 'The effect of mass on heating and cooling loads and on insulation requirements of buildings in different climates', *ASHRAE Transactions*, Vol. 85, No. 1, pp. 869–884.

48 Argiriou, A. (1992). 'A study of thermal mass in Hellenic buildings'. Institute for Technological Applications, Hellenic Productivity Centre, Athens, September.

49 Ruud, M., J. Mitchell and S. Klein (1990). 'Use of building thermal mass to offset cooling loads', *ASHRAE Transactions*, Vol. 96, No. 2, pp. 820–828.

50 Kusuda T. and J. Bean (1981). 'Comparison of calculated hourly cooling load and indoor temperature with measured data for a high mass building tested in an environmental chamber', *ASHRAE Transactions*, Vol. 87, No. 1, pp. 1232–1240.

51 Sunaga N. and N. Ito (1985). 'Experimental analysis of thermal mass effect on the indoor climate in passive solar heating system with test cells', *Proceedings of the International Symposium on Thermal Application of Solar Energy*, pp. 253–258, 7–10 April, Hakone, Japan.

52 Burch, D., W. Johns, T. Jacobsen, G. Walton, and C. Reeve (1984). 'The effect of thermal mass on night temperature setback savings', *ASHRAE Transactions*, Vol. 90, No. 2A, pp. 184–206.

53 Depecker, P., J. Brau and S. Rousseau (1985). *L'Inertie Thermique par le Béton, Economies d'Energie et Confort d'Eté*. Centre d'Information de l'Industrie Cimentière, Paris.

54 *L'Isolation ne Suffit pas. Pour le Comfort et les Economies d'Energie l'Inertie Thermique par le Béton a fait Thermique par le Béton a fait ses Preuve*. (Undated). Publication du Centre d'Information de l'Industrie Cimentière, Paris.

55 Lienhard, J.H. (1981). *A Heat Transfer Textbook*. Prentice-Hall, Englewood Cliffs, NJ.

56 Churchill, S. and H. Chu (1975). 'Correlating equations for laminar and turbulent free convection from a vertical plate, *International Journal of Heat and Mass Transfer*, Vol. 18, pp. 1323–1329.

57 Gebhart, B. (1970). *Heat Transfer*, 2nd ed. McGraw-Hill Book Company, New York.

58 Gryzagoridis, J. (1971). 'Natural convection from a vertical flat plate in the low Grashof number range', *International Journal of Heat and Mass Transfer*, Vol. 14, pp. 162–165.

59 McAdams, W.H. (1954). *Heat Transmission*, 3rd ed. McGraw-Hill Book Company, New York.

60 Bayley, F.J. (1955). 'An analysis of turbulent free convection heat transfer', *Proceedings of the Institution of Mechanical Engineers*, Vol 169, No. 20, p. 361.

61 Eckert, E.R.G. and T.W. Jackson (1951). 'Analysis of turbulent free convection boundary layer on flat plate'. NACA Technical Report 1015.

62 Lewandowski, W.M. and P. Kubski (1984). 'Effects of the use of balance and gradients methods as a result of experimental investigation of natural convection action with regard to the conception and construction of measuring apparatus', *Warme- und Stoffubertragen*, Vol. 18, p. 247.

63 *ASHRAE Fundamentals Handbook* (1981). American Society for Heating Refrigerating and Air Conditioning Engineers, Atlanta.

64 PASSYS Phase-1, Model Validation and Development Subgroup, Final Report, (1990).

65 CIBS (1976). *Guide C3. Heat Transfer*. CIBSE, London.

66 Mitalas, G.P. (1976). 'Calculation of transient flow through wall and roofs', *Proceedings of the ASHRAE Annual Meeting*, Lake Placid, NY, pp. 24–26.

67 Ferries, B. (1980). 'Contribution à l'étude des enveloppes climatiques et aide à leur conception par microordinateur'. Thèse de 3e Cycle, Toulouse, November.

68 Gaignou, A. (1973). 'Régime varié dans les échanges thermiques', *Promoclim E*, April.

69 Alamdari, F. and G. Hammond (1983). 'Improved data correlations for buoyancy-driven convection in rooms', *Building Service Engineering Research Technology*, Vol. 4, pp. 106–112.

70 Nansteel, M. and R. Greif, (1981). 'Natural convection in undivided and partially divided rectangular enclosures, *ASME Journal of Heat Transfer*, Vol. 103, November, pp. 623–629.

71 Bauman, F., A. Gadgil, R. Kammerud and R. Greif, (1980). 'Buoyancy driven convection in rectangular enclosures : experimental results and numerical calculations', ASME Paper 80-HT-66, presented at the 19th National Heat Transfer Conference, 27–30 July, Orlando, FL.

72 Bajorek, S.M. and J.R. (1982). 'Lloyd experimental investigation of natural convection in partitioned enclosures', *ASME Journal of Heat Transfer*, Vol. 104, pp. 527–532.

73 S.M. Bohn, A.T. Kirkpatrick and D.A. Olson, (May 1984). 'Experimental study of three-dimensional natural convection high-Rayleigh number', *ASME Journal of Heat Transfer*, Vol. 106, pp. 339–345.

74 Min, T., L. Schutrum, G. Parmelee and J. Vouris (1956). Natural convection and radiation in a panel-heated room, *ASHRAE Transactions*, 62, pp. 337–358.

75 Allard, F., J. Brau, C. Inard and J.M. Pallier (1987). 'Thermal experiments of full-scale dwelling cells in artificial climatic conditions', *Energy and Buildings*, Vol. 10, pp. 49–58.

76 Khalifa, A.J.N. and R.H. Marshall (1990). Validation of Heat Transfer Coefficients on Interior Building Surfaces using a Real-sized Indoor test cell, *International Journal of Heat and Mass Transfer*, Vol. 33, No.10, pp. 2219–2236

77 Fujii T., and H. Imura 1972. 'Natural convection heat transfer from a plate with arbitrary inclination', *International Journal of Heat and Mass Transfer*, Vol.15, pp. 755–767.

78 Lloyd J.R. and W.R. Moran (1974). 'Natural convection adjacent to horizontal surface of various platforms', *Proceedings of the ASME Annual Meeting*, Paper 74-WA / HT-66.

79 Goldstein, R.J., E.M. Sparrow and D.C. Jones (1973). 'Natural convection mass transfer adjacent to horizontal plates, *International Journal of Heat and Mass Transfer*, Vol. 16, p. 1025.

80 Incropera, F.P. and D.P. Dewitt (1981). *Fundamentals of Heat Transfer*. John Wiley & Sons, New York.

81 Roldan, A. (1985). 'Etude thermique et aeraulique des enveloppes de bâtiment – influence des couplages interieurs et du multizonage'. Thèse de Docteur, Lyon, December.

82 Perez Sanchez, M.M. (1989). 'Typologie et uniformisation syntaxique des modèles de transfert de chaleur dans le contexte de la thermique du bâtiment'. Thèse de Docteur, Lyon.

APPENDIX A
CORRELATIONS FOR NATURAL HEAT TRANSFER COEFFICIENT

Tables 8.A1 to 8.A6 are also available in: Dascalaki, E., M. Santamouris, C.A. Balaras and D. Asimakopoulos (1994). 'Natural convection heat transfer coefficients from vertical and horizontal surfaces for building applications', *Energy & Buildings* Vol. 20, pp. 243–249.

APPENDIX B
COMPARATIVE STATISTICAL ANALYSIS OF AVAILABLE CORRELATIONS

All correlations in Tables 8.B1 to 8.B3 have been modified for air with Prandtl number Pr=0.72 and the results are given for representative Grashof values (ranging between 10,000 and 1E12).

Table 8.A1 Correlations for vertical surfaces

Reference	Correlation	Gr range	Flow condition	Working fluid	Experimental setup
Squire-Eckert [55]	$Nu = 0.505 \, Gr^{1/4}$	$Gr > 10^5$	Laminar	Water, mercury	Theoretical
Churchill, Chu [56]	$Nu = 0.47 \, Gr^{1/4}$	$10^5 < Gr < 10^9$	Laminar		
Churchill, Chu [56]	$Nu = 0.68 + 0.47 \, Gr^{1/4}$	$Gr < 10^9$	Laminar		
Churchill, Chu [56]	$Nu = 0.865 + 0.31 \, Gr^{1/6}$	$0 < Gr < \infty$	Laminar turbulent		
Gebhart [57]	$Nu = 0.511 \, Gr^{1/4}$	$10^5 < Gr < 10^{11}$	Laminar		
Gryzagoridis [58]	$Nu = 0.507 \, Gr^{1/4}$	$10 < Gr < 10^8$		Air	Vertical plates
McAdams [59]	$Nu = 0.544 \, Gr^{1/4}$	$10^4 < Gr < 10^9$	Laminar		
Bayley [60]	$Nu = 0.089 \, Gr^{1/3}$	$2 \times 10^9 < Gr < 10^{12}$			
Eckert, Jackson [61]	$Nu = 0.509 \, Gr^{1/4}$	$Gr < 10^9$	Laminar	Air	
Eckert, Jackson [61]	$Nu = 0.018 \, Gr^{2/5}$	$10^{10} < Gr < 10^{12}$	Turbulent	Air	
Lewandowski [62]	$Nu = 0.564 \, Gr^{1/4}$				Vertical plate
ASHRAE [63]	$Nu = 0.516 \, Gr^{1/4}$	$10^4 < Gr < 10^8$	Laminar	Air	Small plates
ASHRAE [63]	$Nu = 0.117 \, Gr^{1/3}$	$10^8 < Gr < 10^{12}$	Turbulent		

Table 8.A1 (continued)

Reference	Correlation	Gr range	Flow condition	Working fluid	Experimental setup
Fischenden, Saunders [64]	$Nu = 0.514\,Gr^{1/4}$	$Gr < 10^9$	Laminar		Theoretical
Fischenden, Saunders [64]	$Nu = 0.107\,Gr^{1/3}$	$Gr > 10^9$	Turbulent		
Rogers, Mayhew [64]	$Nu = 0.504\,Gr^{1/4}$	$Gr < 10^9$	Laminar		
Wong [64]	$Nu = 0.474\,Gr^{1/4}$	$Gr < 10^9$	Laminar		
CIBS [65]	$Nu = 0.48\,Gr^{1/4}$	$Gr < 10^9$	Laminar		
CIBS [65]	$Nu = 0.119\,Gr^{1/3}$	$Gr > 10^9$	Turbulent		
ASHRAE [63]	$Nu = 0.102\,Gr^{1/3}$	$10^8 < Gr < 10^{10}$	Turbulent	Air	
Mitalas [66]	$Nu = 0.079\,Gr^{1/3}$	$10^8 < Gr < 10^{10}$		Air	
Ferries [67]	$Nu = 0.124\,Gr^{1/3}$	$10^8 < Gr < 10^{10}$		Air	
Gaignou [68]	$Nu = 0.143\,Gr^{1/4}$	$10^8 < Gr < 10^{10}$		Air	
Alamdari, Hammond [69]	$Nu = \{(0.55Gr^{1/4})^6 + (0.095Gr^{1/3})^6\}^{1/6}$	$10^8 < Gr < 10^{10}$		Air	
Nansteel,Greif [70]	$Nu = 0.553\,Gr^{1/4}$	$10^{10} < Gr < 10^{11}$	Laminar	Water	Rectangular enclosure
Bauman et al. [71]					Vertical copper / aluminium walls
Bajorek, Lloyd [72]	$Nu = 0.111\,Gr^{0.3}$	$10^5 < Gr < 10^6$		Air	Rectangular enclosure
Bohn et al. [73]	$Nu = 0.571\,Gr^{1/4}$	$10^9 < Gr < 10^{11}$		Pure water	Cubical enclosure
Min et al. [74]	$Nu = 0.197\,Gr^{0.32}$	$Gr > 10^9$		Air	Rooms
Allard et al. [75]	$Nu = 0.06\,Gr^{0.32}$	$10^9 < Gr < 10^{10}$		Air	MINIBAT Test Facility
Allard et al. [75]	$Nu = 0.146\,Gr^{0.285}$	$10^9 < Gr < 10^{10}$		Air	MINIBAT Test Facility
Khalifa, Marshall [76]	$Nu = 1.53\,Gr^{0.14}$	$0.4 \times 10^8 < Gr < 10^{10}$		Air	Full scale test facility

Table 8.A2 Correlations for horizontal surfaces facing upwards

Reference	Correlation	Gr range	Flow condition	Working fluid	Experimental setup
Fujii, Imura [77]	$Nu = 0.143\ Gr^{1/3}$	$Gr < 2 \times 10^8$	Turbulent	Pure water	5 cm heated plate
Fujii, Imura [77]	$Nu = 0.117\ G^{1/3}$	$Gr > 5 \times 10^8$	Turbulent	Pure water	30 cm heated plate
Lloyd, Moran [78] Goldstein et al. [79]	$Nu = 0.497\ Gr^{1/4}$	$10^5 < Gr < 10^7$			
Lloyd, Moran [78] Goldstein et al. [79]	$Nu = 0.134\ Gr^{1/3}$	$10^7 < Gr < 10^{10}$			
Fischenden & Saunders [64]	$Nu = 0.496\ Gr^{1/4}$	$Gr < 10^9$			
Fischenden & Saunders [64]	$Nu = 0.125\ Gr^{1/3}$	$Gr > 10^9$			
Perez Sanchez [82]	$Nu = 0.195\ Gr^{1/4}$	$10^4 < Gr < 4 \times 10^5$			
Perez Sanchez [82]	$Nu = 0.068\ Gr^{1/3}$	$Gr > 4 \times 10^5$			
CIBS [64]	$Nu = 0.517\ Gr^{1/4}$	$10^8 < Gr < 10^{10}$	Laminar		
CIBS [64]	$Nu = 0.132\ Gr^{1/3}$	$10^8 < Gr < 10^{10}$	Turbulent		
Alamdari & Hammond [69]	$Nu = \{(0.52Gr^{1/4})^6 + (0.126Gr^{1/3})^6\}^{1/6}$	$0 < Gr < \infty$	Laminar turbulent		
ASHRAE [63]	$Nu = 0.487\ Gr^{1/4}$	$10^8 < Gr < 10^{10}$	Laminar		
ASHRAE [63]	$Nu = 0.118\ Gr^{1/3}$	$10^8 < Gr < 10^{10}$	Turbulent		
Min et al. [74]	$Nu = 0.297\ Gr^{0.31}$	$Gr > 2 \times 10^7$		Air	Test rooms
Khalifa, Marshall [76]	$Nu = 1.24\ Gr^{0.24}$	$5{*}10^8 < Gr < 10^{10}$		Air	Full scale test facility

Table 8.A3 Correlations for horizontal surfaces facing downwards

Reference	Correlation	Gr range	Flow condition	Working fluid	Experimental setup
ASHRAE [63]	$Nu = 0.218\,Gr^{1/4}$	$10^8 < Gr < 10^{10}$	Laminar	Air	Small plates
CIBS [64]	$Nu = 0.236\,Gr^{1/4}$	$10^8 < Gr < 10^{10}$		Air	
Alamdari, Hammond [69]	$Nu = 0.56\,Gr^{1/5}$	$10^8 < Gr < 10^{10}$	Laminar	Air	5 cm, 30 cm heated plates
Fujii, Imura [77]	$Nu = 0.936\,Gr^{1/5}$	$10^6 < Gr < 10^{11}$	Laminar	Pure water	
Incropera, Dewitt [80]	$Nu = 0.249\,Gr^{1/4}$	$10^5 < Gr < 10^{10}$			Rectangular plate
Lewandowski [62]	$Nu = 0.717\,Gr^{1/5}$				
McAdams [59]	$Nu = 0.248\,Gr^{1/4}$				
Fischenden & Saunders [64]	$Nu = 0.229\,Gr^{1/4}$				
Min et al. [74]	$Nu = 0.065\,Gr^{0.255}$			Air	Three test rooms (7.35 × 3.6 × 2.7 m, 7.35 × 3.6 × 3.7m, 3.6 × 3.6 × 2.4 m)

Table 8.A4 Correlations for vertical plates

Reference	Correlation	Flow condition	Working fluid	Experimental setup
ASHRAE [63]	$h = 1.42\,(\Delta T/L)^{1/4}$	Laminar	Air	Small plates
ASHRAE [63]	$h = 1.31\,(\Delta T/L)^{1/3}$	Turbulent	Air	Large plates
Mitalas [81]	$h = 1.02\,(\Delta T)^{1/3}$			
Ferries [81]	$h = 1.6\,(\Delta T)^{1/3}$			
Gaignou [81]	$h = 1.845\,(\Delta T)^{1/4}$			
Alamdari, Hammond [69]	$h = \{[1.5(\Delta T/L)^{1/4}]^6 + [1.23(\Delta T)^{1/3}]^6\}^{1/6}$	Laminar turbulent		Isolated plates
Rogers, Mayhew [64]	$h = 1.42\,(\Delta T/L)^{1/4}$	Laminar		
Rogers, Mayhew [64]	$h = 1.31\,(\Delta T/L)^{1/3}$	Turbulent		

Table 8.A5 Correlations for horizontal plates facing upwards

Reference	Correlation	Flow condition	Working fluid	Experimental setup
ASHRAE [63]	$h = 1.32\,(\Delta T/L)^{1/4}$	Laminar	Air	Small plates
ASHRAE [63]	$h = 1.52\,(\Delta T/L)^{1/3}$	Turbulent	Air	Large plates
CIBS [64]	$h = 1.4\,(\Delta T)^{1/4}$	Laminar		
CIBS [64]	$h = 1.7\,(\Delta T)^{1/3}$	Turbulent		
Alamdari, Hammond [69]	$h = \{[1.4(\Delta T/L)^{1/4}]^6 + [1.63(\Delta T)^{1/3}]^6\}^{1/6}$	Laminar turbulent		Isolated plates

Table 8.A6 Correlations for horizontal plates facing downwards

Reference	Correlation	Flow condition	Working fluid	Experimental setup
ASHRAE [63]	$h = 0.59\,(\Delta T/L)^{1/4}$	Laminar	Air	Small plates
CIBS [64]	$h = 0.64\,(\Delta T/L)^{1/4}$	Laminar	Air	Isolated plates
Alamdari, Hammond [69]	$h = 0.6\,(\Delta T)^{1/5}$			

Table 8.B1 Correlations for vertical surfaces

Ref.	Gr=	10,000	100,000	1,000,000	10,000,000	1.00E+08	1.00E+09	1.00E+10	1.00E+11	1.00E+12
55	Nu=		8.980311	15.9695	28.39824	50.5	89.80311	159.695	283.9824	505
56	Nu=		8.357913	14.86271	26.43004	47	83.57913			
56	Nu=		9.037913	15.54271	27.11004	47.68	84.25913			
56	Nu=	5.38	3.059632	4.111098	5.666332	7.96669	11.36917	16.4018	23.84562	34.85582
57	Nu=	2.348753	9.087008	16.15924	28.73564	51.1	90.87008	161.5924	287.3564	
58	Nu=	5.07	9.015877	16.03275	28.51071	50.7				
59	Nu=	5.44	9.67384	17.20279	30.59137	54.4	96.7384			
60	Nu=	5.09	9.051442	16.09599	28.62317	50.9	83.05963	177.5783	379.6558	811.6896
64	Nu=						90.51442	180	452.1396	1135.723
61	Nu=		10.0295	17.83525	31.71605	56.4				
62	Nu=	5.64				51.6				
63	Nu=						91.75922	163.1735	290.1681	516
63	Nu=						109.1908	233.4457	499.098	1067.053
64	Nu=	5.14	9.140356	16.25411	28.90434	51.4	91.40356			
64	Nu=	5.04	8.962528	15.93788	28.342	50.4	89.62528			
64	Nu=	4.74	8.429044	14.9892	26.65498	47.4	84.29044			
64	Nu=	4.8	8.535741	15.17893	26.99238	48	85.35741			
65	Nu=					44.52461	95.19194	237.4362	507.6296	1085.293
65	Nu=					34.48475	73.72709	203.5168		
63	Nu=					54.12796	115.7235	157.6257	456.4401	975.8516
66	Nu=					61.32288	115.5464	247.4125		
67	Nu=		10.24851	18.47747	33.46033			213.4931		
69	Nu=	5.692891						227.9455	471.186	1002.67
70	Nu=							174.874	310.9748	
72	Nu=		3.478505	6.940531						
73	Nu=						101.5398	180.5661	321.0969	
74	Nu=						149.4398	312.224	652.3283	
75	Nu=						45.51465	95.09359		1362.907
75	Nu=						53.62322	103.3601		
76	Nu=					20.16933	27.84142	38.43186		
SD		0.92	2.13	3.92	6.49	12.71	29.15	73.82	159.5	391.98
(%)		18.6	25.5	26.5	23.9	27.4	34.2	42.7	41.4	46.1

Table 8.B2 Correlations for horizontal surfaces facing upwards

Ref.	Gr=	10,000	100,000	1,000,000	10,000,000	1.00E+08	1.00E+09	1.00E+10	1.00E+11	1.00E+12
77	Nu=	2.987693	6.387575	13.65639	29.19685	62.42176				
77	Nu=					51.07235	109.1908	233.4457	499.098	1067.053
78	Nu=		8.838049	15.71652	27.94836					
79	Nu=				27.35929	58.49312	125.0561	267.3652		
64	Nu=	4.96	8.820266	15.6849	27.89213	49.6	88.20266			
64	Nu=						116.6568	249.4078	533.2244	1140.014
82	Nu=	1.95	3.467645	6.49395	13.88382	29.68308	63.46129	135.6778	290.0741	620.1674
82	Nu=		3.037448			51.7	91.93705	163.4898		
64	Nu=					57.62009	123.1896	263.3746		
69	Nu=	5.389593	9.749551	17.81839	33.37709	65.30806	134.1065	284.4095	612.1642	1324.98
63	Nu=					48.7	86.60221	154.0029		
63	Nu=					51.50887	110.124	235.441		
74	Nu=				43.92952	89.69257	183.1287	373.9008	763.4075	1558.678
76	Nu=					103.1387	179.2345	311.4739		
SD		1.62	2.91	4.38	8.9	19.45	35.67	71.1	172.69	348.2
(%)		42.4	43.4	31.6	30.6	32.5	30.3	29.3	32	30.5

Table 8.B3 Correlations for horizontal surfaces facing downwards

Ref.	Gr=	10,000	100,000	1,000,000	10,000,000	1.00E+08	1.00E+09	1.00E+10	1.00E+11	1.00E+12
63	Nu=					21.8	38.76649	68.93765		
65	Nu=					23.6	41.96739	74.62975		
69	Nu=					22.294	35.33361	56	148.346	
77	Nu=			14.8346	23.51126	37.26283	59.05761	93.6		
80	Nu=		4.427916	7.874071	14.0023	24.9	44.27916	78.74071		
62	Nu=					28.54428	45.23964	71.7		
59	Nu=					24.8	44.10133	78.42449		
64	Nu=					22.9	40.7226	72.41616		
74	Nu=					7.127108	12.82075	23.06287		
SD				4.92	6.72	7.83	12.2	19.74		
(%)				43.3	35.8	33.1	30.3	28.8		

9

Natural ventilation

Ventilation of indoor spaces, either natural or mechanical, provides a means to control indoor air quality and achieve thermal comfort. As discussed in Chapter 6, thermal comfort depends on indoor air temperature, humidity and air velocity; indoor air movement can therefore provide appropriate air velocities for thermal comfort, even when the temperature and humidity are not the most appropriate. Moreover, the renewal of the air by ventilation is a powerful means to exhaust air pollutants, generated indoors, into the outdoor environment.

Mechanical ventilation can be used when outdoor conditions are not favourable to provide a healthy and comfortable indoor environment for building occupants. Energy-efficient buildings must be designed to introduce specific amounts of outside air to meet indoor air-quality requirements. However, outdoor air cannot always be regarded as 'fresh' air and its quality is debatable, if both environmental pollution and climatological conditions, which may be far from the desired indoor conditions, are considered. The provision of ventilation air to a mechanically ventilated building usually requires that outdoor air be conditioned before it enters occupied spaces, which may be a significant cost factor in operating the building. During recent decades, energy conservation measures have led to the reduction in the quantity of outside air that is used by air-conditioning systems and this has proved to have a serious impact on indoor air quality.

In cases where the outdoor air conditions allow it, natural ventilation may prove to be an energy-saving way to reduce the internal cooling load, to achieve thermal comfort and to maintain a healthy indoor environment. Natural ventilation may result from air penetration through a variety of unintentional openings in the building envelope, but it also occurs as a result of manual control of a building's openings (doors, windows). In both cases, air is driven in/out of the building as a result of pressure differences across the openings, which are due to the combined action of wind and buoyancy-driven forces.

In the following sections we focus on natural ventilation. Basic principles are analysed, techniques are discussed and simplified methods are presented for the estimation of the ventilation rate, as well as the cooling potential of natural ventilation.

NATURAL VENTILATION

Natural ventilation can be used not only to provide fresh air for the occupants, necessary to maintain acceptable air-quality levels, but also for cooling, in cases where the climatic conditions allow it, because of the direct influence on thermal comfort sensations experienced by occupants. This term is used to describe ventilation processes caused by naturally produced pressure differences due to wind and the stack effect. Natural ventilation is achieved by infiltration and/or by allowing air to flow in and out of a building by opening windows and doors. The term 'infiltration' is used to describe the random flow of outdoor air through leakage paths in the building's envelope. The presence of cracks and a variety of unintentional openings, their sizes and distribution determine the leakage characteristics of a building and its potential for air infiltration. The distribution of air leakage paths in a building determines the magnitude of wind and stack-driven infiltration and the nature of air flow patterns inside the building. Typical air leakage paths in a residential building are shown in Figure 9.1.

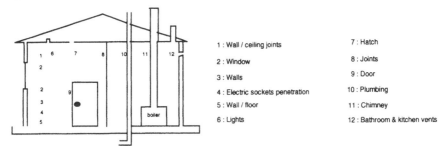

Figure 9.1 Air leakage paths in a building

Modern architecture tends to minimize air infiltration by introducing 'air-tight' buildings, where the cracks in the structure are sealed. Infiltration rates vary seasonally in response to outdoor temperature and wind conditions. Infiltration-associated air-change rates may vary from a low of 0.1–0.2 air changes per hour (ACH) in tight, energy efficient houses to 3.0 ACH in leaky houses under high-infiltration conditions.

The inflow and outflow of air through large building openings, such as windows and doors, in occupant-controlled natural ventilation is mainly due to the effect of the wind. However, thermal forces attributed to temperature differences also play a significant role, especially under conditions of low ambient wind speeds. The effectiveness of natural ventilation also depends on the size of the openings and their orientation to the prevailing wind direction. Parameters influencing the airflow rates through large openings are discussed in the following sections.

Characteristics of the air flow in and around buildings

Successful design of a naturally ventilated building requires a good understanding of the air flow patterns around it and the effect of the neighbouring buildings as well as the existing design strategies to improve ventilation. The objective is to ventilate the largest possible part of the indoor space. Fulfilment of this objective depends on window location, interior design and wind characteristics.

Figure 9.2 illustrates the wind flow pattern around a building with no openings. As the wind flows past the building, a positive pressure is created on the windward facade. The wind is diverted and a negative pressure is created along the side walls due to the high speed of the flow along them. A large, slow-moving eddy on the leeward facade produces a smaller suction.

Figure 9.3 shows the case of a single-zone building which is cross-ventilated as a result of two windows placed on the windward and leeward facades. Cross-ventilation is improved if two outlets of total area equal to the inlet are placed on the building sidewalls (Figure 9.4). In this case, wider recirculation is set up in the room. This design permits more efficient ventilation for a wider range of wind directions.

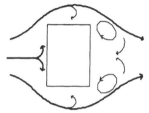

Figure 9.2 Airflow patterns around a building (plan view)

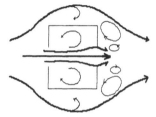

Figure 9.3 Ventilation through windward and leeward openings

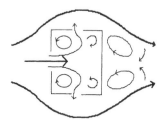

Figure 9.4 Ventilation through windward and side openings

The above airflow patterns vary significantly as a result of the surrounding terrain and the presence of neighbouring buildings. The extent of the wake at the leeward side of a building depends on the building's shape and the wind direction. For a typical house, the average wake length is four times the ground-to-eave height [1]. This is schematically shown in Figure 9.5. Obviously, if the distance between two buildings is shorter than this, the one which falls within the 'wind shade' of the other will be poorly ventilated.

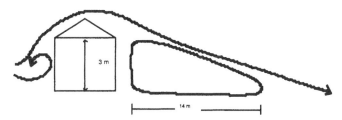

Figure 9.5 Wake of a typical house

In real buildings it is common to have spaces with only one external wall. Ventilation of such spaces will be negligible if there is only one opening, but can be improved by placing two widely spaced openings on the side of the external wall (Figure 9.6). Further improvement may be achieved by use of wing walls, as shown in Figure 9.7. Wing walls are extrusions of the exterior walls (from the ground to the eaves) and their role is to create positive pressure over one opening and negative over another so as to achieve cross-ventilation. Single-sash casement windows act similarly. This technique is effective for wind direction angles from 20 to 160° as shown in Figure 9.7.

Ventilation of spaces with exterior openings in adjacent walls can also benefit from the use of wing walls. An example of this case is shown in Figure 9.8.

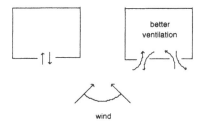

Figure 9.6 Single-sided ventilation

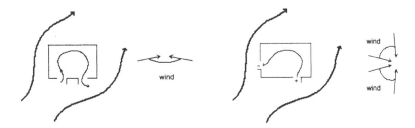

Figure 9.7 The impact of wing walls on single-sided ventilation

Figure 9.8 The impact of wing walls on cross-ventilation

The position of the wing-wall extrusions is critical to ventilation success. Every configuration gives improved airspeeds for a specific band of wind directions. Prevailing wind direction is, therefore, an important characteristic of the microclimate around the building, and it has to be taken into consideration for the design of the wing wall positions. Figures 9.9 and 9.10 show the airflow patterns for two different configurations. In the case shown in Figure 9.10, ventilation is poorer, because positive pressures are created at both openings, which prevents air circulation and confines the flow to the areas near the windows. Comparison of Figures 9.9 and 9.11 shows that air circulation is larger when the openings are widely spaced.

The dimensions of wing-wall extrusions vary according to the exterior opening width. The optimum required dimension for the protrusions is equal to the opening width and the minimum recommended is equal to half the opening width.

Fencing or dense shrubs can act as a barrier and change the wind direction, producing the same effect as wing walls (Figure 9.12).

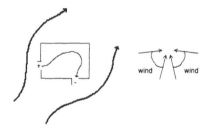

Figure 9.9 Cross-ventilation using wing walls – large air flow

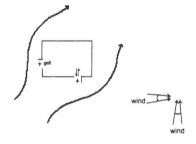

Figure 9.10 Cross-ventilation using wing walls – poor ventilation

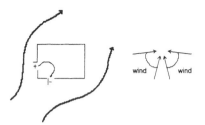

Figure 9.11 Cross-ventilation using wing walls – confined air flow

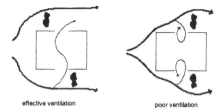

Figure 9.12 Fencing or planting acting as wing walls

Air flow through large openings

In natural ventilation, the driving forces result from pressure differences across openings in the building envelope. The resulting flow rate is given by a power-law equation. There are primarily two distinct cases, depending on the type of the opening, namely:

- cracks (typical dimensions smaller than 10 mm)
- large openings (typical dimensions larger than 10 mm).

For cracks, the air flow rate (Q) can be expressed in $m^3\ s^{-1}$ using the following expression:

$$Q = kL(\Delta P)^n \tag{9.1}$$

where k is the flow coefficient [$m^3\ s^{-1}\ m^{-1}\ Pa^{-n}$], L is the crack length [m] and n is the flow exponent.

The value of the flow exponent n depends on the flow regime; it ranges from 0.5 for fully turbulent flow to 1.0 for laminar flow. In practice, typical values of the flow exponent are set equal to 0.6 and 0.7. The flow coefficient k is a function of the crack geometry. A range of values of k for cracks formed around closed windows are given in Table 9.1 for $n = 0.67$.

Table 9.1 Typical flow coefficient (k) values for windows (n=0.67)

Window type	Average	Range
Sliding	8	2–30
Pivoted	21	6–80
Pivoted (weather stripped)	8	0.5–20

The flow rate (Q) through an opening of relatively large free area is calculated using the common orifice flow equation:

$$Q = C_d A\sqrt{(2\Delta P/\bar{\rho})} \tag{9.2}$$

where C_d is the discharge coefficient of the opening, A is the opening area, [m^2], ΔP is the pressure difference across the opening [Pa] and $\bar{\rho}$ the average air density in the direction of flow [$kg\ m^{-3}$].

The discharge coefficient is a function of the temperature difference, wind speed and opening height. A number of expressions have been proposed for its calculation, especially for internal openings. Interzonal heat and mass flow measurements in a real building [2] have given the following expression for the discharge coefficient in the case of internal openings:

$$C_d = 0.0835 (\Delta T / T)^{-0.31}.$$

For steady state and buoyancy-driven flow the discharge coefficient for internal openings can be calculated from the following expression [3]:

$$C_d = (0.4 + 0.0075 \Delta T).$$

Experimental results have been analysed in order to express the discharge coefficient for internal openings as a function of the temperature difference, air speed and opening height [4]. It was proved that the value of C_d is a strong function of the opening dimensions. For small internal openings, a representative value for the discharge coefficient is 0.65. For large internal openings C_d has a value close to unity. A proposed mean value for a standard opening is $C_d = 0.78$.

An evaluation of the discharge coefficient as a function of the opening height is attempted in [5]. For opening heights $1.5 < H < 2$ m, the proposed relation is: $C_d = 0.21H$. According to [6], the values of C_d can be selected within the range of 0.6–0.75, with reasonable accuracy.

Darliel and Lane Serff [7] have carried out experiments in a $18.6 \times 60 \times 40$ cm box and in a 199×9.4 cm channel using water. They have measured a C_d coefficient close to 0.311.

Measurements of air flow through large openings separating two zones in a test cell, repeated in [6], have shown that the coefficient of discharge (C_d) varies between 0.67 and 0.73, which corresponds to C value between 0.223 and 0.243. This experiment is carried out using cold and hot vertical plates situated at the end of each zone. Therefore, an important boundary layer flow should have been developed.

Khodr Mneimne [8] in a full-scale experiment, using an electrical heater as a heating source, has found that, for openings between 0.9 and 2m, a mean C_d value equal to 0.87 should be used.

Based on the experimental work, carried out in the DESYS Test Cell at CSTB, France [9], it has been concluded that a value of C_d equal to 0.42 for $5 \times 10^8 < \text{Gr} < 3.6 \times 10^9$, can be used with satisfactory accuracy.

Pressure differences result from the combined action of two mechanisms:

- **Wind-induced pressure differences.** Positive pressure is created on the building sides that face the wind (windward sides) whereas suction regions are formed on the opposite sides (leeward sides) and on the sidewalls. This results in a negative pressure inside the building, which is sufficient to introduce large flows through the building openings. In a general case, an inflow of air is induced on the windward side and an outflow on the leeward side.
- **Stack effect.** Air movement due to the stack effect occurs when temperature differences between a zone and the environment adjacent to it, be it another

zone or the exterior, cause light warm air to rise and flow out of the warm zone, while cooler air flows in. The stack effect occurs in tall buildings, particularly at places with vertical passages such as stairwells, elevators or shafts.

For each of the cases described above, it is possible to calculate the resulting pressure difference, as shown in the following discussion.

Wind-induced pressure differences

Airflow through an external opening is mainly attributed to a wind-induced pressure difference across it. The pressure due to wind flow onto or away from a surface is given by:

$$P_w = \frac{C_p \rho V^2}{2} \tag{9.3}$$

where P_w is the wind-induced pressure [Pa], Cp is the pressure coefficient and V is wind speed [m s^{-1}] at a reference height, usually taken as the building height.

The wind speed is calculated as a function of available wind velocity measurements from typical meteorological data. Wind velocity measurements are made available at a fixed height, usually 10 m above the ground level. As a result, the actual wind speed (V) must be properly adjusted for a specific height and account for the building's orientation, topography of the location and roughness of the surrounding terrain in the wind direction.

This can be performed by three wind profiles:

- **Power law wind profile [10].** The actual wind speed V_1 is evaluated by the following expression:

$$\frac{V_1}{V_{10}} = Kz_1^a \tag{9.4}$$

where the coefficient K and the exponent a are constants which depend on terrain roughness. Typical values for K and a are given in Table 9.2.
- **Logarithmic wind profile [11].** Based on this profile, the wind speed is a logarithmic function of height:

$$\frac{V_l}{V_m} = \frac{V_{*,l}}{V_{*,m}} \left[\frac{\ln \dfrac{z_l - d_l}{z_{0,l}}}{\ln \dfrac{z_m - d_m}{z_{0,m}}} \right] \tag{9.5}$$

where

$$\frac{V_{*,l}}{V_{*,m}} = \left[\frac{z_{0,l}}{z_{0,m}}\right]^{0.1};$$

and V_m is the wind speed from meteorological data [m s^{-1}], V_* the atmospheric friction speed [m s^{-1}], z_0 terrain roughness [m] and d terrain displacement length [m].

Typical values for z_0 and d are given in Table 9.2.

• **LBL model wind profile [10].** Another power-law profile is also available:

$$\frac{V_l}{V_m} = \frac{\alpha(z/10)^g}{\alpha_m(z_m/10)^{g_m}} \qquad (9.6)$$

where α and g are terrain-dependent constants; typical values are given in Table 9.2.

Table 9.2 Typical values for terrain dependent parameters (h= building height) [10]

Terrain	K	α	z_0	d	α	g
Open flat country	0.68	0.17	0.03	0.0	1.00	0.15
Country with scattered wind breaks	0.52	0.20	0.1	0.0	1.00	0.15
Rural			0.5	0.7h	0.85	0.20
Urban	0.35	0.25	1.0	0.8h	0.67	0.25
City	0.21	0.33	> 2.0	0.8h	0.47	0.35

The dimensionless pressure coefficient Cp is an empirically derived parameter that accounts for the changes in wind-induced pressure, caused by the influence of surrounding obstructions on the prevailing local wind characteristics. Its value changes according to the wind direction, the building surface orientation and the topography and roughness of the terrain in the wind direction. Typical design data sets based on experimental results are given in Table 9.3 [12]. Every data set comprises Cp values for 16 different wind directions (angle of wind with the normal to the surface: 0°, 22.5°, 45°, 67.5°, 90°, 112.5°, 135°, 157.5°, 180°, 202.5°, 225°, 247.5°, 270°, 292.5°, 315°, 337.5°, progressing clockwise as seen from above). Table 9.3 comprises 29 pressure coefficient data sets corresponding to an equal number of different facade configurations in terms of surface aspect, dimensions and exposure.

The Cp values given in Table 9.3 may be used for low-rise buildings of up to three storeys and they express an average value for each external building surface.

Table 9.3 Pressure coefficient sets [12]

Ref. No.	Facade description	AR*	Exp.†	Cp sets
1	Wall	1:1	E	0.7, 0.525, 0.35, −0.075, −0.5, −0.45, −0.4, −0.3, −0.2, −0.3, −0.4, −0.45, −0.5, −0.075, 0.35, 0.525
2	Roof, pitch > 10 deg	1:1	E	−0.8, −0.75, −0.7, −0.65, −0.6, −0.55, −0.5, −0.45, −0.4, −0.45, −0.5, −0.55, −0.6, −0.65, −0.7, −0.75
3	Roof, pitch > 10–30 deg	1:1	E	−0.4, −0.45, −0.5, −0.55, −0.6, −0.55, −0.5, −0.45, −0.4, −0.45, −0.5, −0.55, −0.6, −0.55, −0.5, −0.45
4	Roof, pitch > 30 deg	1:1	E	−0.3, −0.35, −0.4, −0.5, −0.6, −0.5, −0.4, −0.45, −0.5, −0.45, −0.4, −0.5, −0.6, −0.5, −0.4, −0.35
5	Wall	1:1	SE	0.4, 0.25, 0.1, −0.1, −0.3, −0.325, −0.35, −0.275, −0.2, −0.275, −0.35, −0.325, −0.3, −0.1, 0.1, 0.25
6	Roof, pitch <10 deg	1:1	SE	−0.6, −0.55, −0.5, −0.45, -,4, −0.45, −0.5, −0.55, −0.6, −0.55, −0.5, −0.45, −0.4, −0.45, −0.5, −0.55
7	Roof, pitch 10–30 deg	1:1	SE	−0.35, −0.4, −0.45, −0.5, −0.55, −0.5, −0.45, −0.4, −0.35, −0.4, −0.45, −0.5, −0.55, −0.5, −0.45, −0.4
8	Roof, pitch > 30 deg	1:1	SE	−0.3, −0.4, −0.5, −0.55, −0.6, −0.55, −0.5, −0.5, −0.5, −0.5, −0.5, −0.55, −0.6, −0.55, −0.5, −0.4
9	Wall	1:1	S	0.2, 0.125, 0.05, 0.1, −0.25, −0.275, −0.3, −0.275, −0.25, −0.275, −0.3, −0.275, −0.25, −0.1, 0.05, 0.125
10	Roof, pitch < 10 deg	1:1	S	−0.5, −0.5, −0.5, −0.45, −0.4, −0.45, −0.5, −0.5, −0.5, −0.5, −0.5, −0.45, −0.4, −0.45, −0.5, −0.5
11	Roof, pitch 10-30 deg	1:1	S	−0.3, −0.35, −0.4, −0.45, −0.5, −0.45, −0.4, −0.35, −0.3, −0.35, −0.4, −0.45, −0.5, −0.45, −0.4, −0.35
12	Roof, pitch > 30 deg	1:1	S	0.25, −0.025, −0.3, −0.4, −0.5, −0.4, −0.3, −0.35, −0.4, −0.35, −0.3, −0.4, −0.5, −0.4, −0.3, −0.025
13	Long wall	2:1	E	0.5, 0.375, 0.25, −0.125, −0.5, −0.65, −0.8, −0.75, −0.7, −0.75, −0.8, −0.65, −0.5, −0.125, −0.25, −0.375
14	Short wall	1:2	E	−0.9, −0.35, 0.2, 0.4, 0.6, 0.4, 0.2, −0.35, −0.9, −0.75, −0.6, −0.475, −0.35, −0.475, −0.6, −0.75
15	Roof, pitch < 10 deg	2:1	E	−0.7, −0.7, −0.7, −0.75, −0.8, −0.75, −0.7, −0.7, −0.7, −0.7, −0.7, −0.75, −0.8, −0.75, −0.7, −0.7
16	Roof, pitch 10–30 deg	2:1	E	−0.7, −0.7, −0.7, −0.7, −0.7, −0.65, −0.6, −0.55, −0.5, −0.55, −0.6, −0.65, −0.7, −0.7, −0.7, −0.7
17	Roof, pitch >30 deg	2:1	E	0.25, 0.125, 0, −0.3, −0.6, −0.75, −0.9, −0.85, −0.8, −0.85, −0.9, −0.75, −0.6, −0.3, 0, 0.125
18	Long wall	2:1	SE	0.5, 0.375, 0.25, 0.-125, −0.5, −0.65, −0.8, −0.75, −0.7, −0.75, −0.8, −0.65, −0.5, −0.125, 0.25, 0.375
19	Short wall	1:2	SE	−0.9, −0.35, 0.2, 0.4, 0.6, 0.4, 0.2, −0.35, −0.9, −0.75, −0.6, −0.475, −0.35, −0.475, −0.6, −0.75
20	Roof, pitch <10 deg	2:1	SE	−0.7, −0.7, −0.7, −0.75, −0.8, −0.75, −0.7, −0.7, −0.7, −0.7, −0.7, −0.75, −0.8, −0.75, −0.7, −0.7
21	Roof, pitch 10–30 deg	2:1	SE	−0.7, −0.7, −0.7, −0.7, −0.7, −0.65, −0.6, −0.550. −0.5, −0.55, −0.6, −0.65, −0.7, −0.7, −0.7, −0.7
22	Roof, pitch > 30 deg	2:1	SE	0.25, 0.125, 0, −0.3, −0.6, −0.75, −0.9, −0.85, −0.8, −0.85, −0.9, −0.75, −0.6, −0.3, 0, 0.125
23	Long wall	2:1	S	0.06, −0.03, −0.12, −0.16, −0.2, −0.29, −0.38, −0.34, −0.3, −0.34, −0.38, −0.29, −0.2, −0.16, −0.12, −0.03
24	Short wall	1:2	S	−0.3, −0.075, 0.15, 0.165, 0.18, 0.165, 0.15, −0.075, −0.3, −0.31, −0.32, −0.32, −0.26, −0.2, −0.26, −0.32
25	Roof, pitch < 10 deg	2:1	S	−0.49, −0.475, −0.46, −0.435, −0.41, −0.435, −0.46, −0.475, −0.49, −0.475, −0.46, −0.435, −0.41, −0.435, −0.46, −0.475

Table 9.3 (continued)

Ref. No.	Facade description	AR*	Exp.†	Cp sets
26	Roof, pitch 10–30 deg	2:1	S	−0.49, −0.475, −0.46, −0.435, −0.41, −0.435, −0.46, −0.43, −0.4, −0.43, −0.46, −0.435, −0.41, −0.435, −0.46, −0.475
27	Roof, pitch > 30 deg	2:1	S	0.06, −0.045, −0.15, −0.19, −0.23, −0.42, −0.6, −0.51, −0.42, −0.51, −0.6, −0.42, −0.23, −0.19, −0.15, −0.045
28	Wall	1:1	E	0.9, 0.7, 0.5, 0.2, −0.1, −0.1, −0.2, −0.2, −0.2, −0.2, −0.2, −0.1, −0.1, 0.2, 0.5, 0.7
29	Roof, no pitch	1:1	E	−0.1, −0.1, −0.1, −0.1, −0.1 −0.1 −0.1 −0.1, −0.1, −0.1, −0.1, −0.1, −0.1, −0.1, −0.1, −0.1

* AR: aspect ratio (length-to-width ratio).
† Exp.: exposure, E: exposed, SE: semi-exposed, S: sheltered.

Local (not wall-averaged) evaluation of the Cp parameter is one of the most diffi-cult aspects of air infiltration modelling. In the following section a recently devel-oped Cp calculation model is presented.

A parametrical model for the calculation of the pressure coefficient, Cp [13]

THE MODEL
The model is based on a parametric analysis of results from two wind-tunnel tests, carried out by Hussein and Lee [14] and Akins and Cermak [15]. The model consists of a number of relations between the pressure coefficient on a rectangular-shaped building model and a number of influencing parameters, grouped in three categories:

- **Climate parameters**, involving wind velocity profile exponent (a) and wind incidence angle (anw). Figure 9.13 shows the wind incidence angle for a building facade. The parameter anw is defined as the absolute value of the wind incidence angle. Windward facades have $0° <$ anw $< 90°$ and leeward facades have $90° <$ anw $< 180°$.
- **Environmental parameters**, involving the plan area density (pad) and the relative building height (rbh). The plan area density is defined as the ratio of built area to total area. This ratio has to be calculated within a radius rang-ing from 10 to 25 times the height of the considered building. Figure 9.14 illustrates the plan area density for a building with length L and width W.

  The relative building height is the ratio of the building height to the height of the surrounding buildings, the latter assumed to be regular boxes of the same height.
- **Building parameters**, involving the frontal aspect ratio (far), the side as-pect ratio (sar), the relative vertical position (zh) and the relative horizontal position (xl). The aspect ratio is defined as the ratio of the length to the

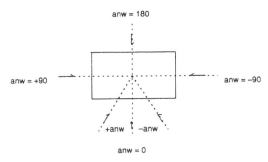

*Figure 9.13 Wind incidence angle (°) in relation to the
front facade (plan view)*

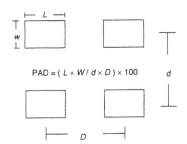

*Figure 9.14 Calculation of the plan
area density (pad)*

height of a building facade. The frontal aspect ratio is related to the considered facade and the side aspect ratio to its adjacent facade, regardless of the angle the wind direction forms with the facades themselves.

The relative horizontal and vertical position of a facade element is defined relative to a reference point on the facade, as illustrated in Figure 9.15.

The range of parameters common to both of the wind-tunnel tests mentioned above was used as a reference. It corresponds to the profile of Cp in the vertical centreline of the windward and leeward facades of a model with approaching wind normal to the facades, in a boundary layer typical of a suburban area. The model assumes that the reference horizontal distribution of Cp does not change with plan area density, relative building height, aspect ratio and wind velocity profile exponent.

The reference profiles as functions of the relative vertical position zh are third or fifth degree polynomials:

$$Cp_{ref}(zh) = a_0 + a_1(zh)^2 + \ldots + a_n(zh)^n \tag{9.7}$$

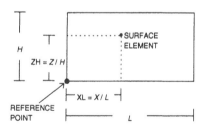

Figure 9.15 Facade element positioning

where $n = 3$ for the windward facade and $n = 5$ for the leeward facade.

The reference profiles were defined for the following parameters:

$a = 0.22$,
$pad = 0.0$,
$rbh = 1.0$,
$far = 1.0$,
$sar = 1.0$,
$anw = 0°$ (windward facade),
$anw = 180°$ (leeward facade).

The rest of the Cp data were normalized for each parameter, with respect to the Cp corresponding to the reference value of the parameter. Thus, the normalized Cp value for the n, m and t values of the parameters i, j and d respectively is

$$Cp_{norm_{i_n, j_m, d_t}} = \frac{Cp_{i_n, j_m(d_t)}}{Cp_{i_n, j_m(d_{ref})}}. \tag{9.8}$$

The normalized Cp values as functions of the various parameters are given by first-to fifth-degree polynomials for the leeward side and by first-to-third degree for the windward side. Polynomial coefficients are given in Appendix A at the end of the chapter in Tables 9.A1–A6 for the windward facades ($0° < anw < 90°$) and in Tables 9.A9–A14 for the leeward side ($90° < anw < 180°$). Non-polynomial functions relate the normalized Cp values to the parameters far and sar, for far > 1.0 and sar > 1.0 (Tables 9.A7, 9.A8, 9.A15 and 9.A16 in Appendix A).

The pressure coefficient of an element k with coordinates xl and zh on the facade of a building with shape defined by specific values of far and sar and in environmental conditions defined by specific values of a, pad, rbh and anw is calculated by:

$$Cp_k = Cp_{ref}(zh) \times CF \tag{9.9}$$

where CF is the global correction factor:

$$CF = Cf_{zh}(a) \times Cf_{zh}(pad) \times Cf_{zh,pad}(rbh)$$

$$\times Cf_{zh,pad}(far) \times Cf_{zh,pad}(sar) \times Cf_{zh,anw}(xl) \tag{9.10}$$

where

$$Cf_{i_n,j_m(d_l)} = Cp_{norm i_n,j_m,d_l}. \tag{9.11}$$

If n and m values of the i and j parameters are different from those given in the tables, then the correction factor is calculated for the closest lower and higher values and the results are linearly interpolated.

Application of this method is restricted because of the variation range defined for each parameter. In particular, the model cannot be applied to:

- high terrain roughness ($a > 0.33$) and/or high density of the immediate surrounding buildings (pad > 50);
- immediate surrounding buildings with staggered or irregular pattern layout;
- immediate surrounding buildings with pad > 12.5, when the considered building has a different height from its surroundings or a shape other than a cube;
- buildings four times higher or half the height of the surroundings;
- buildings with irregular shape or overhangs;
- regular block-shaped buildings with aspect ratios less than 0.5 or greater than 4.

A case study of the method is presented in Appendix A at the end of the chapter.

Stack effect

An additional component which controls the air motion through the building envelope is buoyancy. Buoyancy forces are attributed to temperature differences between a zone and the zones adjacent to it or the outside environment. If the zone temperature is higher than that of its surrounding environment, then warm air rises and flows out of the zone near the top, while cooler air flows in by infiltration near the bottom, as shown in Figure 9.16. The magnitude of air flow associated with the infiltration process grows with the temperature difference.

If P_0 is the static pressure at the bottom of a zone, then the pressure, due to stack effect only, at a height z of the zone is given by:

$$P_s = P_0 - \rho gz \tag{9.12}$$

where P_0, P_s are the pressures at the bottom of the zone and at a height z respectively [Pa], g is the gravitational acceleration [m s^{-2}], ρ is the density of the air at a temperature T equal to the indoor air temperature [kg m^{-3}]

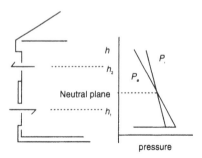

Figure 9.16 Buoyancy driven airflow
through two vertical openings

Assuming that the air behaves as an ideal gas, the density ρ can be calculated by the following expression:

$$\rho = \rho_0 \frac{T_0}{T}$$

where T is the absolute temperature (°K), ρ_0, T_0 are the reference density and temperature of the air, i.e. for $T_0 = 273.15$ °K, $\rho_0 = 1.29$ kg m^{-3}).

From equation (9.12) it is clear that the stack pressure decreases with height. In the case of two isothermal zones that are interconnected by a component (door or window), the pressure difference at a height z (m), across the component will be:

$$\Delta P_s = P_{1,0} - P_{2,0} + (\rho_1 - \rho_2)gz \tag{9.13}$$

where $P_{1,0}$, $P_{2,0}$ are the static pressures at a reference height (i.e. the bottom of the zones) and ρ_1, ρ_2 are the air densities in zones 1 and 2 respectively.

This theory assumes that the temperature inside the zones does not change with height (isothermal zones). A more complex model has been proposed [16] to represent, in more detail, the behaviour of large openings. The model accounts for temperature stratification and turbulence effects by assuming:

- steady flow, inviscid and incompressible fluid;
- linear density stratification on both sides of the opening;
- turbulence effects represented by an equivalent pressure-difference profile.

Thus, on each side of the opening, a linear density stratification is assumed:

$$\rho_i(z) = \rho_{0i} + b_i z \tag{9.14}$$

and a linear pressure difference is introduced to simulate the turbulence effect:

$$\Delta P_t = P_{t0} + b_t z. \tag{9.15}$$

Introducing these terms into equation (9.13) gives for the case of gravitational flow (no wind effect):

$$\Delta P = P_{1,0} - P_{2,0} - g\left[\left(\rho_{01}z + \frac{b_1 z^2}{2}\right) - \left(\rho_{02}z + \frac{b_2 z^2}{2}\right)\right] + (P_{t0} + b_t z). \tag{9.16}$$

The air velocity at any level z is given by:

$$V(z) = \sqrt{\frac{2[P_1(z) - P_2(z)]}{\rho}} \tag{9.17}$$

where ρ represents the air density transported by the velocity.

Combined action of wind and temperature difference

For the calculation of the total pressure difference across the opening, the terms of the dynamic pressure must be added to those representing the stack effect. Thus, combining equations (9.3) and (9.13):

$$\Delta P = P_{1,0} - P_{2,0} + \frac{\rho_1 \text{Cp} V_1^2}{2} - \frac{\rho_2 \text{Cp} V_2^2}{2} + (\rho_1 - \rho_2)gz \tag{9.18}$$

where V_1 and V_2 are the air velocities at the two sides of the opening and at height z.

Neutral level

Air flow through large openings is usually bi-directional. In the general case, cold air flows in through the lower part of the opening, while warmer air flows out from the upper part. This is shown schematically in Figure 9.17 where the vertical velocity profile along the opening is presented.

As shown in this figure, the air velocity decreases and becomes zero at a height H_n from the bottom of the opening. At that level, the so-called neutral level, the pressure difference across the opening is zero.

The position of the neutral level can be determined with equation (9.18), for $\Delta P = 0$. Thus, for an opening between two zones with $T_1 \neq T_1$, the height of the neutral level from the reference level (usually taken at the bottom of a zone) is given by:

$$H_n = \frac{P_{1,0} - P_{2,0} + \frac{\rho_1 \text{Cp} V_1^2}{2} - \frac{\rho_2 \text{Cp} V_2^2}{2}}{\rho_2 - \rho_1} \tag{9.19}$$

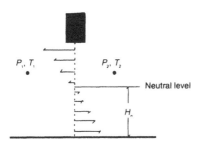

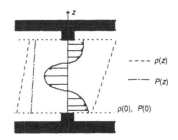

Figure 9.17 Bi-directional air flow across a doorway

Figure 9.18 A model for airflow through a large opening assuming temperature stratification and turbulence

Figure 9.18 shows the case of gravitational flow as presented in the previous section. The locations of the two possible neutral levels can be determined by solving equation (9.16) for $\Delta P = 0$.

EMPIRICAL/SIMPLIFIED METHODS FOR ESTIMATING VENTILATION RATES IN A SINGLE-ZONE BUILDING

Several simplified procedures based on empirical data have been developed to produce estimates of ventilation rates in essentially single-zone buildings. Representative methods are described below.

The British Standard Method [17]

The British Standard Method proposes formulae for the calculation of the air infiltration and ventilation in single side and cross-ventilation configurations. The method assumes two-dimensional flow through a building and ignores all internal partitions. Tables 9.4 and 9.5 give schematically the proposed formulae for different air flow patterns and for different conditions.

NORMA – a simplified theoretical model [18]

The method calculates the air ventilation rates and is only applicable to a single-zone building.

NORMA provides methods to calculate the air flow in single-sided and cross-ventilation configurations. In single-sided ventilation stack flow is dominant while the influence of wind is not so important. In cross ventilation the flow depends directly on the differences in pressure at the openings and therefore the difficulty of predicting the flow is mainly due to the uncertainty in the pressure coefficients at the openings. The scientific basis, including the forms to complete as well as examples, for each case, is given in the following sections.

Table 9.4 Formulae for single-sided ventilation

Ventilation

Due to wind

$Q = 0.025AV$

where A is the opening surface and V is the wind velocity.

Due to temperature difference with two openings

$$Q = C_d A \left[\frac{\varepsilon\sqrt{2}}{(1+\varepsilon)(1+\varepsilon^2)^{1/2}} \right] \left(\frac{\Delta T g H_1}{T} \right)^{1/2}$$

$\varepsilon = A_1 / A_2, \quad A = A_1 + A_2$

where C_d is the discharge coefficient

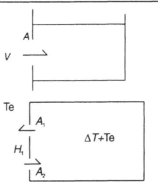

Due to temperature difference with one opening:

$$Q = C_d \frac{A}{3} \sqrt{\frac{\Delta T g H_2}{T}}$$

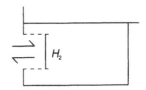

SINGLE-SIDED VENTILATION

Problems related to single-sided ventilation are classified in two groups:

- single-sided ventilation with openings at the same height;
- single-sided ventilation with openings at different levels.

In both cases, the air flow is due to the temperature difference between the indoor and the outdoor environment.

Ventilation due to temperature difference with openings at the same heights.
The temperature difference between the indoor and the outdoor environment creates pressure differences that promote the air flow through the opening. When the indoor temperature is higher than the outdoor, cooler air enters from the lower part of the opening while warm air escapes through the higher level. The flow direction is reversed when the external temperature is higher than the internal.

Table 9.5 Formulae for cross-ventilation

Ventilation

Due to wind only

$$Q_w = C_d A_w V \sqrt{\Delta Cp}$$

$$\frac{1}{A_w^2} = \frac{1}{(A_1 + A_2)^2} + \frac{1}{(A_3 + A_4)^2}$$

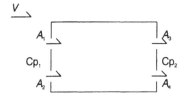

Due to temperature difference only:

$$Q_b = C_d A_b (2\Delta\theta g H_1 / T)^{0.5}$$

$$\frac{1}{A_b^2} = \frac{1}{(A_1 + A_3)^2} + \frac{1}{(A_2 + A_4)^2}$$

$$T = 0.5(Te + Ti)$$

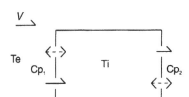

Due to wind and temperature difference:

$$Q = Q_b$$

for $V / (\Delta T)^{0.5} < 0.26(A_b / A_w)^{0.5}(H_1 / \Delta Cp)^{0.5}$

$$Q = Q_w$$

for $V / (\Delta T)^{0.5} > 0.26(A_b / A_w)^{0.5}(H_1 / \Delta Cp)^{0.5}$

$$\Delta T = Ti - Te$$

The parameters which determine the levels of air flow are mainly the following:

- the area of the opening A;
- The temperature difference DT; and
- The vertical height of the opening h.

The method of NORMA, designed to calculate the air flow Q due to temperature difference in single-sided ventilation configurations with openings of the same height, is based on an analytical expression coupling the air flow Q, the height of the opening h, the width of the opening w, the temperature difference between indoors and outdoors DT and the mean indoor/outdoor temperature MT. The expression proposed is:

$$Q = 790h^{1.5}w\sqrt{DT/MT}.\qquad\qquad (9.20)$$

Values of Q are in cubic metres per hour. The height and width are in metres and the temperature in degrees Celsius.

The above expression has been derived by correlating data obtained from simulations of hundreds of single-opening configurations. Simulations have been performed using the monozone computerized tool NORMA. The accuracy of the expression has been tested by comparing the predicted values with the predictions of various multizone computational tools. It is found that the values predicted by NORMA are very close to the predictions of the other tools.

In order to use the method the user has to complete Form 1 given in the Appendix B at the end of this chapter. The necessary input includes all the parameters introduced in equation (9.20). In addition, by specifying the volume of the ventilated space, the user can calculate the air changes per hour.

The method consists of 13 steps. Information on each step is given below:

1 **Opening height h.** Give the vertical distance between the top and bottom of the window.
2 **Opening width w.** Give the width of the opening, defined as the horizontal distance between the sides of the window.
3 **Volume of the ventilated space.** Specify the volume of the space to be ventilated.
4 **Design month.** The monthly ambient climatic data is used. Therefore, specify the summer month for which calculations will be performed.
5 **Mean monthly ambient temperature DTA.** Enter, from local climatological tables, the mean monthly ambient temperature.
6 **Indoor air temperature TA.** Specify the design indoor air temperature.
7 **Mean temperature MT.** Calculate the mean value of the indoor and outdoor temperature. Add DTA and TA and divide by two.
8 **Mean temperature difference DT.** Calculate the difference between the indoor and the outdoor temperature. Enter the absolute value of the difference. This means that if TA is lower than DTA, the difference is equal to DTA – TA. If TA is higher than DTA, DT = TA – DTA.
9 **Temperature ratio RT.** Divide DT by MT.
10 **Square root of RT.** Calculate the square root of RT.
11 **Determine $h^{1.5}$.** Calculate the value of $h^{1.5}$.
12 **Air flow rate Q.** Calculate the air flow rate, in cubic metres per hour, using the formula given above. The air flow rate can be calculated as the product of the values calculated in Step 2 × Step 10 × Step 11 × 790.
13 **Air changes per hour.** If there is more than one window at almost the same height and on the same side, then calculations should be repeated for each window. The total air flow QT is the sum of the flow calculated for

each window. The air changes per hour can be calculated as the ratio of the total air flow QT to the volume of the ventilated space as given in Step 3.

To help the user, two examples of single-sided ventilation in a building are given below: The first example is a one-opening configuration, while the second is a three-openings configuration.

Example 1. Calculate the air changes per hour in a 50 cubic metre zone having an opening with the following characteristics: height = 1.2 m and width = 1.4 m. The mean ambient temperature is equal to 29.4°C and the design indoor temperature is 27°C.
   The results are given in Appendix B, Form 2. The calculated air changes per hour are close to 8.5.

Example 2. Calculate the air changes per hour in a 65 cubic metres building with three openings. The characteristics of the openings are the following:

opening 1: height = 0.8 m , width = 1 m,
opening 2: height = 1.2 m , width = 0.8 m,
opening 3: height = 1.3 m , width = 1.2 m,
outdoor temperature = 29.4°C, indoor temperature = 28°C.

The results of this example are given in Appendix B, Form 3. The calculated air changes per hour are close to 9.5.

Ventilation due to temperature difference with openings at different heights. The temperature difference between the indoor and the outdoor environments creates density differences which produce pressure differences and promote the air flow through the openings. As previously mentioned, when the interior temperature is higher than the exterior, cooler air enters through the lower openings and warm air escapes through the higher openings. A reverse flow is observed when the external temperature is higher than the internal.
   The main parameters influencing ventilation processes due to temperature difference with two openings are the following:

- the temperature difference DT between the interior and the exterior of the building;
- the vertical distance between the two openings H;
- the area of the two openings, A_1 and A_2.

The method proposed to calculate the air flow Q, due to the stack effect in single-sided ventilation configurations with openings at two different heights, is based on an analytical expression coupling the air flow Q, the vertical distance between the two openings H, the areas of the lower A_1 and higher A_2 openings, the temperature difference between indoors and outdoors DT, and the mean indoor–outdoor temperature MT. The expression proposed is:

$$Q = 1590K(A_1 + A_2)\sqrt{DT/MT}\sqrt{H} \qquad (9.21)$$

where K is a correction factor given as a function of A_1 and A_2 in Figure 9.19. For A_2 > 6 the K value that corresponds to $A_2 = 6$ should be used.

The values of Q are in cubic metres per hour. The height H is in metres and the temperature in degrees Celsius.

In order to use the method the user has to complete Form 4, given in Appendix B at the end of this chapter. The necessary inputs include all the parameters introduced in equation (9.21). In addition, by specifying the volume of the ventilated space, the user can calculate the air changes per hour.

The method consists of 16 steps. Information on each step is given below:

1   **Area of the lower opening A_1.** Give the area of the lower opening.
2   **Area of the higher opening A_2.** Give the area of the higher opening.
3   **Volume of the ventilated space.** Specify the volume of the space to be ventilated.
4   **Total area of openings A.** The total area of the openings is equal to the sum of the higher and lower openings, as specified in steps 1 and 2.
5   **Vertical distance H.** Give the vertical distance between the mid-levels of

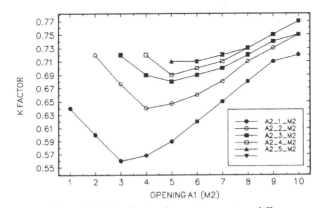

Figure 9.19 Ventilation due to temperature difference with openings at different heights: The correction factor K as a function of A₁ and A₂

the lower and higher openings.

6   **Design month.** The mean monthly ambient climatic data is used. Therefore, specify the summer month for which calculations will be performed.

7   **Mean monthly ambient temperature DTA.** Enter the mean monthly ambient temperature from local climatological tables.

8   **Indoor air temperature TA.** Specify the design indoor air temperature.

9   **Mean temperature MT.** Calculate the mean value of the indoor and outdoor temperature. Add DTA and TA and divide by two.

10 **Mean temperature difference DT.** Calculate the difference between the
indoor and the outdoor temperatures. Introduce the absolute value of the
difference. This means that, if TA is lower than DTA, the difference is
equal to DTA – TA. If TA is higher than DTA, DT = TA – DTA.

11 **Temperature ratio RT.** Divide DT by MT.

12 **Square root of RT.** Calculate the square root of RT.

13 **Determine $H^{0.5}$.** Calculate the value of $H^{0.5}$.

14 **K correction factor.** From Figure 9.19 determine the K correction factor.

15 **Air flow rate Q.** Calculate the air flow rate, in cubic metres per hour,
using equation (9.21). The air flow rate can be calculated as the product of
the values calculated in Step 4 × Step 12 × Step 13 × Step 14 × 1590.

16 **Air changes per hour.** The air changes per hour can be calculated as the
ratio of the total air flow Q to the volume of the ventilated space as given
in Step 3.

To help the user, an example of a building having a lower opening of 5 m² and a
higher opening of 7 m² is presented. The volume of the building is 90 m³. The outdoor temperature is 29°C, while the indoor temperature is 26°C. The vertical distance between the openings is $H = 2.5$ m.

The results of the example are given in Appendix B, Form 5. It is calculated that
the air flow is close to 6900 cubic metres per hour which corresponds to 77 air
changes per hour.

CROSS-VENTILATION

In cross-ventilation the air flow depends directly on the difference in pressure at the
openings. The main parameters influencing the air-flow levels are:

•   the inlet and outlet surfaces of the openings;
•   the wind velocity and direction;
•   the temperature difference between the indoor and the outdoor environ-
    ments;
•   the relative position of the openings; and
•   the relative wind shadowing of the building.

The possible combinations of these parameters are almost infinite and therefore it is virtually impossible to classify and present in an simple way results of all possible configurations. However, it is possible to study the impact of some decisive architectural design parameters under certain urban and climatic conditions.

The method is based on the use of calculation forms (Form 6 in Appendix B at the end of the chapter), where the air flow rate is manually calculated after specifying all the architectural, urban and climatic data. The method is designed for cross-ventilation configurations with openings in two facades. It includes algorithms for the calculation of the air flow due to wind as well as to the stack effect.

Air flow due to wind. The method is designed for an inlet-to-outlet openings ratio between 0 and 2:

  $0 <$ air inlet openings / air outlet openings ≤ 2.

For all the configurations with openings in three or more facades, or with inlet-to-outlet aspect ratios higher than two, computerized tools should be used.

For the prediction of the air flow due to wind Q_w in cubic metres per hour, the following expressions are used:

for $0 <$ inlet openings/outlet openings ≤ 1

$$Q_w = 1620 B^{-1.02} V \sqrt{\Delta Cp} \qquad (9.22)$$

for $1 <$ inlet openings/outlet openings ≤ 2

$$Q_w = 1512 B^{-1.07} V \sqrt{\Delta Cp} \qquad (9.23)$$

where:

$$B = 1 / \sqrt{A_1^2 + A_2^2} ; \qquad (9.24)$$

V is the wind speed, (m s^{-1}), ΔCp the difference of the wind pressure coefficients between the windward and the leeward facades, A_1 the surface area of the windward opening and A_2 the surface area of the leeward opening.

Air flow due to stack effect. For the calculation of the air flow due to the stack effect Q_s, equation (9.21) is again used:

$$Q_s = 1590 K (A_1 + A_2) \sqrt{DT / MT} \sqrt{H} \qquad (9.25)$$

where K is again determined from Figure 9.19. For $A_2 > 6$ the K value that corresponds to $A_2 = 6$ should be used. All units are the same as above.

Total air flow. The total air flow QT is calculated from:

$$QT = \sqrt{Q_w^2 + Q_s^2},$$
(9.26)

where Q_s and Q_w are as defined previously.

The method consists of 35 steps (Appendix B, Form 6).

1 **Design month.** The mean monthly ambient climatic data is used. Therefore, specify the summer month for which calculations will be performed.

2 **Wind speed and wind direction.** From the local climatological tables give the mean monthly wind speed and direction which correspond to the design month.

3 **Mean monthly ambient temperature DTA.** From local climatological tables enter the mean monthly ambient temperature.

4 **Indoor air temperature TA.** Specify the design indoor air temperature.

5 **Mean temperature difference DT.** Calculate the difference between the indoor and the outdoor temperature. Introduce the absolute value of the difference. This means that, if TA is lower than DTA, the difference is equal to DTA – TA. If TA is higher than DTA, DT = TA – DTA.

6 **Mean temperature MT.** Calculate the mean value of the indoor and outdoor temperatures. Add DTA and TA and divide by two.

7 **Temperature ratio RT.** Divide DT by MT.

8 **Square root of RT.** Calculate the square root of RT.

9 **Incidence angle f on the windward wall.** The incidence angle on the windward wall is defined as the angle between the vertical on the windward window and the wind direction (Figure 9.20).

10 **Wind pressure coefficient Cp_1 on the windward facade.** Select from the Tables given in Appendix D at the end of the chapter the form of the building which is closest to the studied case. Then select the wind angle which is closest to the incidence angle calculated in Step 9. Read from the table the value of the pressure coefficient which corresponds to the windward face. For the figure given in the tables of Appendix D, the windward facade corresponds to face 1.

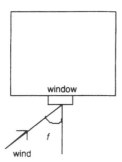

Figure 9.20 Wind incidence angle f

11 **Wind pressure coefficient Cp₂ on the leeward facade.** From the same table used in Step 10, select which is the leeward facade (face 2, 3 or 4), that corresponds to the studied configuration. Then read from the table the value of the pressure coefficient which corresponds to your leeward facade.

12 **Difference of wind pressure coefficients DCp.** Calculate the difference $Cp_1 - Cp_2$.

13 **Determine DCp$^{0.5}$.** Calculate the square root of DCp.

14 **Volume of the ventilated space.** Specify the volume of the space to be ventilated.

15 **Surface of the windward openings A_1.** Specify the total surface area of the openings in the windward facade.

16 **Surface of the leeward openings A_2.** Specify the total surface area of the openings in the leeward facade.

17 **Vertical distance between the openings H.** If the windward and the leeward openings are not at the same height from the ground, specify the mean vertical distance between the openings. The vertical distance is calculated as the distance between the mid-levels of the windward and the leeward windows.

18 **Total area of openings AC.** The total area of the openings is equal to the sum of the windward and leeward openings, as specified in Steps 1 and 2.

19 **Square root of A_1.** Calculate the square root of A_1.

20 **Square root of A_2.** Calculate the square root of A_2.

21 **Sum of $A = A_1^2 + A_2^2$.** Calculate the sum of the squares of A_1 and A_2.

22 **Square root of A, B.** Calculate the square root of A.

23 **$C = 1/B$.** Calculate $1/B$.

24 **$D = A_1/A_2$.** Calculate the ratio A_1/A_2.

25 **Square root of H.** Calculate h, the square root of H.

26 **K_1 and K_2.** Determine the coefficients K_1 and K_2.
   (a) For $0 < D \leq 1$ then $K_1 = 0.45$, $K_2 = -1.02$.
   (b) For $1 < D \leq 2$ then $K_1 = 0.42$, $K_2 = -1.07$.

27 $E = C^{K_2}$. Calculate the coefficient $E = C^{K_2}$.

28 **Air flow rate due to wind Q_w.** Calculate the air flow rate due to wind, in cubic metres per hour, using equation (9.22) or equation (9.23). The air flow rate can be calculated as the product of the values calculated in Step 26(a) × Step 27 × Step 13 × Step 2 × 3600.

29 **K correction factor.** From Figure 9.19 determine the K correction factor.

30 **Air flow rate due to stack effect Q_s.** Calculate the air flow rate due to the stack effect, in cubic metres per hour, using equation (9.25). The air flow rate can be calculated as the product of the values calculated in Step 29 × Step 18 × Step 8 × Step 25 × 1590.

31 **The square of Q_w, Q_{w1}.** Calculate the square of Q_w.

32 **The square of Q_s, Q_{s1}.** Calculate the square of Q_s.

33 **The sum of $Q_{s1} + Q_{w1}$, Q_{t1}.** Calculate the sum of $Q_{s1} + Q_{w1}$.

34 **The square root of Q_{t1}, QT.** Calculate the square root of Q_{t1}.

35 **Air changes per hour.** The air changes per hour can be calculated as the ratio of the total air flow QT to the volume of the ventilated space, as given in Step 14.

Example: Consider a building characterized by the following data:

wind speed = 2.2 m/sec,
indoor temperature = 27°C, outdoor temperature = 32°C,
CP_1 = 0.7, CP_2 = −0.5, volume= 500 m³, A_1=6 m², A_2=4 m²,
vertical distance = 5 m.

Calculate the air changes per hour.

The results are given in Appendix B, Form 7. The calculated air flow due to wind is equal to 30,166 m³ h⁻¹, while the air flow due to the stack effect is equal to 10,271 m³ h⁻¹. The total air flow is equal to 31,843 m³ h⁻¹ and the corresponding air changes per hour are close to 64.

The PHAFF method [9]

Most of the existing correlations for natural ventilation fail to predict the air flow observed in cases where wind or buoyant effects are absent. Experimental results have shown that fluctuating effects are responsible for the air flow in the case of single-sided ventilation or when the wind direction is parallel to openings in two parallel facades [16]. Fluctuating flows are attributed to the turbulence characteristics of the incoming wind and/or to turbulence induced by the building itself. Turbulence in the air flow along an opening causes simultaneous positive and negative pressure fluctuations of the inside air. An empirical correlation that integrates the turbulence effect in a more general air flow model is presented in [9].

A general expression is given for the ventilation rate Q through an open window as a function of temperature difference, wind velocity and fluctuating terms. For the case of single-sided ventilation, an effective velocity, U_{eff} was defined and refers to the flow through half a window opening. In a general form the effective velocity is defined as:

$$U_{eff} = \frac{Q}{A/2} = \sqrt{\frac{2}{g}(\Delta p_{wind} + \Delta p_{stack} + \Delta p_{turb})} \qquad (9.27)$$

leading to the form:

$$U_{eff} = \frac{Q}{A/2} = \sqrt{C_1 U_{met}^2 + C_2 H\Delta T + C_3} \qquad (9.28)$$

where U_{met} is the meteorological wind velocity, H is the vertical size of the opening, C_1 is a dimensionless coefficient depending on the wind, C_2 is a boundary constant and C_3 is a turbulence constant. The term C_3 is equivalent to an effective turbulence pressure that provides ventilation in the absence of stack effect or steady wind. Comparison between measured and calculated values has led to the following values for the fitting parameters: $C_1 = 0.001$, $C_2 = 0.0035$ and $C_3 = 0.01$.

NATURAL CONVECTION HEAT AND MASS TRANSFER THROUGH LARGE INTERNAL OPENINGS

Heat and mass transfer between zones in buildings is caused, almost entirely, by natural convection when the area of the flow connection, or doorway, is much smaller than the overall cross-sectional area of the zone and when there is no supply in one zone.

There are two major mechanisms which are responsible for the natural convection flow between the building zones:

- gravitational flow, which is created by air temperature differences between hot and cold zones, and
- flows driven by boundary layer pumping, which occurs next to heated and cooled surfaces.

Natural convection flows in buildings generally result from a combination of boundary-layer flows and bulk density differences.

Important research has been carried out to cover specific scientific aspects of the topic. However, the use of existing knowledge in passive cooling applications should be carefully examined mainly because of the experimental methodology which has been used. On the other hand, it is widely accepted that existing knowledge about specific problems is not sufficient and therefore further research should be planned and carried out.

The aim of this section is:

- to report the existing scientific work on air flow through large vertical and horizontal openings and to review the solutions already developed in the literature;
- to compare the existing algorithms and investigate the limits of their validity; and
- to identify the gaps in scientific knowledge.

Gravitational flow

ONE-DIMENSIONAL FLOW

Neutral plane at the mid-height of the door – isothermal zone. If inviscid and incompressible air flow between two zones of a building separated by a large opening is assumed, the application of the Bernoulli equation (equation (9.17) results in the following equation for the air velocity at different heights:

$$V_{12} = C_d \sqrt{\frac{2(P_1(y) - P_2(y))}{\rho}}$$

(9.29)

where C_d is the discharge coefficient and $P_1(y)$ and $P_2(y)$ are the pressures at y of zones 1 and 2 respectively (Figure 9.21).

Taking

$$P_i(y) = P_{0i} - \int_0^y \rho_i(y)g \, dy$$

(9.30)

$$\frac{\rho_1 - \rho_2}{\rho_1 + \rho_2} = \frac{T_2 - T_1}{T_2 + T_1}$$

(9.31)

where P_{0i} is the pressure at the floor level of the zone and ρ_i and T_i are the mean density and temperature of zone i. Substituting equations (9.30) and (9.31) into (9.29), we obtain

$$V_{12} = C_d \sqrt{2gy \frac{\Delta T}{T}}.$$

(9.32)

If the neutral level is at a height Y_x equal to the mid-height of the opening, then the mass transfer is determined from the continuity equation applied at the aperture:

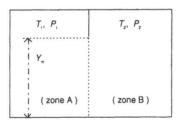

Figure 9.21 Isothermal zones communicating through a large opening

$$m = \int_0^{Y_x} \rho V \, dA \qquad\qquad\qquad (9.33)$$

where $Y_x = Y_m/2$ and A is the surface area, V the velocity and Y_m the height of the opening (Figure 9.22). Also $m_{12} = m_{21} = m$. After integration of equation (9.33):

$$\dot{m} = \frac{2}{3} C_d \bar{\rho} W Y_x \left[g \frac{\Delta T}{T} Y_x \right]^{0.5} \qquad\qquad (9.34)$$

where W and Y_x are the width and the height of the opening, respectively.

However, equation (9.34) cannot be exact for real flow situations, because of the following effects:

- The air flow in the opening is not one-dimensional and the effects of viscosity may not be negligible.
- ΔT may not be independent of Y. The neutral plane may not coincide with the mid-height of the opening.

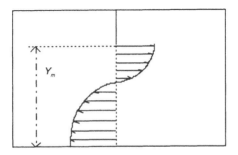

Figure 9.22: Isothermal flow between two zones

Neutral plane not at the mid-height of the door – isothermal zone. If the distance between the neutral plane and the mid-height of the door is $\Delta Y \left(\Delta Y = Y_x - \dfrac{Y_m}{2} \right)$, it has been shown [19] that the velocity in the inflow stream is:

$$V_1 = C_1 \left[\left(2g \frac{\Delta T}{\overline{T}} \right)(Y - \Delta Y) \right]^{0.5}, \quad Y = \frac{Y_x}{2} \text{ for } \frac{\Delta Y}{Y_x} < \frac{Y}{Y_x} < 1 \tag{9.35}$$

$$V_2 = C_2 \left[\left(2g \frac{\Delta T}{\overline{T}} \right)(Y - \Delta Y) \right]^{0.5}, \quad \text{for } -1 \leq \frac{Y}{Y_x} < \frac{\Delta Y}{Y_x}. \tag{9.36}$$

Integration of equation (9.33), using equations (9.35) and (9.36) gives:

$$\dot{m} = \frac{2}{3} \overline{\rho}_1 C_1 W \left(2g \frac{\Delta T}{\overline{T}} \right)^{0.5} [Y_x - \Delta Y]^{0.5}$$

$$= \frac{2}{3} \overline{\rho}_2 C_2 W \left(2g \frac{\Delta T}{\overline{T}} \right)^{0.5} [Y_x + \Delta Y]^{1.5}. \tag{9.37}$$

From equation (9.37)

$$\Delta Y = Y_x \left[\frac{1 - \alpha}{1 + \alpha} \right] \tag{9.38}$$

where

$$\alpha = \left(\frac{\rho_2 C_2}{\rho_1 C_1} \right)^{2/3} = \left(\frac{T_1 C_2}{T_2 C_1} \right)^{2/3}. \tag{9.39}$$

Neutral plane not at the mid-height of the door – stratified zone. If it is assumed that the temperature distribution in the zone is linear and is defined as:

$$T_i(y) = T_{0i} + b_i y \tag{9.40}$$

where T_{0i} is the zone temperature at $y = 0$ and the zone density distribution is given by:

$$\rho_i(y) = \rho_0 - \overline{\rho}\beta(T_i(y) - T_0) \quad \text{where } \beta = 1/\overline{T}. \tag{9.41}$$

It has been calculated that the aperture velocity at height y for flow from zone 1 to zone 2 is given by [20]:

$$V_{12}(y) = C_d \left\{ \frac{2}{\rho}(P_{01} - P_{02}) + 2\beta g \left[(T_{01} - T_{02})y + (b_1 - b_2)\frac{y^2}{2} \right] \right\}^{0.5} \tag{9.42}$$

where P_{0i} is the zone pressure at $y = 0$.

The neutral height Y_x is that at which the zone pressures are equal and the aperture velocity is zero. Equation (9.42) is thus set equal to zero and solved for $y = Y_x$ to determine the neutral height:

$$Y_x = \frac{-(T_{01} - T_{02})}{b_1 - b_2} \pm \left[\left(\frac{T_{01} - T_{02}}{b_1 - b_2} \right)^2 - \frac{2(P_{01} - P_{02})}{\rho \beta g(b_1 - b_2)} \right]^{0.5}. \tag{9.43}$$

Integration of equation (9.33) using the velocity expression given in equation (9.36), and using the neutral height given in equation (9.43) as one of the limits, gives the total aperture mass flow rate.

The base pressure difference, which is contained in the integral limits as well as in the integral, has been calculated numerically in [21], using a modified secant profile [20].

For isothermal zones, the base pressure difference required in equation (9.43) is given by [20]:

$$P_{01} - P_{02} = \bar{\rho}\beta g[T_2 - T_1] \frac{Y_m}{\left(1 + \frac{\rho_1}{\rho_2}\right)^{2/3}}. \tag{9.44}$$

TWO-DIMENSIONAL FLOW

In this case, the air velocity is characterized by two velocity components, u and v. If θ is the local inclination of a streamline with respect to the X axis, then (Figure 9.23):

$$u = V \cos\theta \quad \text{and} \quad v = V \sin\theta. \tag{9.45}$$

If ΔY is the distance between the mid-height of the opening and the neutral plane $\left(\Delta Y = \frac{Y_m}{2} - Y_x \right)$, then for the inflow stream velocity components we obtain [19]:

$$u_1 = C_1 \left[2g \frac{\Delta T}{T}(Y - \Delta Y) \right]^{0.5} \cos\theta, \quad \text{for} \quad \frac{\Delta Y}{Y_x} < \frac{Y}{Y_x} < 1, \quad Y = \frac{Y_m}{2} \tag{9.46}$$

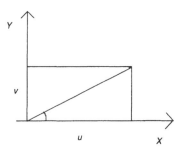

Figure 9.23 Two-dimensional flow

$$v_1 = C_1\left[2g\frac{\Delta T}{\overline{T}}(Y-\Delta Y)\right]^{0.5}\sin\theta, \quad \text{for} \quad \frac{\Delta Y}{Y_x} < \frac{Y}{Y_x} < 1, \quad Y = \frac{Y_m}{2} \tag{9.47}$$

while the outflow velocity components are given by:

$$u_2 = C_2\left[2g\frac{\Delta T}{\overline{T}}(Y+\Delta Y)\right]^{0.5}\cos\theta, \quad \text{for} \quad -1 \le \frac{Y}{Y_x} < \frac{\Delta Y}{Y_x} \tag{9.48}$$

$$v_2 = C_2\left[2g\frac{\Delta T}{\overline{T}}(Y+\Delta Y)\right]^{0.5}\sin\theta, \quad \text{for} \quad -1 \le \frac{Y}{Y_x} < \frac{\Delta Y}{Y_x}. \tag{9.49}$$

Integration of equation (9.33) using equations (9.46) to (9.49) gives:

$$M = \frac{2}{3}\overline{\rho}_1 C_1 W\left(2g\frac{\Delta T}{\overline{T}}\right)^{0.5}[Y_x - \Delta Y]^{0.5}\cos\theta$$

$$= \frac{2}{3}\overline{\rho}_2 C_2 W\left(2g\frac{\Delta T}{\overline{T}}\right)^{0.5}[Y_x + \Delta Y]^{1.5}\cos\theta. \tag{9.50}$$

It has been found experimentally [19] that

$$\frac{\Delta Y}{Y_x} = 0.06 \quad \text{and} \quad \theta = 25°. \tag{9.51}$$

Experimental results reported in [19] give $C_1 = 0.5$ and $C_2 = 0.48$. Those values are not in agreement with the discharge coefficient given in [22–23]. In [22] it is reported that $C_1 = 0.73$ and $C_2 = 0.68$, while in [23] it is found that $C_1 = C_2 = 0.68$. A further discussion on the discharge coefficients is given in the next section.

HEAT TRANSFER BY NATURAL CONVECTION

The heat flow below and above the neutral level Z_n can be calculated from the expression:

$$Q_{0,z_n} = c_p C_d W \int_0^{z_n} \rho_0(y) u(y) T_0(y) \, dy. \tag{9.52}$$

In [24] it is shown that, when the heat flow is expressed in terms of an average air temperature difference ΔT and the coefficients are defined in equation (9.40), then

$$Q = mc_p [\Delta T + 0.34(b_h + b_c)]. \tag{9.53}$$

This approximation is valid for small temperature gradients.

In a simpler approach, defined in [25], the associated rate of heat transfer per unit door width through the opening may be given by:

$$Q = \dot{m} c_p \Delta T = hA\Delta T. \tag{9.54}$$

Thus:

$$h = \frac{\dot{m} c_p}{A}. \tag{9.55}$$

If the neutral plane is situated at the mid-height of the opening and equation (9.34) is substituted into (9.55), one obtains:

$$Nu = C \, Gr^{0.5} \, Pr \tag{9.56}$$

where $C = C_d/3$, C_d is defined in equation (9.32) and Gr is the Grashof number and Pr is the Prandtl number.

The dimensionless numbers, Pr and Gr, are given by the expressions:

$$Pr = c_p \mu / k \quad \text{and} \quad Gr = \frac{g\beta H^3 \Delta T}{v^2}. \tag{9.57}$$

In equation (9.57), ΔT is the temperature difference between the zones and H is the characteristic length. The approximation that the neutral plane is situated at the mid-height of the opening does not really introduce an important inaccuracy in the overall calculations. Estimation of ΔY using equation (9.38) gives that $\Delta Y \approx 5$ mm for an opening of $Y_m = 2$ m and for a temperature difference of 10°C.

A review of the C values used in equation (9.56) [26–27] indicates that the values of C can be as low as 0.09 and exceed a value of 0.25 for large openings. A method to calculate the C values, as a function of the temperature difference, air speed and height of the opening, is proposed in [28]. According to this, the value of C does not vary significantly as a function of the temperature difference and air speed, but is a strong function of the dimensions of the opening. A proposed mean value for a standard opening is $C = 0.26$.

A second expression for the prediction of the Nu number is proposed by Brown et al. [29]:

$$Nu = Pr\ 1.16(10 - 0.6t\ /\ Y_m)Gr^{0.432} - 215 \qquad (9.58)$$

for $10^6 \le Gr < 10^8$ and $0.19 \le t/Y_m \le 0.75$, where t is the thickness of the partition and

$$Nu = 0.343\ Pr\ Gr^{0.5}(1 - 0.498t\ /\ Y_m)\ \text{ for } Gr \ge 10^8. \qquad (9.59)$$

Wray and Weber [30] and Weber et al. [31] have performed experiments in a small two-zone scaled model enclosure with Freon as a working fluid. They have proposed the following expression.

$$Nu = C\ Pr\ Gr^{0.46} \qquad (9.60)$$

where the coefficient C would have the value of 0.29 and 0.41 for Y_m/H values of 0.82 and 1 respectively.

Cockroft [32] has proposed the following expression:

$$Nu = \frac{2}{3}\frac{\rho c_p}{k}Y_m\left[C_d\left(\frac{2}{\rho}\right)^{1/2}\frac{[C_a^{3/2} - C_b^{3/2}]}{C_t}\right] \qquad (9.61)$$

where

$$C_a = (1 - r_p)C_t + (P_1 - P_2),\ \ C_b = (P_n - P_m) - r_p C_t,\ \ r_p = \frac{H_r}{Y_m} \qquad (9.62)$$

$$C_t = \frac{gP_0 Y_m}{R\left(\dfrac{1}{T_1} - \dfrac{1}{T_2}\right)}. \qquad (9.63)$$

H_r is the reference height at which P_1 and P_2 are measured and P_0 is the atmospheric pressure.

Flows driven by boundary layer pumping

In the case where there are heated and cooled surfaces inside the two zones, thermosyphon pumping is produced by the boundary layers that are formed next to the heated and cooled surfaces.

In this case, there are three significant resistances to natural convection heat transfer: the hot-wall boundary layer resistance (R_{hw}), the cold-wall boundary layer resistance (R_{cw}) and the aperture resistance (R) (Figure 9.24).

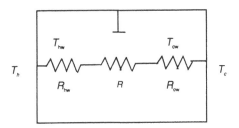

Figure 9.24 Schematic representation of the three resistances to natural convection heat transfer

If the resistances are assumed to be connected in series, then the Nu number is given by [24]

$$Nu = \frac{1}{kW[R_{hw} + R + R_{cw}]} \tag{9.64}$$

where k is the thermal conductivity of the surface.

According to [24], R_{hw} and R_{cw} can be calculated from the following expressions:

$$R_{hw} = \frac{1}{kWA \, Ra_{Y_m}^{1/3}} \tag{9.65}$$

$$R_{cw} = \frac{1}{kWA \, Ra_H^{1/3}} \tag{9.66}$$

where A is a constant $(A = 0.107)$, Ra is the Rayleigh number, Y_m is the height of the opening and H the height of the zone.

The two wall-resistance terms can be combined to give:

$$R = R_{hw} + R_{cw} = \frac{(1 + (Y_m / H)^{3/4})^{4/3}}{kWA(Y_m / H)(1 - \Delta T / DT)^{1/3} Ra^{1/3}} \tag{9.67}$$

where ΔT is the temperature difference between the two zones and DT is the wall-to-wall temperature difference.

According to [24], the aperture resistance term is given by:

$$R = \cfrac{1}{kW \, 0.2\left(\cfrac{w}{W}\right)\left(\cfrac{Y_m}{H}\right)^{3/2}\left(\cfrac{\mathrm{Pr}}{\mathrm{Ra}}\right)^{1/2}\left(\cfrac{\Delta T}{\mathrm{DT}}\right)^{1/2}\left(1+0.3F\left(\cfrac{\mathrm{DT}}{\Delta T}\right)\right)\left(1+\left(\cfrac{Y_m}{H}\right)\right)}$$

(9.68)

where w is the width of the aperture, W is the width of the enclosure and F is a constant defined as follows:

$$F = \frac{b_h + b_c}{\dfrac{\mathrm{DT}}{Y_m} + \dfrac{\mathrm{DT}}{H}}.$$

(9.69)

In equation (9.69) b_h and b_c are the coefficients of equation (9.40) for the hot and cold zones respectively. Using experimental results [24] it is found that

$$Nu = \frac{1}{kW(0.83R_w + 0.69R)}.$$

(9.70)

Neymark et al. [33] have conducted similar experiments to those reported in [24] and have proposed the following expression:

$$Nu = \frac{1}{12\mathrm{Ra}^{-0.25} + 5.08\left(\dfrac{W}{W_1}\right)^{-2/3}(\mathrm{Pr}\,\mathrm{Ra}^*)^{-1/3}}$$

(9.71)

where W_1 is the width of the zone and Ra^* is the modified Rayleigh number.

Numerical studies of natural convection in a partially divided rectangular enclosure with an opening in the partition plate and isoflux side walls [34] led to the following expression for the Nu number:

$$Nu = 0.176(\mathrm{Ra}^*)^{-0.258}(1 - A_o)^{-0.328}, \quad \text{for } 10^4 \leq \mathrm{Ra}^* < 10^8$$

(9.72)

where A_o is the opening ratio, Y_m/H, $0 \leq A_o \leq 0.5$, and $\mathrm{Pr} = 7$.

Nansteel and Greif [35] carried out experiments in a small 2D enclosure with water at $2.3 \times 10^{10} \leq \mathrm{Ra}_L \leq 1.1 \times 10^{11}$ and aperture ratios ($A_p = Y_m/H$) of 1, 3/4, 1/2 and 1/4.

The results were correlated using the following relationships:

$$Nu = 0.748 A_p^{0.256} \mathrm{Ra}_L^{0.226} \text{ for a conducting partition}$$

(9.73)

$$Nu = 0.762 A_p^{0.473} \mathrm{Ra}_L^{0.226} \text{ for a non-conducting partition}$$

(9.74)

where L is the sum of the lengths of the two zones.

In an extension to their work, Nansteel and Greif [36] carried out similar experiments using silicon oil over the range Ra = 1.55×10^9 and using a low-conductivity partition. The following correlation was developed:

$$\text{Nu} = 0.28 A_p^{0.271} \text{Ra}_L^{0.276}. \tag{9.75}$$

In further work, the same authors [37], using water, found that the Nu number can be obtained for $2.4 \times 10^{10} < \text{Ra} < 1.1 \times 10^{10}$ from the following expression:

$$\text{Nu} = 1.19 A_p^{0.401} \text{Ra}_L^{0.207}. \tag{9.76}$$

Liu and Bejan [38] performed experiments similar to those of Nansteel and Greif, but for opening size $1/16 \le A_p \le 1$. The heat transfer data was correlated using the equation:

$$\text{Nu} = 0.336 \frac{\text{Ra}_H^{0.25}}{A_p^{-0.75} + 0.5}. \tag{9.77}$$

Bauman et al. [39] proposed the following expressions for the convection heat transfer coefficient h:

$$h = 1.03 \left(\frac{Y_m}{H} \right)^{0.47} \left[\frac{T_h - T_c}{2H} \right]^{0.22} \quad \text{for a 2D flow} \tag{9.78}$$

and

$$h = 1.95 \left(\frac{Y_m}{H} \right)^{0.25} \left[\frac{T_h - T_c}{2H} \right]^{0.2} \quad \text{for a 3D flow.} \tag{9.79}$$

Bajorek and Lloyd [40] conducted experiments for a single value of the aperture ratio $A_p = 0.5$. The experiments were performed at Ra $\approx 10^6$. Their heat transfer results were correlated by the relationship:

$$\text{Nu}_H = 0.063 \text{Gr}_H^{0.33} \tag{9.80}$$

where Gr_H was based on $(T_h - T_c)$.

Numerical calculation on the free-convection heat transfer in a partially divided vertical enclosure with conducting end walls [41] has indicated that for moderate temperature differences, the prediction for perfectly conducting end walls given by equation (9.73) fits well with calculation, while results with an adiabatic end wall shows a significant difference.

Classification of the existing information – sensitivity analysis

A classification of the existing information, including papers published up to 1986, is given in [42–43]. Table 9.C1, in Appendix C at the end of the chapter, reports the information on heat and mass transfer through large openings, published up to 1992.

Plots of 20 expressions for the prediction of the Nu number due to the gravitational flow created by air-temperature differences between hot and cold zones are given in [4]. A list of the plotted expressions is given in Table 9.6. It is deduced that for Gr $< 10^8$ the values of the Nu number predicted by the proposed expression are of the same order. However, For Gr $> 10^9$ an important difference between the calculated values is obtained.

Table 9.6 Gravitational flow between hot and cold zones – expressions for the Nu number [4]

1	$Nu = 0.2\,Gr^{0.5}\,Pr$	12	$Nu = 0.22\,Pr\,Gr^{0.5}$
2	$Nu = 0.33\,Pr\,Gr^{0.5}$	13	$Nu = 0.2[1 + (1 + \beta(T_2 - T_1)^{2/3}/2]^{3/2} \times Pr\,Gr^{0.5}$
3	$Nu = 0.3\,Gr^{0.5}\,Pr$	14	$Nu = 0.33[1 + (1 + \beta(T_2 - T_1)^{2/3}/2]^{3/2} \times Pr\,Gr^{0.5}$
4	$Nu = 0.29\,Gr^{0.46}\,Pr$	15	$Nu = (C_T C_V/3)\,Gr^{0.5}\,Pr$
5	$Nu = 0.41\,Gr^{0.46}\,Pr$	16	$Nu = 0.26\,Gr^{0.5}\,Pr$
6	$Nu = 0.15\,Gr^{0.5}\,Pr$	17	$Nu = 0.223\,Gr^{0.5}\,Pr$
7	$Nu = (0.0835(\Delta T/T)^{-0.313})/3\ Pr\,Gr^{0.5}$	18	$Nu = 0.243\,Gr^{0.5}\,Pr$
8	$Nu = (0.4 + 0.0075\Delta T)/3\ Pr\,Gr^{0.5}$	19	$Nu = 0.103\,Gr^{0.5}\,Pr$
9	$Nu = (1.16(1 - 0.6(t/D))\,Gr^{0.432} - 215)\,Pr$	20	$Nu = 0.29\,Gr^{0.5}\,Pr$
10	$Nu = 0.343\,Pr\,Gr^{0.5}(1 - 0.498t/D)$	21	$Nu = 0.14\,Gr^{0.5}\,Pr$
11	$Nu = (44/k)\,Pr\,(Y^3\Delta T)^{0.5}$		

Comparison of the predicted Nu number for $10^8 < Gr < 10^{10}$ shows that the ratio between two predicted Nu numbers can be of the order of 1/3 to 1/5. This can be explained by the variability of the experimental conditions and of the procedure which is followed in each experiment.

Most of the experiments have been carried out in test cells where electrical emitters or heaters and cooling walls and plates were used. The use of the equipment has created in each case a specific but different air movement in the heated and cooled room and this is one of the sources of the observed differences. The variable accuracy of the measurements can also be a source of error. However, a general remark that should be made is that almost all the experiments correspond more to air-flow patterns observed in buildings where there is heating or air conditioning rather than to the air-flow processes corresponding to natural ventilation techniques.

If we take into account that for real buildings the Grashof number Gr is between 10^9 and 10^{10}, it is clear that a further examination of the applicability of the proposed correlations in actual buildings is necessary.

For natural ventilation purposes it is also essential to study the inter-zone convection under combined natural- and forced-convection situations. For example, the combined effect of forced convection due to air distribution in the building, together with the natural convection due to temperature differences, should be investigated.

The variation of the Nu number, for the case of flows driven by boundary-layer pumping as a function of the Ra $\times A_p$ product, is given in [4]. The parameter A_p is the ratio of the opening height Y_m to the room height H. The proposed expressions for the Nu number have been classified into two groups based on the characteristic length. The first group uses the length of the two zones L and the second group uses the height of the room.

The variation of the Nu number as a function of the Ra $\times A_p$ product is illustrated in [4]. The values of the Nu number are estimated using the four expressions, defined in Table 9.7.

Table 9.7 Boundary-layer pumping
– expressions for Nu as a function of
$Ra_L \times A_p$ [4]

1	$Nu_L = 0.748 A_p^{0.256} Ra_L^{0.226}$
2	$Nu_L = 0.762 A_p^{0.473} Ra_L^{0.226}$
3	$Nu_L = 0.28 A_p^{0.271} Ra_L^{0.276}$
4	$Nu_L = 1.19 A_p^{0.401} Ra_L^{0.207}$

It has been shown that, for the same A_p and Ra_L number the maximum difference between the predicted values of the Nu number is close to 40% of the lower predicted value. If the variability of the conditions under which the equations have been derived is taken into account, this deviation is not unexpected.

The variation of the Nu number, as predicted by the equations having as characteristic length the height of the room, is also given in [4], using the equations in Table 9.8.

For the same Ra number and A_p value the difference between the minimum and maximum predicted Nu number does not exceed 20%. This difference is acceptable, if we also take into account that in the case of boundary-layer pumping flows the Nu number is low compared with the Nu of the gravitational air flow due to temperature differences.

Table 9.8 Boundary-layer pumping –
Nu as a function of $Ra_H \times A_p$ [4]

1	$Nu = 0.336 \dfrac{Ra_H^{0.25}}{A_p^{-0.75} + 0.5}$
2	$Nu = \dfrac{1.03}{k}(A_p)^{0.47} H\left[\dfrac{T_h - T_c}{2H}\right]^{0.22}$
3	$Nu = \dfrac{1.95}{k}(A_p)^{0.25} H\left[\dfrac{T_h - T_c}{2H}\right]^{0.22}$

Boundary-layer pumping flows, although they occur in natural ventilation processes, are not as important as the air-flow processes due to air distribution in the building and to temperature differences.

Natural convection through horizontal openings

Knowledge of the air-flow characteristics, as well as of the corresponding natural convection through horizontal openings, is essential for air-flow modelling in buildings. The most interesting aspect is that where the fluid above the opening has a greater density than the fluid below, an exchange occurs between the heavier fluid flowing downward and the lighter flowing upward.

Very limited information is available on this topic. Brown [44] studied natural convection through square openings in a horizontal partition, using air as the fluid medium. He concluded that natural convection phenomena can be described by the following equation:

$$Nu_H = 0.0546 Gr_H^{0.55} Pr\left(\frac{L}{H}\right)^{2/3} \tag{9.81}$$

where H is the thickness of the partition and L the characteristic width of the opening. The range of validity of the equation is $3 \times 10^4 \leq Gr_H < 4 \times 10^7$ and $0.0825 \leq H/L < 0.66$.

In a more recent paper, Epstein [45] has studied the buoyancy exchange flow through small openings in horizontal partitions using brine and water as working fluids. He presents results as a function of the ratio L/H. The main conclusions are [46]:

- at very small L/H, Taylor instability leads to the intrusion of each liquid into the other one and an oscillating exchange is observed;
- a Bernoulli flow is subsequently found, while for higher values of L/H a turbulent diffusion regime appears.

An empirical formulation of the flow for a wide range of aspect ratios was also proposed: for $0.01 \leq L/H \leq 20$ and $0.025 \leq \Delta\rho/\rho < 0.17$

$$\frac{\dot{m}}{(H^5 g\Delta\rho/\rho)^{1/2}}$$

$$= \frac{0.0055[1+400(L/H)^3]^{1/6}}{[1+0.00527(1+400(L/H)^3)^{1/2}((L/H)^6+117(L/H)^2)^{3/4}]^{1/3}}. \quad (9.82)$$

For the case of a staircase, however, it is difficult to interpret the flow in terms of the parameter L/H. Experiments reported in [3] conclude that the discharge coefficient decreases strongly with temperature difference. This represents the combined effect of the counterflow in the staircase and the air flow through a door in the presence of strong stratification.

In a second study [47] on the buoyancy driver flow in a stairwell, it was found that this is more much more significant than a two-way boundary-layer flow, while it is difficult to identify a horizontal partition.

Modelling of the air flow through horizontal openings is discussed in [9], where formulae are given to calculate the downward and upward volume flows.

The overall knowledge on the mass and heat flow through horizontal openings is not sufficient and it is evident that much more research is needed to describe accurately the behaviour of horizontal openings in natural or mixed convection configurations.

MULTIZONE MODELLING

Fluid flow analysis is an important aspect of building simulation, because knowledge of the results is necessary for the calculation of heating/cooling load as well as for the thermal-comfort assessment of a building. However, owing to computational difficulties and lack of sufficient data, progress on air-flow modelling has been rather slow. Two approaches for fluid flow simulation have been developed: the computational fluid dynamics modelling (CFD) and the mass balance/flow network approach (zonal method).

Computational fluid dynamics modelling (CFD)

CFD modelling is based on the solution of Navier–Stokes equations, namely the mass, momentum and thermal-energy conservation equations at all points on a two- or three-dimensional grid that represents the building under investigation and its surroundings.

MASS CONSERVATION

The mass conservation equation expresses the requirement that the rate of increase in the density ρ within a control volume $dxdydz$ is equal to the net rate of influx of mass to the control volume:

$$\frac{\partial \rho}{\partial t} + \frac{\partial}{\partial x}(\rho U) + \frac{\partial}{\partial y}(\rho V) + \frac{\partial}{\partial z}(\rho W) = 0 \qquad (9.83)$$

where ρ is the air density and U, V and W the air velocity components in the x, y and z directions respectively.

MOMENTUM CONSERVATION

According to the law of momentum conservation, the net force on the control volume in any direction is equal to the outlet momentum flux minus the inlet momentum flux in the same direction. This is conveniently summarized in vector notation:

$$\rho \frac{D\mathbf{V}}{Dt} = -\nabla p + \mathbf{B} + \nabla \times [\mu(\nabla \times \mathbf{V})] + \nabla[(\zeta + \frac{4}{3}\mu)\nabla.\nabla] \qquad (9.84)$$

where $\mathbf{V}$ is the vector of velocity, p the pressure, $\mathbf{B}$ the body force, λ and μ the coefficients of viscosity and $\zeta = \lambda + (2/3)\mu$.

THERMAL ENERGY CONSERVATION

According to the law of the thermal-energy conservation, the net increase in internal energy in a control volume is equal to the net flow of energy by convection plus the net inflow by thermal and mass diffusion. In vector notation this is expressed:

$$\rho c_v \frac{DT}{Dt} = -p\nabla.\mathbf{V} + k\nabla^2 T - \nabla.\mathbf{q}_r + q''' + \Phi \qquad (9.85)$$

where $\mathbf{q} = -k\nabla T$, $\mathbf{q}_r$ is the radiation heat-flux vector, q''' is the internal heat generation rate per unit volume, Φ is the dissipation function, k is the thermal conductivity and c_v the specific heat.

CFD models are mainly used for steady-state problems to predict the temperature and velocity fields inside and the pressure field outside a building.

Figures 9.25 and 9.26 illustrate the temperature and velocity fields inside a two-zone test cell, as simulated by a well known CFD program, PHOENICS [48].

The flow network approach

The zonal method is based on the concept that each zone of a building can be represented by a pressure node. Boundary nodes are also used to represent the environment outside the building. Nodes are interconnected by flow paths, such as cracks, windows or doors to form a network. Figure 9.27 shows a network representation of a multizone building.

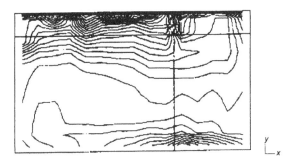

Figure 9.25 Contours of air velocity U in a test cell
(U = velocity parallel to the x axis)

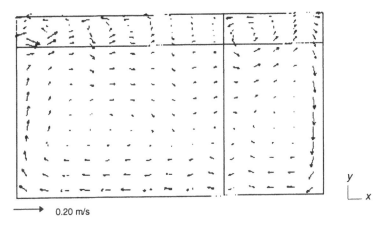

0.20 m/s

Figure 9.26 Contours of temperature in a test cell

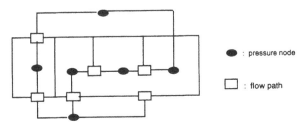

Figure 9.27 Network representation of a multi-zone building

Pressure at boundary nodes is known and it has to be determined for internal nodes. Application of the mass balance equation in a zone i with j flow paths gives:

$$\sum_{i=1}^{J} \rho_i Q_i = 0. \tag{9.86}$$

A multi-zone problem involving N zones is therefore described by a set of N equations with an equal number of unknown pressures. Application of mass balance at each internal node of the network leads to a set of simultaneous non-linear equations, the solution of which gives the internal node pressures. The flow rates through the building openings can then be determined as earlier in this chapter.

Computational tools based on the concept of the zonal method are fast and easy to use. The output comprises air-flow predictions for user-specified building structures and instantaneous climatic conditions. A recent comparative study of five existing computational tools (AIRNET, BREEZE, ESP, PASSPORT-AIR and COMIS) has shown that their predictions are in good agreement for a large number of configurations tested [49]. A representative example of this comparison follows.

Figure 9.28 shows a plan view of the building that was simulated. Only three zones were used in the simulation namely zone A, zone B and zone C. The dimensions of each zone are shown in the figure (zone height is 4.5m).

The case of ventilation with all windows and doors open was simulated using five computational tools. The wind incidence angle on window W1 is 292.5° (clockwise from north) and the wind speed is 1.5 m s^{-1}. The ambient temperature is equal to 26.8°C while indoor temperatures are 26.4°C for zone A, 25.8°C for zone B and 25°C for zone C. Simulation results for each of the computational tools are given in Tables 9.9 and 9.10.

Table 9.9 Flow characteristics of internal and external openings; values in kg s^{-1}

	W1	W2	W3	W4 inflow	W4 outflow	P1	P2
AIRNET	2.54	1.32	2.21	0.24	1.90	2.54	1.66
ESP	2.48	1.18	2.30	0.33	1.77	2.50	1.43
BREEZE	2.23	1.01	1.93	0.10	1.40	2.23	1.30
PASSPORT	2.19	1.24	2.00	0.22	1.65	2.18	1.43
COMIS	2.38	1.04	2.06	0.82	1.95	2.38	1.14
S.DEVIATION	0.15	0.13	0.15	0.28	0.22	0.16	0.19
S.ERROR	0.07	0.06	0.07	0.12	0.10	0.07	0.09

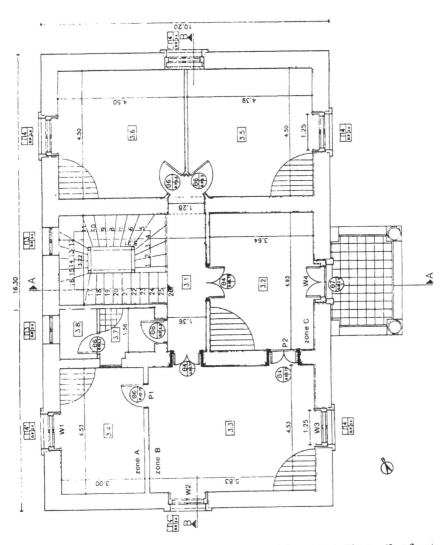

Figure 9.28 Plan view of the Institute of Meteorology and Atmospheric Physics (first floor) of the National Observatory in Athens, Greece

Table 9.10 Flow characteristics of zones, in kg s⁻¹. The maximum difference is defined as the ratio of the difference between the maximum and minimum values divided by the maximum value (%)

	Zone A (outflow, kg s⁻¹)	Zone B (outflow, kg s⁻¹)	Zone C (outflow, kg s⁻¹)
AIRNET	2.54	3.87	1.89
ESP	2.50	3.73	1.77
BREEZE	2.23	3.23	1.40
PASSPORT	2.19	3.42	1.65
COMIS	2.38	3.43	1.95
Max. Differ. (%)	13.9	16.5	28.2

NIGHT VENTILATION

Introduction

Daytime ventilation introduces outdoor fresh air, which is necessary in order to maintain acceptable indoor air quality. In addition to these amounts of outdoor air, ventilation can also provide appropriate means for passive cooling, once the outdoor air is at a lower temperature.

Alternatively, ventilation can also continue into the night time, with positive results on reducing the cooling load of the building. During the night, outdoor temperatures are lower than the indoor ones. Consequently, it is possible to ventilate the building by allowing the outdoor air to enter the spaces and remove the stored heat that has been trapped during the day. The air movement increases the heat dissipation from the building materials and the warmer air is then exhausted into the low temperature atmospheric heat sink. This process continues during the night and, as a result, the indoor air temperature and mass of the building are at lower levels when the temperature-increase cycle starts again the following day. Consequently, in the morning, occupants enter a cooler environment, which means that even in air-conditioned buildings, one could have substantial energy savings from the reduced operation of the mechanical system.

The success of night ventilation depends on the relative difference of indoor and outdoor air temperatures. The lower the outdoor night-time temperatures, the higher the effectiveness of night-time ventilation. It is also necessary to achieve the best possible air movement of the outdoor air through the indoor spaces. The heat convected from the mass of the building is increased by the relative velocity of the air passing over the various surfaces. This process can be facilitated by the use of ceiling fans, which will increase indoor air movement and, as a result, raise the convective heat transfer coefficient between the various surfaces and the passing air. This technique can be used to increase night-time ventilation rates to 1.5 m^3 min^{-1} m^{-2} [50].

Night-time ventilation for cooling purposes can be successful in heavy buildings, with large diurnal outdoor temperature variations. In buildings of large thermal capacity the indoor air temperature can be close to the average outdoor temperature. Buildings where night-ventilation techniques are applied are often equipped with specially designed windows with top openings. For reasons of safety and privacy, ground-floor windows remain closed and safety window screens are used.

A range of night-cooling applications is described in [51] and [52]. In a case study [53] a building located in Athens was ventilated from 21:00 until 07:00. Ventilation rates ranged from 2 to 8 ACH in steps of 2 ACH. Results showed that the maximum observed reduction in peak indoor temperature was 1°K. The decrease in indoor air temperature was found to be more significant in June and August than in July because the night ambient air temperature is higher then. A study on the energy savings potential in office buildings in Athens [54] has shown that night-ventilation techniques can reduce the total cooling load of air-conditioned offices by up to 30%.

Simulations have shown that the use of night ventilation techniques can provide part of the cooling load required. However, to maintain acceptable daytime temperature levels, the use of additional cooling systems is often unavoidable [53].

Table 9.11 Monthly cooling loads [GJ] of a building using night ventilation techniques

1 ACH	$T_{in} = 26°C$	$T_{in} = 27°C$	$T_{in} = 28°C$	$T_{in} = 29°C$
June	9.37	7.12	6.12	4.77
July	24.28	19.32	15.85	11.86
August	14.21	10.96	9.09	7.39
Total	48.36	37.40	31.10	24.10
5 ACH	$T_{in} = 26°C$	$T_{in} = 27°C$	$T_{in} = 28°C$	$T_{in} = 29°C$
June	6.21	4.74	3.44	2.06
July	20.45	15.76	11.88	8.09
August	11.12	8.62	6.25	4.49
Total	38.10	29.12	21.57	14.60
10 ACH	$T_{in} = 26°C$	$T_{in} = 27°C$	$T_{in} = 28°C$	$T_{in} = 29°C$
June	4.93	3.17	2.08	1.46
July	18.99	13.87	10.09	6.13
August	9.29	8.86	4.76	3.37
Total	33.21	25.90	16.93	10.96
20 ACH	$T_{in} = 26°C$	$T_{in} = 27°C$	$T_{in} = 28°C$	$T_{in} = 29°C$
June	3.94	2.15	1.07	0.79
July	17.51	12.68	8.12	5.17
August	8.54	5.84	3.62	2.24
Total	30.00	20.67	12.81	2.00

Night ventilation potential – case studies

A single-zone building (20 × 20 × 3.2 m) of a 400 m² total floor area has been simulated for various air change rates during the night. Monthly cooling loads have been calculated for 1, 5, 10 and 20 ACH and for four different values of desired indoor air temperature: 26°C, 27°C, 28°C and 29°C. Table 9.11 summarizes the results for June, July and August.

As shown in Figures 9.29 to 9.32, air change rates up to 10 ACH reduce the cooling load significantly, whereas greater air change rates only affect the cooling load to a minor extent.

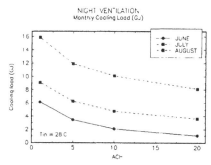

Figure 9.29 Monthly cooling loads of a single-zone building (400 m²). Desired indoor air temperature 26°C

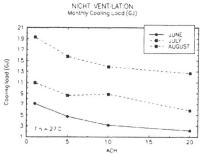

Figure 9.30 Monthly cooling loads of a single-zone building (400 m²). Desired indoor air temperature 27°C

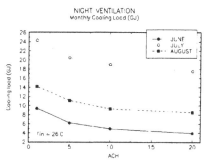

Figure 9.31 Monthly cooling loads of a single-zone building (400 m²). Desired indoor air temperature 28°C

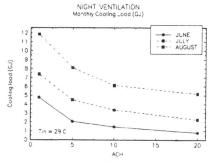

Figure 9.32 Monthly cooling loads of a single-zone building (400 m²). Desired indoor air temperature 29°C

Figure 9.33 illustrates the impact of night ventilation on the day-time indoor air temperatures. The values shown, are temperature differences for the cases of 5, 10, 15, 20, 25 and 30 ACH from the base case of 1ACH. For ventilation rates up to 10 ACH the indoor air temperatures are significantly reduced. For greater air change rates no further reduction is observed.

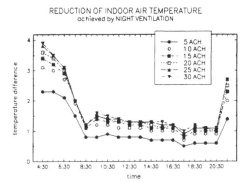

*Figure 9.33 Reduction of indoor air temperature
achieved by night ventilation*

Potential of night-time ventilation for cooling

A simple model for the estimation of potential cooling due to night-time ventilation arising from stack-effect is presented in [56]. The algorithm couples the ventilative heat loss rate with the heat transfer between the interior air and the walls of a building. The model can be used for simple energy calculations of stack-ventilative night cooling to predict the influence of ventilation opening dimensions, building materials and climate. In the following the basic assumptions of the model are presented and the algorithm for calculating the temperature of the exhaust air is described. The algorithm is a combination of a ventilation and a thermal model, which, as shown in Figure 9.34, form a four-node network.

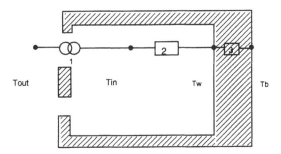

Figure 9.34 A four-node model for ventilative cooling

VENTILATION MODEL

Consider a volume with two openings, an inlet at the bottom and an outlet at a distance H above it. If the inlet and the outlet area are A_{in} and A_{out} respectively and the inside/outside temperature difference is $T_{in} - T_{out}$, the exhaust velocity is [55]:

$$V_{out} = C_1 \sqrt{\frac{2g(T_{in} - T_{out})}{T}} \sqrt{\frac{H}{(1 + (A_{out} / A_{in}))^2}} \qquad (9.87)$$

where C_1 is a single discharge coefficient for the inlet and exhaust openings – a typical value of C_1 for sharp-edged orifices is about 0.6 – and T is the average absolute temperature of the indoor and outdoor air [°K].

A single neutral level is assumed to be situated between the inlet and the outlet openings. The model implies that the vertical size of the openings must be much smaller than the distance between them (H) and that A_{in} and A_{out} are of the same order of magnitude.

The volume flow rate V is, then, given by:

$$V = A_{out}V_{out}. \qquad (9.88)$$

For a given volume flow rate, the heat removal capacity F is proportional to the temperature difference between supply and exhaust air:

$$F = C_p dV(T_{in} - T_{out}). \qquad (9.89)$$

THERMAL MODEL

The building is assumed to be in quasi-thermal equilibrium before the ventilation starts. Radiant heat loss through the ventilation openings is neglected. If S_i is the total internal surface, the heat balance equation is

$$C_{in} \frac{dT_{in}}{dt} = -F + h_c S_i (T_{wall} - T_{in}) \qquad (9.90)$$

where C_{in} is the thermal capacity of the heated air volume taking into consideration the effect of furnishings and h_c is the average convective heat transfer coefficient (typical value for low-velocity forced convection h_c=6 W m^{-2}K^{-1}).

If the cooling rate is F, then, for a total interior wall surface S_i, the density of the heat flow rate is $q = F/S_i$ and the difference between the wall surface temperature and the air temperature is

$$T_{wall} - T_{in} = q / h_c = F / (S_i h_c). \qquad (9.91)$$

In this equation the heat capacity is neglected. When cooling starts, the air–wall temperature difference rises nearly exponentially to the value given by:

$$T_{\text{wall}} - T_{\text{in}} = \left(1 - \exp\left(-\frac{t}{\tau}\right)\right) F / (C_2 S_i h_c) \tag{9.92}$$

where C_2 is a coefficient that accounts for the fraction of wall surface that is active in the heat transfer process, $\tau (= C_{\text{in}}/h_c S_i)$ is a time constant, with a value that is typically a few minutes for unfurnished rooms. For long time periods the term $\exp(-t/\tau)$ can be neglected.

The wall surface temperature as a function of time is given by

$$\Delta T_{\text{wall}}(t) = T_{\text{wall}}(t) - T_{\text{wall}}(0) = \frac{2q}{b}\sqrt{\frac{t}{\pi}} = R_{\text{dyn}}q \tag{9.93}$$

where b is the thermal effusivity of the wall material, defined as the square root of the product of thermal conductivity c, density ρ and specific heat λ: $b = (\rho c \lambda)^{1/2}$ [56]. The thermal effusivity accounts for the response of a surface temperature to a change in the heat flow density at the surface. The lower the thermal effusivity of a material, the more sensitive the surface temperature is to changes in the heat flow at it. If the walls of the enclosure are homogenous, a theoretical value for b can easily be found; for composite walls, however, b has to be defined experimentally.

The term R_{dyn} is the dynamic thermal resistance per wall surface area, which is

zero at $t = 0$ and increases with the square root of time: $R_{\text{dyn}} = \frac{2}{b}\sqrt{\frac{t}{\pi}}$.

REFERENCES

1 Evans, B.H. (1957). 'Natural air flow around buildings', Research Report No 59, Texas Engineering Experiment Station, Texas A&M College System, March.

2 Riffat, S.B. (1989). 'A study of heat and mass transfer through a doorway in a traditionally built house', *ASHRAE Transactions*, pp. 584–589.

3 Kiel, D.E. and D.J. Wilson (1989). 'Combining door swing pumping with density driven flow', *ASHRAE Transactions*, pp. 590–599.

4 Santamouris, M. (1992). 'Natural convection heat and mass transfer through large openings', Model Development Subgroup Meeting, 23–27 October, Lyon, France.

5 Pelletret, R., F. Allard, F. Haghighat and J. van der Maas (1991). 'Modelling of large openings', presented at the 12th AIVC Conference, Canada.

6 Limam, K., C. Innard and F. Allard (1991). 'Etude experimentale des transferts de masse et de chaleur à travers les grandes ouvertures verticales', *Conference Groupe d'Etude de la Ventilation et du Renouvellement d'Air*, Institut National des Sciences Appliqués, Lyon, pp. 98–111.

7 Darliel, S. B. and G.F. Lane-Serff (1991). 'The hydraulics of doorway exchange flows', *Building and Environment*, Vol. 26, No. 2, pp. 121–135.

8 Khodr Mneimne, H. (1990). 'Transferts thermo-aerouliques entre pieces à travers les grandes ouvertures', PhD Thesis, Nice University.

9 Annex 20: 'Air flow patterns within buildings – Air flow through large openings in buildings' (1992). International Energy Agency, Energy Conservation in Buildings and Community Systems Programme, June.

10 Awbi, H.B. (1991). *Ventilation of Buildings*, 1st ed. Chapman and Hall, London.

11 Liddament, M.W. (1986). *International Energy Agency Air Infiltration Calculation Techniques – An Applications Guide*. Air Infiltration and Ventilation Centre (AIVC).

12 Clarke, J., J. Hand and P. Strachan (1990). *ESP– A Building and Plant Energy Simulation System*. ESRU Manual U90/1, February.

13 Grosso, M. (1992). 'Wind pressure distribution around buildings: a parametrical model', *Energy and Buildings*, Vol. 18, pp. 101–131.

14 Hussein, M. and B.E. Lee (1980). 'An investigation of wind forces on three dimensional roughness elements in a simulated atmospheric boundary layer', BS 55. Department of Building Science, University of Sheffield, UK, July.

15 Akins, R.E. and J.E. Cermak (1976). 'Wind pressures on buildings'. CER76-77EA-JEC15. Fluid Dynamic and Diffusion Laboratory, Colorado State University, CO.

16 Technical Note AIVC 29 (Related Project) – 'Fundamentals of the multizone air flow model' (1990). COMIS, Air Infiltration and Ventilation Centre (AIVC), May.

17 BS 5925 (1980). *Code of Practice for Design of Buildings: Ventilation Principles and Designing for Natural Ventilation*. British Standards Institution, London.

18 Santamouris M. (1993). 'NORMA – A method to calculate the thermal performance of passively cooled buildings', *Cooling Load of Buildings*, Vol. 5, Version 1.1. School of Architecture, University College Dublin, Dublin, Ireland.

19 Mahajan, B.M. (1987). 'Measurement of air velocity components of natural convective interzonal airflow', *Journal of Solar Energy Engineering*, Vol. 109, pp. 267–273.

20 Hill, D., A. Kirkpatrick and P. Burns (1986). 'Analysis and measurements of interzonal natural convection heat transfer in buildings', *Journal of Solar Energy Engineering*, Vol. 108, pp. 178–184.

21 Hill, D., A. Kirkpatrick and Burns P. (1985) 'Interzonal natural convection heat transfer in a passive building'. *Heat Transfer in Buildings and Structures*, HTD – Vol 41, *23rd ASME/AICHE National Heat Transfer Conference*, Denver, CO, August, pp. 61–66.

22 Steckler, K.D, H.R. Baum and J.A. Quintiere (1984). 'Five induced flows through room openings – Flow coefficients'. National Bureau of Standards Internal Report NBSIR-83-2801.

23 Prahl, J. and H.W. Emmons (1975). 'Fire induced flow through an opening', *Combustion and Flame*, Vol. 25, pp. 369–385.

24 Boardman, C.R., A. Kirkpatrick and R. Anderson (1989). 'Influence of aperture height and width on interzonal natural convection in a full scale air filled enclosure' *Journal of Solar Energy Engineering*, Vol. 111, pp. 278–285.

25 Brown, W.G. and K.R. Solvason (1962). 'Natural convection through rectangular openings in partitions', *International Journal of Heat and Mass Transfer*, pp. 859–868.

26 Kiel, D.E. (1985). 'Measuring and modelling air flow through doorways'. MSc Thesis, Univeristy of Alberta, Edmonton, Canada.

27 Lane-Serff, G.F., J.F. Linden and J.E. Simpson (1987). 'Transient flow through doorways produced by temperature differences', *Proceedings on Room Ventilation*, Vol. 87, Session 22.

28 Pelletret, R. (1987). 'Les transferts internes en thermique du bâtiment'. CSTB, DPE / 87-500 / RP / JM.

29 Brown, W.G., A.G. Wilson and K.R. Solvason (1963). 'Heat and moisture flow through openings by convection', *ASHRAE Journal*, Vol. 5, No. 9, pp. 49–54.

30 Wray, N.O. and D.D. Weber (1979). 'LASL similarity studies: Part I. Hot zone/cold zone: A quantitative study of natural heat distribution mechanisms in passive solar building', *Proceedings of the 4th National Passive Solar Conference*, Kansas City, Vol. 4, pp. 226–230.

31 Weber, D.D., W.O. Wray and R. Kearney (1979). 'LASL similarity studies, Part II. Similitude modelling of interzone heat transfer by natural convection', *Proceedings of the 4th National Passive Solar Conference*, Kansas City, Vol. 4, pp. 231–234.

32 Cockroft, J.P. (1979). 'Heat transfer and air flow in buildings'. PhD Thesis, University of Glasgow, Scotland.

33 Neymark, J., C.R. Boardman, A. Kirkpatrick and R. Anderson (1989). 'High Rayleigh number natural convection in partically divided air and water filled enclosures', *International Journal of Heat and Mass Transfer*, Vol. 32, No. 9, pp. 1671–1679.

34 Chen, K.S. and P.W. Ko (1991). 'Natural convection in a partially divided rectagular enclosure with an opening in the partition plate and isoflux side walls', *International Journal of Heat and Mass Transfer*, Vol. 34, No. 1, pp. 237–246.

35 Nansteel, M.W and R. Greif (1981). 'Natural convection in undivided and partially divided rectagular enclosures', *Transactions, ASME Journal of Heat and Transfer*, Vol. 103, pp. 623–629.

36 Nansteel, M.W and R. Greif (1983). 'Natural convection heat transfer in complex enclosure at large Prandtl numbers', *ASME Journal of Heat Transfer*, Vol. 105, pp. 912–915.

37 Nansteel, M.W and R. Greif (1984).'An investigation of natural convection in enclosures with two and three dimensional partitions', *International Journal of Heat and Mass Transfer*, Vol. 27, No. 4, pp. 561–571.

38 Liu, N.N. and A. Bejan (1983). 'Natural convection in a partially divided enclosure' *International Journal of Heat and Mass Transfer*, Vol. 26, No. 12, pp. 1867–1878.

39 Bauman, F., A. Gadgil, R. Kammerud, E. Altmayer and M.W. Nansteel (1983). 'Convection heat transfer in buildings. Recent research results' *ASHRAE Transactions*, Vol. 89.

40 Bajorek, S. and J.R. Mand Lloyd (1982). 'Experimental investigation of natural convection in partitioned enclosures', *ASME Journal of Heat Transfer*, Vol. 104, pp. 527–532.

41 Zimmerman, E. and S. Acharya (1987). 'Free convection heat transfer in a partially divided vertical enclosure with conducting end walls', *International Journal of Heat and Mass Transfer*, Vol. 30, No. 2, pp. 319–331.

42 Anderson, R. (1986). 'Natural convection research and solar buildings applications' *Passive Solar Journal*, Vol. 3, No. 1, pp. 33–76.

43 Barakat, S.A. (1987). 'Inter-zone convective heat transfer in buildings: A review', *Journal of Solar Energy Engineering*, Vol. 109, pp. 71–78.

44 Brown, W.G. (1962). 'Natural convection through rectangular openings in partitions – 2', *International Journal of Heat and Mass Transfer*, Vol. 5, pp. 869–879.

45 Esptein, M. (1988). 'Buoyancy-driven exchange flow through small openings in horizontal partitions', *Journal of Heat Transfer*, Vol. 110, pp. 885–893.

46 Allard, F. and Y. Ustumi (1992). 'Airflow through large openings', *Energy and Building*, Vol. 18, pp. 133–145.

47 Zohrabian, A.S., M.R. Mokhtarradeh-Dehgau and A.J. Reynolds (1989). 'Buoyancy-driven airflow in stairwell model with through flow', Brunel University, UK.

48 Santamouris, M., E. Dascalaki, A. Argiriou, C. Helmis and D.N. Asimakopoulos (1992). 'Analysis of heat and mass flow due to temperature difference in a two-zone test cell'. Model Development Subgroup PASCOOL Meeting, 22–24 May, Florence, Italy.

49 Dascalaki, E., P. Droutsa and M. Santamouris (1992). 'Interzonal comparison of five multizone air flow prediction tools', Model Development Subgroup PASCOOL Meeting, 22–24 May, Florence, Italy.

50 Balaras, C. A. (1992). 'The role of thermal mass on the cooling load of buildings. An overview of computational methods and proposed experimental work within PASCOOL', Model Development Subgroup PASCOOL Meeting, 23–27 October, Lyon, France.

51 Santamouris, M. (1988). 'Passive and hybrid cooling projects in Greece', *Proceedings of Building 2000 Workshop*, Barcelona, Spain, pp.155–193.

52 Hoffman, M.E. and M. Gideon (1984). 'Window design practical directions for passive heating and cooling in heavy and light buildings', *Proceedings: Windows in Building Design and Maintenance*, Sweden, pp. 277–286.

53 Goulding, J.R., J. Owen Lewis and T.C. Steemers (1993). *Energy in Architecture. The European Passive Solar Handbook*. Commission of the European Communities.

54 Santamouris, M., A. Argiriou, E. Dascalaki, C. Balaras and A. Gaglia (1994). 'Energy characteristics and savings potential in office buildings', *Solar Energy*, Vol. 52, pp. 59–66.

55 *ASHRAE Handbook* (1985). 'Fundamentals'. ASHRAE, Atlanta.

56 Koos van der Maas, J. and C.-A. Roulet (1991). 'Nighttime ventilation by stack effect', *ASHRAE Technical Data Bulletin*, Vol. 7, No. 1, *Ventilation and Infiltration*, pp. 32–40.

57 Weber, D.D and R. Kearney (1980). 'Natural convection heat transfer through an aperture in passive solar heated buildings', *Proceedings of the 5th Natural Passive Solar Conference*, Amherst, MA, pp. 1037–1041.

58 Shaw, B.H. (1972). 'Heat and mass transfer by natural convection and combined natural and forced air flow through large rectangular openings in a vertical partition', *Proceedings of the International Mechanical Engineering Conference on Heat and Mass Transfer by Combined Forced and Natural Convection*, Manchester, Vol. 819, pp. 31–39.

59 Shaw, B.H. and W. Whyte (1974). 'Air movement through doorways: the influence of temperature and its control by forced air flow', *Building Services Engineering*, Vol. 42, pp. 210–218.

60 Mahajan, B.M. (1987). 'Measurement of interzonal heat and mass transfer by natural convection', *Solar Energy*, Vol. 38, No. 6, pp. 437–446.

61 Balcomb, V.D. and K. Yamaguchi (1983). 'Heat distribution by natural convection', *Proceedings of the 8th Natural Passive Solar Conference*, Santa Fe, New Mexico.

62 Bauman, F., A. Gadgil, R. Kammerud and R. Greif (1980). 'Buoyancy driven convection in a rectangular enclosure experimental results and numerical calculations', ASME Paper No 80-HT-66, Transfer Conference., Orlando, FL.

APPENDIX A
PARAMETRICAL MODEL FOR THE CALCULATION OF THE PRESSURE COEFFICIENT

A case study

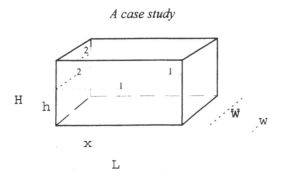

The pressure coefficient is to be calculated for two points of building envelope : point (1) on facade (1) and point (2) on facade (2).

The environmental and geometrical parameters have the following values:

pad = 0.0, rbh = 1, a = 0.22, L = 6 m, H = 3 m, W = 3 m,
x = 3 m, w = 1.5 m, h = 1.5 m, anw(1) = –45°, anw(2) = –135°.

The frontal aspect ratio for facades (1) and (2) is

far(1) = L/H = 2, far(2) = W/H = 1.

The side aspect ratio for facades (1) and (2) is

sar(1) = W/H = 1, sar(2) = L/H = 2.

The relative position of points (1) and (2) is

xl(1) = x/L = 0.5, xl(2) = w/W = 0.5

zh(1) = zh(2) = h/H = 0.5.

SIDE 1 – WINDWARD SIDE

Cp_{ref}(zh = 0.5) = 1.21   (from equation 9.7)

$Cp_{norm0.5}$(a = 0.22) = 0.999   (from Table 9.A1)

$Cp_{norm0.5}$(pad = 0) = 0.979   (from Table 9.A2)

$Cp_{norm0.5,\,0}$(rbh = 1) = 0.957   (from Table 9.A3)

$Cp_{norm0,\,0.5}(far = 2) = 0.874$ (from Table 9.A7, by linear interpolation between $Cp_{norm0,\,0.4}(far = 2)$ and $Cp_{norm0,\,0.6}(far = 2)$)

$Cp_{norm0,\,0.5}(sar = 1) = 1$ (from Table 9.A8)

$Cp_{norm0.5,\,45}(xl = 0.5) = 0.464$ (from Table 9.A6)

$$Cp(1) = Cp_{ref}(zh = 0.5) \times Cp_{norm0.5}(a = 0.22) \times Cp_{norm0.5}(pad = 0)$$
$$\times Cp_{norm0.5,0}(rbh = 1) \times Cp_{norm0,\,0.5}(far = 2) \times Cp_{norm0,\,0.5}(sar = 1)$$
$$\times Cp_{norm0.5,\,45}(xl = 0.5) = 0.459.$$

Thus, $Cp(1) = 0.459.$

SIDE 2 – LEEWARD SIDE

$Cp_{ref}(zh = 0.5) = -0.454$ (from equation 9.7)

$Cp_{norm0.5}(a = 0.22) = 0.954$ (from Table 9.A9)

$Cp_{norm0.5}(pad = 0) = 1.034$ (from Table 9.A10)

$Cp_{norm0.5,\,0}(rbh = 1) = 0.998$ (from Table 9.A11)

$Cp_{norm0.5,\,0}(far = 1) = 1$ (from Table 9.A15)

$Cp_{norm0,\,0.5}(sar = 2) = 0.645$ (from Table 9.A16, by linear interpolation between $Cp_{norm0,\,0.4}(sar = 2)$ and $Cp_{norm0,\,0.6}(sar = 2)$)

$Cp_{norm0.5,\,135}(xl = 0.5) = 1.707$ (from Table 9.A14)

$$Cp(2) = Cp_{ref}(zh = 0.5) \times Cp_{norm0.5}(a = 0.22) \times Cp_{norm0.5}(pad = 0)$$
$$\times Cp_{norm0.5,\,0}(rbh = 1) \times Cp_{norm0,\,0.5}(far = 1) \times Cp_{norm0,\,0.5}(sar = 2)$$
$$\times Cp_{norm0.5,\,135}(xl = 0.5) = -0.492.$$

Thus, $Cp(2) = -0.492.$

Coefficients for the curve-fitting equations

WINDWARD SIDE
Polynomial function for the reference Cp.

$$Cp_{ref}(zh) = -2.381082(zh)^3 + 2.89756(zh)^2 - 0.774649(zh) + 0.745543.$$

Coefficients for the equations of the normalized Cp as a function of environmental and geometrical parameters are given in Tables 9.A1 to 9.A8.

Table 9.A1 Coefficients for the equations of the normalized Cp as a function of terrain roughness: $Cp_{norm_{zh}}(a) = a_0 + a_1(a) + a_2(a)^2$

zh	a_2	a_1	a_0
0.1	−10.820106	+2.312434	+1.014958
0.3	−10.42328	+1.268783	+1.225354
0.5	−8.531746	+0.688492	+1.261468
0.7	−0.939153	−1.691138	+1.417505
0.9	5.10582	−3.350529	+1.489995

Table 9.A2 Coefficients for the equations of the normalized Cp as a function of density of surrounding buildings: $Cp_{norm_{zh}}(pad) = a_0 + a_1(pad) + a_2(pad)^2 + a_3(pad)^3$

zh	a_3	a_2	a_1	a_0
0.0–0.65	−2.14966e−05	+2.37444e−03	−0.089797	+0.979603
0.66–0.75	−1.775637e−05	+2.034996e−03	−0.081741	+0.995399
0.76–0.85	−1.523628e−05	+1.788998e−03	−0.074881	+1.00378
0.86–0.95	−1.571837e−05	+1.693211e−03	−0.06647	+0.994355
0.96–1.0	−1.987115e−05	+1.968606e−03	−0.067063	+0.966038

Table 9.A3 Coefficients for the equations of the normalized Cp as a function of height of surrounding buildings: $Cp_{norm_{zh,pad}}(rbh) = a_0 + a_1(rbh) + a_2(rbh)^2 + a_3(rbh)^3$

zh	pad	a_3	a_2	a_1	a_0
0.07	0.0	0.0	0.0	0.111687	0.848151
	5.5	0.0	0.0	0.303608	0.693641
	12.5	0.0	0.0	0.665827	0.450229
	25.0	−0.354662	1.416299	3.925792	−3.814382
0.20	0.0	0.0	0.0	0.152862	0.78183
	5.5	0.0	0.0	0.35057	0.60962
	12.5	0.0	0.0	0.691757	0.407027
	25.0	0.0	1.534332	−17.32797	14.40045

Table 9.A3 (continued)

zh	pad	a_3	a_2	a_1	a_0
0.50	0.0	0.0	0.0	0.251497	0.705467
	5.5	0.0	0.0	0.661656	0.348851
	12.5	0.0	0.0	1.601127	−0.4244487
	25.0	2.743878	−18.09787	13.731616	2.08857
0.70	0.0	0.0	0.0	0.280233	0.697339
	5.5	0.0	0.0	0.693236	0.3469922
	12.5	0.0	0.0	1.566717	−0.325088
	25.0	−1.2113787	6.301881	4.370901	−6.988637
0.80	0.0	0.0	0.0	0.338131	0.637794
	5.5	0.0	0.0	0.719554	0.349286
	12.5	0.0	0.0	1.373569	−0.175915
	25.5	−0.403791	1.579764	5.205654	−4.533334
0.90	0.0	0.0	0.0	0.436478	0.555708
	5.5	0.0	−0.155809	1.523391	−0.266623
	12.5	0.0	−0.217166	2.2467	−0.855572
	25.0	0.0	−0.733177	6.203364	−3.94136
0.93	0.0	0.0	0.0	0.464299	0.535423
	5.5	0.0	−0.17031	1.579231	−0.294406
	12.5	0.0	−0.235091	2.28368	−0.853961
	25.0	0.0	−0.62338	5.154261	−3.165345

Table 9.A4 Coefficients for the equations of the normalized Cp as a function of frontal aspect ratio far < 1.0:
$$Cp_{norm_{pad.zh}}(far) = a_0 + a_1(far)$$

pad	zh	a_1	a_0
0.0	0.07	0.21	0.79
	0.20	0.166	0.834
	0.40	0.102	0.898
	0.60	0.066	0.934
	0.80	−0.04	1.04
	0.93	−0.292	1.292
5.0	0.07	0.286	0.714
	0.20	0.21	0.79
	0.40	0.148	0.852
	0.60	0.156	0.844
	0.80	0.028	0.972
	0.93	−0.364	1.364

Table 9.A4 (continued)

pad	zh	a_1	a_0
7.5	0.20	0.12	0.88
	0.40	0.054	0.946
	0.60	0.6245004e−17	1.0
	0.80	0.038	0.962
	0.93	−0.352	1.352
10.0	0.07	0.182	0.818
	0.20	0.046	0.954
	0.40	−0.12	1.12
	0.60	−0.166	1.166
	0.80	−0.052	1.052
	0.93	−0.428	1.428
12.5	0.07	0.1	0.9
	0.20	−0.068	1.068
	0.40	−0.058	1.058
	0.60	−0.044	1.044
	0.80	0.032	0.968
	0.93	−0.334	1.334

Table 9.A5 Coefficients for the equations of the normalized Cp as a function of side aspect ratio sar < 1.0: $Cp_{norm_{pad,zh}}(sar) = a_0 + a_1(sar)$

pad	zh	a_1	a_0
0.0	0.07	−0.022	1.022
	0.20	0.056	0.944
	0.40	−0.03	1.03
	0.6	6.245004e−17	0.1
	0.80	−0.02	1.02
	0.93	−0.166	1.166
5.0	0.07	0.172	0.828
	0.20	0.19	0.81
	0.40	0.334	0.666
	0.60	0.438	0.562
	0.80	0.31	0.69
	0.93	−0.09	1.09
7.5	0.07	0.266	0.734
	0.20	0.298	0.702
	0.40	0.46	0.54
	0.60	0.436	0.564
	0.80	0.324	0.676
	0.93	−0.118	1.118

Table 9.A5 (continued)

pad	zh	a_1	a_0
10.0	0.07	0.328	0.672
	0.20	0.318	0.682
	0.40	0.8	0.2
	0.60	0.66	0.334
	0.80	0.206	0.794
	0.93	−0.286	1.286
12.5	0.07	0.75	0.25
	0.20	1.104	−0.104
	0.40	1.428	−0.428
	0.60	1.2	−0.2
	0.80	0.634	0.366
	0.93	6.245004e−17	1.0

Table 9.A6 Coefficients for the equations of the normalized Cp: horizontal distribution versus wind direction: $Cp_{norm_{zh,anw}}(xl) = a_0 + a_1(xl) + a_2(xl)^2 + a_3(xl)^3$

zh	anw (°)	a_3	a_2	a_1	a_0
0.50	0.0	0.0	−3.04662	3.04662	0.268462
	10.0	0.0	−3.142447	2.873329	0.38632
	20.0	0.0	−2.001162	1.398438	0.693916
	30.0	0.0	−1.275862	0.278803	0.935081
	40.0	0.0	−1.058275	−0.01627	0.871259
	50.0	0.0	−0.891626	0.247508	0.428414
	60.0	0.0	−1.560755	1.496049	−0.257573
	70.0	0.0	−1.990676	2.614312	−0.994965
	80.0	0.0	−1.651067	2.530479	−1.359928
	90.0	−5.984848	10.036713	−3.883683	−0.778811
0.70	0.0	0.0	−2.501166	2.501166	0.401189
	10.0	0.0	−2.665435	2.355141	0.523287
	20.0	0.0	−1.674825	1.008462	0.802867
	30.0	0.0	−0.869048	−0.176541	1.051723
	40.0	0.0	−0.635198	−0.467520	0.973357
	50.0	0.0	−0.667077	3.841881e−03	0.485571
	60.0	0.0	−1.415846	1.367316	−0.231142
	70.0	0.0	−2.064103	2.719557	−1.005524
	80.0	0.0	−1.842775	2.788363	−1.37687
	90.0	−4.015152	6.670746	−2.319231	−0.836434

Table 9.A6 (continued)

zh	anw (°)	a_3	a_2	a_1	a_0
0.90	0.0	0.0	−2.456876	2.456876	0.451469
	10.0	0.0	−2.681034	2.335446	0.581156
	20.0	0.0	−1.724942	0.981305	0.888531
	30.0	0.0	−0.832512	−0.270429	1.118564
	40.0	0.0	−0.547786	−0.547786	0.992378
	50.0	0.0	−0.88711	0.279757	0.426546
	60.0	0.0	−1.85509	1.935973	−0.375921
	70.0	0.0	−2.815851	3.659487	−1.236923
	80.0	0.0	−2.449507	3.577449	−1.585214
	90.0	−6.959984	10.745338	−3.502826	−0.877273

Table 9.A7 Coefficients for the equations of the normalized Cp as a function of frontal aspect ratios far > 1.0: $Cp_{norm_{pad,zh}}(far) = [a_1 far + a_2/far + a_3]^{1/2}$

pad	zh	a_1	a_2	a_3
0.0	0.07	−0.070887	0.335565	0.741492
	0.20	−0.061746	0.39232	0.670057
	0.40	−0.071734	0.370249	0.700161
	0.60	−0.075213	0.280472	0.799646
	0.80	−0.081452	0.261036	0.821341
	0.93	−0.05991	0.441293	0.620374
5.0	0.07	−0.625867	−3.31499	4.938818
	0.20	−0.700802	−3.691923	5.39902
	0.40	−0.551417	−2.657088	4.2088561
	0.60	−0.394759	−1.857109	3.243966
	0.80	−0.384892	−1.582766	2.964682
	0.93	−0.471534	−1.938719	3.408053
7.5	0.07	−0.464735	−4.370468	5.827134
	0.20	−0.484764	−4.700937	6.175447
	0.40	−0.357666	−3.421083	4.761667
	0.60	−0.430568	−3.272576	4.686477
	0.80	−0.538978	−3.080677	4.608249
	0.93	−0.295157	−2.106807	3.39147
10.0	0.07	−0.445623	−5.965503	7.414155
	0.20	−0.562911	−8.352512	9.919405
	0.40	−0.303556	−5.104654	6.409214
	0.60	−0.396287	−4.685712	6.096834
	0.80	−0.326486	−3.146084	4.485651
	0.93	−0.491857	−3.607476	5.109896
12.5	0.07	0.39952	−6.357705	6.938206
	0.20	0.560605	−10.512008	10.939653
	0.40	0.460531	−5.146305	5.668398
	0.60	0.052937	−4.346084	5.273574
	0.80	−0.17023	−3.285382	4.448491
	0.93	−0.489256	−4.363034	5.840238

Table 9.A8 Coefficients for the equations of the normalized Cp as a function of side aspect ratio sar > 1.0: $Cp_{norm_{pad,zh}}(sar)=[a_1 sar + a_2/sar + a_3]^{1/2}$

pad	zh	a_1	a_2	a_3
0.0	0.07	0.102648	0.307944	0.589408
	0.20	−0.044242	−0.132726	1.176968
	0.40	−0.02005	−0.06025	1.0802
	0.60	−2.751206e−10	−5.399712e−10	1.0
	0.80	−0.127266	−0.101574	1.22884
	0.93	0.175931	0.527814	0.296255
5.0	0.07	−0.61983	−2.745612	4.364542
	0.20	−0.455586	−2.714454	4.17004
	0.40	0.01539	−1.522998	2.507608
	0.60	8.495999e−03	−1.108008	2.099512
	0.80	0.03363	−0.665862	1.632232
	0.93	−0.83599	−2.639028	4.475018
7.5	0.07	−0.672534	−4.465068	6.137602
	0.20	−0.589638	−4.571604	6.161242
	0.40	0.44127	−2.377428	2.935258
	0.60	0.313214	−2.334822	3.021608
	0.80	0.53643	−1.011222	1.474792
	0.93	−0.32829	−2.984262	4.312552
10.0	0.07	−1.31805	−7.924662	10.242712
	0.20	−2.14576	−11.416512	14.562272
	0.40	0.0608	−6.2016	7.1408
	0.60	0.699422	−3.950934	4.251512
	0.80	0.51795	−2.521878	3.003928
	0.93	−1.627836	−6.191754	8.81959
12.5	0.07	1.15625	−5.8125	5.65625
	0.20	0.811914	−10.848372	11.036458
	0.40	3.144588	−2.954106	0.809518
	0.60	3.525422	−0.048534	−2.476888
	0.80	1.802288	−0.832296	0.030008
	0.93	−0.384444	−4.326666	5.71111

LEEWARD SIDE

Polynomial function for the reference Cp:

$$Cp_{ref}(zh) = -0.079239(zh)^3 + 0.542317(zh)^2 - 0.496769(zh) + 0.331533.$$

Coefficients for the equations of the normalized Cp as a function of environmental and geometrical parameters are given in Tables 9.A9 to 9.A16.

Table 9.A9 Coefficients for the equations of the normalized Cp as a function of terrain roughness: $Cp_{norm_{zh}}(a) = a_0 + a_1(a) + a_2(a)^2$

zh	a_2	a_1	a_0
0.1	−14.368685	4.520431	0.0667639
0.3	−13.490491	4.101437	0.706052
0.5	−8.775919	1.322245	1.088822
0.7	−4.662405	−0.929782	1.395398
0.9	2.382908	−4.837467	1.940878

Table 9.A10 Coefficients for the equations of the normalized Cp as a function of density of surrounding buildings: $Cp_{norm_{zh}}(pad) = a_0 + a_1(pad) + a_2(pad)^2 + a_3(pad)^3 + a_4(pad)^4 + a_5(pad)^5$

zh	a_5	a_4	a_3	a_2	a_1	a_0
0.07	9.118209e−08	−1.050363e−05	3.932533e−04	−4.734698e−03	−0.015304	1.047295
0.20	5.934754e−08	−6.708652e−06	2.340744e−04	−1.943067e−03	−0.031483	1.043295
0.40	5.052791e−08	−5.537346e−06	1.722449e−04	−3.926684e−04	−0.046517	1.034663
0.60	5.595805e−08	−6.121612e−06	1.8897e−04	−3.177597e−04	−0.051446	1.032759
0.80	5.553558e−08	−5.931215e−06	1.719758e−04	3.013991e−04	−0.059971	1.037969
0.93	6.211419e−08	−6.759794e−06	2.024378e−04	1.182029e−04	−0.065764	1.033975

Table 9.A11 Coefficients for the equations of the normalized Cp as a function of height of surrounding buildings: $Cp_{norm_{zh,pad}}(rbh) = a_0 + a_1(rbh) + a_2(rbh)^2$

zh	pad	a_2	a_1	a_0
0.07	0.00	0.0	0.547959	0.465538
	5.00	0.0	0.625743	0.308268
	6.25	0.0	0.859533	0.107587
	12.50	0.0	1.710552	−0.681624
0.20	0.00	0.0	0.473757	0.527487
	5.00	0.0	0.636732	0.294108
	6.25	0.123639	0.432008	0.44064
	12.50	0.080203	1.471191	−0.547645
0.40	0.00	−0.043739	0.599345	0.427938
	5.00	0.054539	0.299349	0.645489
	6.25	0.100427	0.35117	0.483096
	12.50	0.175853	0.568029	0.223168
0.60	0.00	−0.069086	0.793503	0.287883
	5.00	0.029377	0.402683	0.594877
	6.25	0.066082	0.524015	0.376383
	12.50	0.145046	0.567979	0.264523

Table 9.A11 (continued)

zh	pad	a_2	a_1	a_0
0.80	0.00	−0.036376	0.781825	0.258777
	5.00	0.011009	0.55164	0.435343
	6.25	−1.58012e−03	1.127839	−0.084281
	12.50	0.09395	1.114736	−0.111437
0.93	0.00	2.138076e−03	0.655048	0.38064
	5.00	0.03126	0.526521	0.418668
	6.25	0.102993	0.946754	−0.122071
	12.50	0.202243	1.119405	−0.353569

Table 9.A12 Coefficients for the equations of the normalized Cp as a function of frontal aspect ratio far < 1.0:
$$Cp_{norm_{pad,zh}}(far) = a_0 + a_1(far)$$

pad	zh	a_1	a_0
0.0	0.07	0.77	0.23
	0.20	0.694	0.306
	0.40	0.624	0.376
	0.60	0.6	0.4
	0.80	0.666	0.334
	0.93	0.55	0.45
5.0	0.07	1.31	−0.31
	0.20	1.096	−0.096
	0.40	1.048	−0.048
	0.60	1.096	−0.096
	0.80	1.142	−0.142
	0.93	1.042	−0.042
7.5	0.07	1.32	−0.32
	0.20	1.17	−0.17
	0.40	1.142	−0.142
	0.60	1.17	−0.17
	0.80	1.292	−0.292
	0.93	1.25	−0.25
10.0	0.07	1.302	−0.302
	0.20	1.166	−0.166
	0.40	1.12	−0.12
	0.60	1.25	−0.25
	0.80	1.428	−0.428
	0.93	1.428	−0.428

Table 9.A12 (continued)

pad	zh	a_1	a_0
12.5	0.07	1.336	−0.366
	0.20	1.174	−0.174
	0.40	1.166	−0.166
	0.60	1.244	−0.244
	0.80	1.4	−0.4
	0.93	1.412	−0.412

Table 9.A13 Coefficients for the equations of the normalized Cp as a function of side aspect ratio sar < 1.0: $Cp_{norm_{pad,zh}}(sar)=a_0+a_1(sar)$

pad	zh	a_1	a_0
0.0	0.07	−0.462	1.462
	0.20	−0.444	1.444
	0.40	−0.5	1.5
	0.60	−0.6	1.6
	0.80	−0.666	1.666
	0.93	−0.986	1.986
5.0	0.07	0.62	0.38
	0.20	0.484	0.516
	0.40	0.286	0.714
	0.60	0.322	0.678
	0.80	0.358	0.642
	0.93	0.124	0.876
7.5	0.07	0.56	0.44
	0.20	0.416	0.584
	0.40	0.358	0.642
	0.60	0.378	0.622
	0.80	0.416	0.584
	0.93	6.245004e−17	1.0
10.0	0.07	0.418	0.582
	0.20	0.374	0.626
	0.40	0.28	0.72
	0.60	0.334	0.666
	0.80	0.286	0.714
	0.93	0.058	0.942
12.5	0.07	0.586	0.414
	0.20	0.392	0.608
	0.40	0.208	0.792
	0.60	0.088	0.912
	0.80	0.2	0.8
	0.93	−0.118	1.118

Table 9.A14 Coefficients for the equations of the normalized Cp: horizontal distribution versus wind direction: $Cp_{norm_{zh,anw}}(xl)=a_0+a_1(xl)+a_2(xl)^2+a_3(xl)^3+a_4(xl)^4$

zh	anw (°)	a_4	a_3	a_2	a_1	a_0
0.50	90.0	0.0	9.325952	−16.031002	6.08061	2.162909
	110.0	0.0	2.526807	−5.145221	3.28289	1.400238
	130.0	0.0	0.200855	−1.520047	1.734472	1.275364
	160.0	0.0	0.861888	−1.966841	1.561282	0.923007
	180.0	0.0	4.145989e−16	−0.107692	0.107692	0.975846
0.70	90.0	0.0	11.862859	−19.086364	6.79763	2.204853
	110.0	0.0	1.79934	−2.526981	1.326103	1.631755
	130.0	0.0	−0.069542	0.404196	0.124611	1.506259
	160.0	0.0	1.003108	−0.873077	0.398465	1.093671
	180.0	0.0	3.88578e−16	0.449883	−0.449883	1.102028
0.90	90.0	−13.234266	47.482906	−48.637238	13.933178	2.493133
	110.0	−18.269231	38.486402	−24.083741	4.338003	1.973497
	130.0	−9.985431	17.831974	−8.056789	0.346156	1.844014
	160.0	−8.458625	17.902681	−10.191521	1.433689	1.232881
	180.0	−6.555944	13.106061	−7.364394	0.809767	1.244049

Table 9.A15 Coefficients for the equations of the normalized Cp as a function of frontal aspect ratios far > 1.0: $Cp_{norm_{pad,zh}}(far)=[a_1 far+a_2/far+a_3]^{1/2}$

pad	zh	a_1	a_2	a_3
0.0	0.07	0.391319	0.275277	0.305879
	0.20	0.208852	0.045117	0.727577
	0.40	0.176644	0.135403	0.657545
	0.60	0.222872	0.219437	0.5177
	0.80	0.352525	0.51124	0.095033
	0.93	0.409298	0.101415	0.461285
5.0	0.07	0.313066	1.29096	−0.679717
	0.20	0.262845	1.187068	−0.511316
	0.40	0.198393	0.852449	−0.107538
	0.60	0.202255	0.824728	−0.109405
	0.80	0.266436	0.989084	−0.34636
	0.93	0.378433	0.831703	−0.27258
7.5	0.07	0.355636	1.865418	−1.293254
	0.20	0.256393	1.501845	−0.83996
	0.40	0.195066	1.248485	−0.513001
	0.60	0.179345	1.132885	−0.406631
	0.80	0.248347	1.426085	−0.79038
	0.93	0.286457	1.200878	−0.562477

Table 9.A15 (continued)

pad	zh	a_1	a_2	a_3
10.0	0.07	0.162696	1.401255	−0.650645
	0.20	0.14259	1.382313	−0.611037
	0.40	0.072493	1.036706	−0.199349
	0.60	0.062272	0.956828	−0.131138
	0.80	0.116832	1.191314	−0.445541
	0.93	0.111723	0.959598	−0.190495
12.5	0.07	0.187639	1.532033	−0.830662
	0.20	0.113114	1.30869	−0.518821
	0.40	0.090391	1.096843	−0.281639
	0.60	0.058215	0.921987	−0.086177
	0.80	0.138563	1.304438	−0.561468
	0.93	0.115601	1.108345	−0.337801

Table 9.A16 Coefficients for the equations of the normalized Cp as a function of side aspect ratio sar > 1.0: $Cp_{norm_{pad,zh}}(sar) = [a_1 sar + a_2/sar + a_3]^{1/2}$

pad	zh	a_1	a_2	a_3
0.0	0.07	1.549121	4.008955	−4.558076
	0.20	1.293432	3.376296	−3.669728
	0.40	0.818276	2.757414	−2.575691
	0.60	0.622491	2.463733	−2.086225
	0.80	0.431822	2.206986	−1.638808
	0.93	1.15475	3.567738	−3.722488
5.0	0.07	1.234668	3.821814	−4.056482
	0.20	1.086419	3.381557	−3.467976
	0.40	1.110227	3.330677	−3.440903
	0.60	1.248462	3.745386	−3.993848
	0.80	1.158504	3.817008	−3.975512
	0.93	0.924129	3.214321	−3.13845
7.5	0.07	1.6176	4.7352	−5.352801
	0.20	1.405914	4.082196	−4.48811
	0.40	1.39227	4.047642	−4.439912
	0.60	1.446764	4.209078	−4.655842
	0.80	1.541118	4.623354	−5.164472
	0.93	1.395	4.185	−4.58
10.0	0.07	1.728091	5.065453	−5.793544
	0.20	1.675056	4.762584	−5.437641
	0.40	1.632	4.6368	−5.2688
	0.60	1.623354	4.746708	−5.370063
	0.80	2.133661	5.767382	−6.900996
	0.93	2.099225	5.670099	−6.769291

Table 9.A16 (continued)

pad	zh	a_1	a_2	a_3
12.5	0.07	2.249115	6.239501	−7.488376
	0.20	2.121972	5.826368	−6.948252
	0.40	1.99874	5.578012	−6.576709
	0.60	2.373076	6.268238	−7.641063
	0.80	2.133851	5.94792	−7.081692
	0.93	2.204859	6.021059	−7.225708

APPENDIX B
NORMA – A SIMPLIFIED THEORETICAL MODEL
CALCULATION FORMS AND EXAMPLES

Form 1: Form to calculate air flow in naturally ventilated buildings. Single-sided ventilation – openings at the same height

Form 2: Form to calculate air flow in naturally ventilated buildings. Single-sided ventilation – openings at the same height (example)

Form 3: Form to calculate air flow in naturally ventilated buildings. Single-sided ventilation – openings at the same height (example)

Form 4: Form to calculate air flow in naturally ventilated buildings. Single-sided ventilation – openings at different heights – stack effect

Form 5: Form to calculate air flow in naturally ventilated buildings. Single-sided ventilation – openings at different heights – stack effect (example)

Form 6: Form to calculate air flow in naturally ventilated buildings. Cross ventilation

Form 7: Form to calculate air flow in naturally ventilated buildings. Cross ventilation (example)

Form 1:   Form to calculate air flow in naturally ventilated buildings. Single-sided ventilation – openings at the same height

1	Opening height – *h*:	_____ (m)
2	Opening width – *w*:	_____ (m)
3	Give the volume of the ventilated space – *V*:	_____ (m³)
4	Design month:	_____
5	From local weather data determine ambient temperature for design month – DTA:	_____ (C)
6	Specify the design indoor air temperature – TA:	_____ (C)

7   Determine mean temperature – MT:

MT = (DTA + TA)/2   [(Step 5 + Step 6)/2]   _____ (C)

8   Determine mean temperature difference – DT:

DT = TA – DTA. If TA < DTA, then DT = DTA – TA.

[DT = (Step 6 – Step 5. If DT < 0, then Step 5 – Step 6]   _____ (C)

9   Determine temperature ratio – RT:

RT = DT/MT [RT = Step 8/Step 7]   _____

10   Determine the square root of RT: $(RT)^{0.5}$

[(Step 9)$^{0.5}$]   _____

11   Determine $h^{1.5}$: [(Step 1)$^{1.5}$]   _____ (m)

12   Determine air flow rate – *Q*:

$Q = 790 \times h^{1.5} \times w \times (RT)^{0.5}$

[Q = 790 × (Step 11) × (Step 2) × (Step 10)]   _____ (m³ h⁻¹)

If there is more than one opening in the zone, then calculate the air flow Q_i that corresponds to each opening. Calculate the total air-flow rate as the sum of the air-flow rate from each opening, QT.

13   Determine air changes per hour – ACH:

QT/*V*   _____

Form 2: *Form to calculate air flow in naturally ventilated buildings. Single-sided ventilation – openings at the same height (example)*

1 Opening height – h: 1.2 ___ (m)
2 Opening width – w: 1.4 ___ (m)
3 Give the volume of the ventilated space
   – V: 50 ___ (m³)
4 Design month: July ___
5 From local weather data determine
   ambient temperature for design month –
   DTA: 29.4 ___ (C)
6 Specify the design indoor air
   temperature – TA: 27 ___ (C)
7 Determine mean temperature – MT:

   MT = (DTA + TA)/2 [(Step 5 + Step 6)/2]   28.2 ___ (C)

8 Determine mean temperature
   difference – DT:

   DT = TA – DTA. If TA < DTA, then DT = DTA – TA.
   [DT = (Step 6 – Step 5. If DT < 0, then Step 5 – Step 6] 2.4 ___ (C)

9 Determine temperature ratio – RT:

   RT = DT/MT [RT = Step 8/Step 7]   0.085 ___

10 Determine the square root of RT: $(RT)^{0.5}$

   [(Step 9)$^{0.5}$]   0.292 ___

11 Determine $h^{1.5}$: [(Step 1)$^{1.5}$]   1.314 ___ (m)
12 Determine air flow rate – Q:

   $Q = 790 \times h^{1.5} \times w \times (RT)^{0.5}$

   [Q = 790 × (Step 11) × (Step 2) × (Step 10)]   424 ___ (m³ h⁻¹)

If there is more than one opening in the zone, then calculate the air
flow Q, that corresponds to each opening. Calculate the total air flow
rate as the sum of the air flow rate from each opening, QT.

13 Determine air changes per hour – ACH:

   QT/V 8.5 ___

Form 3:   Form to calculate air flow in naturally ventilated buildings. Single-sided ventilation – openings at the same height (example)

		1st	2nd	3rd	Total	
1	Opening height – h:	0.8	1.2	1.3		(m)
2	Opening width – w:	1.0	0.8	1.2		(m)
3	Give the volume of the ventilated space – V:	65				(m³)
4	Design month:	July				
5	From local weather data determine ambient temperature for design month – DTA:	29.4				(C)
6	Specify the design indoor air temperature – TA:	28				(C)
7	Determine mean temperature – MT: $MT = (DTA + TA)/2$ [(Step 5 + Step 6)/2]	28.7				(C)
8	Determine mean temperature difference – DT: $DT = TA - DTA$. If $TA < DTA$, then $DT = DTA - TA$. [DT = (Step 6 – Step 5. If DT < 0, then Step 5 – Step 6]	1.4				(C)
9	Determine temperature ratio – RT: $RT = DT/MT$ [RT = Step 8/Step 7]	0.049				
10	Determine the square root of RT: $(RT)^{0.5}$ [(Step 9)$^{0.5}$]	0.22				
11	Determine $h^{1.5}$: [(Step 1)$^{1.5}$]	0.715	1.31	1.48		(m)
12	Determine air flow rate – Q: $Q = 790 \times h^{1.5} \times w \times (RT)^{0.5}$ [$Q = 790 \times$ (Step 11) × (Step 2) × (Step 10)]	124.3	182.1	308.6	615	(m³ h⁻¹)
	If there is more than one opening in the zone, then calculate the air flow Q_i that corresponds to each opening. Calculate the total air flow rate as the sum of the air flow rate from each opening, QT.					
13	Determine air changes per hour – ACH: QT/V				9.5	

*Form 4:   Form to calculate air flow in naturally ventilated buildings. Single-sided
ventilation – openings at different heights – stack effect*

1	Give the area of the lower openings – A_1:	_____ (m²)
2	Give the area of the higher openings – A_2:	_____ (m²)
3	Give the volume of the ventilated space – V:	_____ (m³)
4	Determine the total area of the openings – A: $A = A_1 + A_2$ [(Step 1) + (Step 2)]	_____ (m²)
5	Give the vertical distance between the lower and higher openings – H:	_____ (m)
6	Design month:	_____
7	From local weather data determine ambient temperature for design month – DTA:	_____ (C)
8	Specify the design indoor air temperature – TA:	_____ (C)
9	Determine mean temperature – MT: MT = (DTA + TA)/2 [(Step 7 + Step 8)/2]	_____ (C)
10	Determine mean temperature difference – DT: DT = TA – DTA. If TA < DTA, then DT = DTA – TA. [(Step 8 – Step 7. If DT < 0, then Step 7 – Step 8]	_____ (C)
11	Determine temperature ratio – RT: RT = DT/MT [Step 10/Step 9]	_____
12	Determine the square root of RT: $(RT)^{0.5}$ [(Step 11)$^{0.5}$]	_____
13	Determine $H^{0.5}$: [(Step 5)$^{0.5}$]	_____
14	From Figure 9.19 determine the K correction factor	_____
15	Determine air flow rate – Q: $1590 \times H^{0.5} \times A \times K \times (RT)^{0.5}$ [1590 × (Step 13) × (Step 4) × (Step 14) × (Step 12)]	_____ (m³ h⁻¹)
16	Determine air changes per hour – ACH: Q/V [(Step 15)/(Step 3)]	_____

Form 5:   Form to calculate air flow in naturally ventilated buildings. Single-sided ventilation – openings at different heights – stack effect (example)

1   Give the area of the lower openings –
    A_1:   5____   (m²)
2   Give the area of the higher openings –
    A_2:   7____   (m²)
3   Give the volume of the ventilated space
    – V: 90____   (m³)
4   Determine the total area of the openings
    – A: $A = A_1 + A_2$ [(Step 1) + (Step 2)]   12____   (m²)
5   Give the vertical distance between the
    lower and higher openings – H: 2.5____   (m)
6   Design month: July____
7   From local weather data determine
    ambient temperature for design month –
    DTA: 29____   (C)
8   Specify the design indoor air
    temperature – TA: 26____   (C)
9   Determine mean temperature – MT:
    MT = (DTA + TA)/2 [(Step 7 + Step 8)/2]   27.5____   (C)
10 Determine mean temperature
    difference – DT:
    DT = TA – DTA. If TA < DTA, then DT = DTA – TA.
    [(Step 8 – Step 7. If DT < 0, then Step 7 – Step 8]   3____   (C)
11 Determine temperature ratio – RT:
    RT = DT/MT [Step 10/Step 9]   0.109____
12 Determine the square root of RT: $(RT)^{0.5}$
    [(Step 11)$^{0.5}$]   0.33____
13 Determine $H^{0.5}$: [(Step 5)$^{0.5}$]   1.58____
14 From Figure 9.19 determine the K
    correction factor   0.7____
15 Determine air flow rate – Q:
    $1590 \times H^{0.5} \times A \times K \times (RT)^{0.5}$
    [1590 × (Step 13) × (Step 4) × (Step 14) × (Step 12)]   6934____   (m³ h⁻¹)
16 Determine air changes per hour – ACH:
    Q/V [(Step 15)/(Step 3)]   77.3____

Form 6:   Form to calculate air flow in naturally ventilated buildings. Cross ventilation

1   Design month:

2   From weather data tables determine wind speed (WS) and wind direction (WD) for the design month
     WS (m s^{-1}):
     WD (for example, N for north, SE for south-east, etc.):

3   From local weather data determine ambient temperature for design month – DTA (C):

4   Design indoor air temperature – TA (C):

5   Determine mean temperature difference – DT:
     DT = TA – DTA. If TA < DTA, then DT = DTA – TA.
     [(Step 4 – Step 3. If DT < 0, then Step 3 – Step 4]

6   Determine mean temperature – MT: MT=(DTA + TA)/2
     [(Step 4 + Step 3)/2]

7   Determine temperature ratio – RT: RT = DT/MT [Step 5/Step 6]

8   Determine the square root of RT: (RT)$^{0.5}$ [(Step 7)$^{0.5}$]

9   From prevailing wind direction and building orientation determine the incidence angle on the windward wall (degrees)

10   From tables in Appendix D determine the wind pressure coefficient on the windward facade – Cp(1)

11   From tables in Appendix D determine the wind pressure coefficient on the leeward facade – Cp(2)

12   Determine the difference of wind pressure coefficents – DCP:
     DCP = Cp(1) – Cp(2) [Step 10 – Step 11]

13   Determine the square root of DCP – SDCP: DCP$^{0.5}$ [(Step 12)$^{0.5}$]

14   Building conditioned volume – V (m^3):

15   Give the total surface area of the windward facade openings – A_1 (m^2):

16   Give the total surface area of the leeward facade openings – A_2 (m^2):

17   Give the vertical distance between the leeward and windward openings – H:

18   Give the total surface area of all openings – AC: $A_1 + A_2$
     [(Step 15) + (Step 16)]

19   Determine the square of A_1: $(A_1)^2$ [(Step 15)^2]

20   Determine the square of A_2: $(A_2)^2$ [(Step 16)^2]

21   Determine $A = (A_1)^2 + (A_2)^2$ [(Step 19) + (Step 20)]

22   Determine B, the square root of A: $B = (A)^{0.5}$ [(Step 21)$^{0.5}$]

23   Determine $C = 1/B$: [1/(Step 22)]

Form 6 (continued)

24   Determine the ratio $D = A_1/A_2$ [(Step 15)/(Step 16)]   _____

25   Determine h, the square root of H: [(Step 17)$^{0.5}$]   _____

26   From the value of D determine the coefficients K_1, K_2:

For $0 < D \leq 1$ $K_1 = 0.45$, $K_2 = -1.02$; For $1 < D \leq 2$ $K_1 = 0.42$, $K_2 = -1.07$

(a) K_1:   _____

(b) K_2:   _____

27   Determine $E = E^{K_2}$ [(Step 23)$^{(\text{Step 26(b)})}$]   _____

28   Determine the air flow Q_w, due to wind (m^3 h^{-1}):

$K_1 \times E \times$ SDCP $\times$ WS $\times$ 3600 [(Step 26(a)) $\times$ (Step 27) $\times$ (Step 13) $\times$ (Step 2) $\times$ 3600]   _____

29   From Figure 9.19 determine the coefficient K: _____

30   Determine air flow due to stack effect Q_s (m^3 h^{-1}):

1590 $\times K \times$ AC $\times$ SRT $\times h$ [1590 $\times$ (Step 29) $\times$ (Step 18) $\times$ (Step 8) $\times$ (Step 25)]   _____

31   Determine $Q_w(1)$, the square of Q_w: $(Q_w)^2$ [(Step 28)^2]   _____

32   Determine $Q_s(1)$, the square of Q_s: $(Q_s)^2$ [(Step 30)^2]   _____

33   Determine $Q_t(1) = Q_w(1) + Q_s(1)$: [Step 32) + (Step 31)]   _____

34   Determine total air flow – QT = : $(Q_t(1))^{0.5}$ [(Step 33)$^{0.5}$]   _____

35   Determine air changes per hour – ACH: QT/V [(Step 34)/(Step 14)]   _____

*Form 7: Form to calculate air flow in naturally ventilated buildings. Cross
ventilation (example)*

1	Design month:	July
2	From weather data tables determine wind speed (WS) and wind direction (WD) for the design month	
	WS (m s^{-1}):	2.2
	WD (for example, N for north, SE for south-east, etc.):	NE
3	From local weather data determine ambient temperature for design month – DTA (C):	32
4	Design indoor air temperature – TA (C):	27
5	Determine mean temperature difference – DT:	
	DT = TA – DTA. If TA < DTA, then DT = DTA – TA.	
	[(Step 4 – Step 3. If DT < 0, then Step 3 – Step 4]	5
6	Determine mean temperature – MT: MT=(DTA + TA)/2	
	[(Step 4 + Step 3)/2]	29.5
7	Determine temperature ratio – RT: RT = DT/MT [Step 5/Step 6]	0.17
8	Determine the square root of RT: (RT)$^{0.5}$ [(Step 7)$^{0.5}$]	0.412
9	From prevailing wind direction and building orientation determine the incidence angle on the windward wall (degrees)	0.0
10	From tables in Appendix D determine the wind pressure coefficient on the windward facade – Cp(1)	0.7
11	From tables in Appendix D determine the wind pressure coefficient on the leeward facade – Cp(2)	-0.5
12	Determine the difference of wind pressure coefficents – DCP:	
	DCP = Cp(1) – Cp(2) [Step 10 – Step 11]	1.2
13	Determine the square root of DCP – SDCP: DCP$^{0.5}$ [(Step 12)$^{0.5}$]	1.09
14	Building conditioned volume – V (m^3):	500
15	Give the total surface area of the windward facade openings – A_1 (m^2):	6
16	Give the total surface area of the leeward facade openings – A_2 (m^2):	4
17	Give the vertical distance between the leeward and windward openings – H:	5
18	Give the total surface area of all openings – AC: $A_1 + A_2$	
	[(Step 15) + (Step 16)]	10
19	Determine the square of A_1: $(A_1)^2$ [(Step 15)^2]	36
20	Determine the square of A_2: $(A_2)^2$ [(Step 16)^2]	16
21	Determine $A = (A_1)^2 + (A_2)^2$ [(Step 19) + (Step 20)]	52
22	Determine B, the square root of A: $B = (A)^{0.5}$ [(Step 21)$^{0.5}$]	7.2
23	Determine $C = 1/B$: [1/(Step 22)]	0.138

Form 7 (continued)

24   Determine the ratio $D = A_1/A_2$ [(Step 15)/(Step 16)]   1.5

25   Determine h, the square root of H: [(Step 17)$^{0.5}$]   2.24

26   From the value of D determine the coefficients K_1, K_2:

     For $0 < D \le 1$ $K_1 = 0.45$, $K_2 = -1.02$; For $1 < D \le 2$ $K_1 = 0.42$, $K_2 = -1.07$

     (a) K_1:   0.42

     (b) K_2:   -1.07

27   Determine $E = E^{K_2}$ [(Step 23)$^{\text{(Step 26(b))}}$]   8.32

28   Determine the air flow Q_w, due to wind (m^3 h^{-1}):

     $K_1 \times E \times$ SDCP $\times$ WS $\times$ 3600 [(Step 26(a)) $\times$ (Step 27) $\times$ (Step 13) $\times$ (Step 2) $\times$ 3600]   30166

29   From Figure 9.19 determine the coefficient K: 0.7

30   Determine air flow due to stack effect Q_s (m^3 h^{-1}):

     $1590 \times K \times$ AC $\times$ SRT $\times h$ [1590 $\times$ (Step 29) $\times$ (Step 18) $\times$ (Step 8) $\times$ (Step 25)]   10271

31   Determine $Q_w(1)$, the square of Q_w: $(Q_w)^2$ [(Step 28)^2] 0.909×10^9

32   Determine $Q_s(1)$, the square of Q_s: $(Q_s)^2$ [(Step 30)^2] 0.105×10^9

33   Determine $Q_t(1) = Q_w(1) + Q_s(1)$: [Step 32) + (Step 31)] 1.014×10^9

34   Determine total air flow – QT = : $(Q_t(1))^{0.5}$ [(Step 33)$^{0.5}$]   31843

35   Determine air changes per hour – ACH: QT/V [(Step 34)/(Step 14)]   63.7

APPENDIX C

SUMMARY OF INTER-ZONE NATURAL CONVECTIVE HEAT TRANSFER IN BUILDINGS

Table 9.C1 Summary of inter-zone natural convective heat transfer in buildings

Source	Experimental set-up/Fluid	Aspect ratio (height/length)	A_p (Y_m/H)	Temperature-difference location	Gr	Pr	Characteristic length	Coefficient C
Brown and Solvason [25]	Large test chambers/Air	0.44	0.062–0.125	Air near partition	10^6-10^8	0.7	Y_m	0.2–0.33
Brown et al [29]	Large test chambers/Air	0.44	0.062–0.125	Air near partition	10^6-10^8 $\geq 10^8$	0.7	Y_m	–
Shaw [58]	Full-size rooms /Air	–	0.84	Air at opening top and bottom	$10^8-1.3\times10^{10}$	0.7	Y_m	–
Shaw and Whyte [59]	Full-size rooms/ Air	–	0.84	Centre room air	$10^8-1.3\times10^{10}$	0.7	Y_m	ΔT dependent 0.22 $3 \leq \Delta T \leq 10$
Weber et al. [31]	Model/Freon	0.3	0.82–1.0	Average air	10^9-10^{10}	0.7	H	ΔT dependent 0.27 $\Delta T \leq 10$
Weber and Keaney [57]	Model/Freon	0.3	0.46–1.0	1. AVG at partition 2. AVG near partition	10^7-10^9	0.7	Y_m	0.29–0.41
Mahajan [19]	Test building/Air	0.35	0.75	Centre room air	$\sim 10^{10}$	0.7	Y_m	$C_0 = 0.58/3$, $C_1 = 0.48/3$, $C = (0.4 + 0.0075\,\Delta T)/3$
Kiel and Wilson [56]	Real buildings	–	–	–	10^6-10^{10}	–	Y_m	–
Mahajan B.M. [60] Riffat [2]	Test building/Air Real building	0.35 –	0.75 –	Centre room air Centre room air	$\sim 10^{10}$ $3\times10^8-$ 3×10^{10}	0.7 0.7	Y_m Y_m	0.45/3 $(0.0835[\Delta T/T]^{-0.313})/3$
Cockroft [32]	–	–	–	–	–	–	H	–

Table 9.C1 (continued)

Source	Experimental set-up/Fluid	Aspect ratio (height/length)	A_p (Y_m/H)	Temperature-difference location	Gr	Pr	Characteristic length	Coefficient C
Hill et al. [20]	Test building/Air	–	–	Mean zone temperature	10^9–10^{10}	0.7	Y_m	$C_d[1/2+(\rho_1/\rho_1)^{2/3}/2]^{3/2}$
Balcomb and Yamaguchi [61]	Real building/Air	0.7	0.058	Mean zone temperature	10^9–10^{10}	0.7	Y_m	–
Pelletret et al. [5]	Test building	0.5	0.66	Mean zone temperature	10^8–10^{10}	0.7	Y_m	0.26
Darliel and Lane-Serff [6]	Test box	0.35	–	Mean zone temperature	10^6–10^9	3–4	Y_m	0.103
Limam et al. [7]	Test room	0.5	0.6	Mean zone temperature	10^9–10^{11}	0.7	Y_m	0.22–0.24
Khodr Mneimne [8]	Full-scale test	–	–	Mean zone temperature	10^8–10^{10}	0.7	Y_m	0.29
Pelletret et al. [5]	Test building	0.5	0.66	Mean zone temperature	10^8–10^9	0.7	Y_m	0.14
Bauman et al. [62]	2D model/Water	0.5	0.55–1	Hot and cold surface	$Ra_L = 1.6 \times 10^6$ – 5.4×10^{10}	2.6–6.8	Length of 2 zones	–
Nansteel and Greif [35]	2D model/Water	0.5	0.55–1	Hot and cold surface	$Ra_L = 2.3 \times 10^{10}$ – 1.1×10^{11}	3–4.3	Length of 2 zones	–
Liu and Bejan [38]	3D model/Water	0.305	0.25–1	Hot and cold surface	$Ra_{y_m} = 10^9$ – 10^{10}	3–4	Y_m	–
Nansteel and Greif [35]	2D model/Silicon oil	0.5	0.25–1	Hot and cold surface	$Ra_L = 1.55 \times 10^9$ – 5.9×10^9	620–910	Length of 2 zones	–
Nansteel and Greif [36]	3D model/Water	0.5	0.25–1	Hot and cold surface	$Ra_L = 2.4 \times 10^{10}$ – 1.1×10^{11}	3–4.3	Length	–

Table 9.C1 (continued)

Source	Experimental set-up/Fluid	Aspect ratio (height/length)	A_ρ (Y_m/H)	Temperature-difference location	Gr	Pr	Characteristic length	Coefficient C
Bauman et al. [39]	Summary of LBL Work/Water–Air	0.5	0.25–1	Hot and cold surface	$Ra_{Ym} = 1.5 \times 10^9 - 6 \times 10^9$	0.7	Y_m	–
Bajorek and Lloyd [40]	Air	1	0.5	Hot and cold surface	$\sim 10^6$	0.7	Y_m	–
Boardman et al. [24]	Test cell /Air	1	0.125–1	Hot and cold surface	$Ra_w = 5 \times 10^{11} - 5 \times 10^{12}$	0.7	H	–
Neymark et al. [33]	Test cell /Air	1	0.125–1	Hot and cold surface	$5 \times 10^{11} - 5 \times 10^{12}$	0.7	H	–

APPENDIX D

WIND PRESSURE COEFFICIENT DATA [11]

Low-rise buildings (up to three storeys)

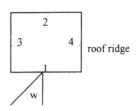

Length-to-width ratio: 1:1
Shielding condition: Exposed
Wind speed reference level: Building height

Table 9.D1 Wind pressure coefficient data

Location		Wind angle (°)							
		0	45	90	135	180	225	270	315
Face 1		0.7	0.35	−0.5	−0.4	−0.2	−0.4	−0.5	0.35
Face 2		−0.2	−0.4	−0.5	0.35	0.7	0.35	−0.5	−0.4
Face 3		−0.5	0.35	0.7	0.35	−0.5	−0.4	−0.2	−0.4
Face 4		−0.5	−0.4	−0.2	−0.4	−0.5	0.35	0.7	0.35
Roof	Front	−0.8	−0.7	−0.6	−0.5	−0.4	−0.5	−0.6	−0.7
(< 10° pitch)	Rear	−0.4	−0.5	−0.6	−0.7	−0.8	−0.7	−0.7	−0.5
	Average	−0.6	−0.6	−0.6	−0.6	−0.6	−0.6	−0.6	−0.6
Roof	Front	−0.4	−0.5	−0.6	−0.5	−0.4	−0.5	−0.6	−0.5
(11–30° pitch)	Rear	−0.4	−0.5	−0.6	−0.5	−0.4	−0.5	−0.6	−0.5
	Average	−0.4	−0.5	−0.6	−0.5	−0.4	−0.5	−0.6	−0.5
Roof	Front	0.3	−0.4	−0.6	−0.4	−0.5	−0.4	−0.6	−0.4
(> 30° pitch)	Rear	−0.5	−0.4	−0.6	−0.4	0.3	−0.4	−0.6	−0.4
	Average	−0.1	−0.4	−0.6	−0.4	−0.1	−0.4	−0.6	−0.4

Low-rise buildings (up to three storeys)

Length to width ratio: 1:1

Shielding condition: Surrounded by obstructions equivalent to half the height of the building

Wind speed reference level: Building height

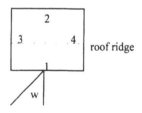

Table 9.D2 Wind pressure coefficient data

		Wind angle (°)							
Location		**0**	**45**	**90**	**135**	**180**	**225**	**270**	**315**
Face 1		0.4	0.1	−0.3	−0.35	−0.2	−0.35	−0.3	−0.1
Face 2		−0.2	−0.35	−0.3	0.1	0.4	0.1	−0.3	−0.35
Face 3		−0.3	0.1	0.4	0.1	−0.3	−0.35	−0.2	−0.35
Face 4		−0.3	−0.35	−0.2	−0.35	−0.3	0.1	0.4	0.1
Roof	Front	−0.6	−0.5	−0.4	−0.5	−0.6	−0.5	−0.4	−0.5
(< 10° pitch)	Rear	−0.6	−0.5	−0.4	−0.5	−0.6	−0.5	−0.4	−0.5
	Average	−0.6	−0.5	−0.4	−0.5	−0.6	−0.5	−0.4	−0.5
Roof	Front	−0.35	−0.45	−0.55	−0.45	−0.35	−0.45	−0.55	−0.45
(11–30° pitch)	Rear	−0.35	−0.45	−0.55	−0.45	−0.35	−0.45	−0.55	−0.45
	Average	−0.35	−0.45	−0.55	−0.45	−0.35	−0.45	−0.55	−0.45
Roof	Front	0.3	−0.5	−0.6	−0.5	−0.5	−0.5	−0.6	−0.5
(> 30° pitch)	Rear	−0.5	−0.5	−0.6	−0.5	0.3	−0.5	−0.6	−0.5
	Average	−0.1	−0.5	−0.6	−0.5	−0.1	−0.5	−0.6	−0.5

Low-rise buildings (up to three storeys)

Length to width ratio: 1:1
Shielding condition: Surrounded by obstructions
equivalent to the height of
the building
Wind speed reference level: Building height

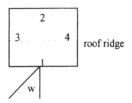

Table 9.D3 Wind pressure coefficient data

Location		Wind angle (°)							
		0	45	90	135	180	225	270	315
Face 1		0.2	0.05	−0.25	−0.3	−0.25	−0.3	−0.25	0.05
Face 2		−0.25	−0.3	−0.25	0.05	0.2	0.05	−0.25	−0.3
Face 3		−0.25	0.05	0.2	0.05	−0.25	−0.3	−0.25	−0.3
Face 4		−0.25	−0.3	−0.25	−0.3	−0.25	0.05	0.2	0.05
Roof	Front	−0.5	−0.5	−0.4	−0.5	−0.5	−0.5	−0.4	−0.5
(< 10° pitch)	Rear	−0.5	−0.5	−0.4	−0.5	−0.5	−0.5	−0.4	−0.5
	Average	−0.5	−0.5	−0.4	−0.5	−0.5	−0.5	−0.4	−0.5
Roof	Front	−0.3	−0.4	−0.5	−0.4	−0.3	−0.4	−0.5	−0.4
(11–30° pitch)	Rear	−0.3	−0.4	−0.5	−0.4	−0.3	−0.4	−0.5	−0.4
	Average	−0.3	−0.4	−0.5	−0.4	−0.3	−0.4	−0.5	−0.4
Roof	Front	0.25	−0.3	−0.5	−0.3	−0.4	−0.3	−0.5	−0.3
(> 30° pitch)	Rear	−0.4	−0.3	−0.5	−0.3	0.25	−0.3	−0.5	−0.3
	Average	−0.08	−0.3	−0.5	−0.3	−0.08	−0.3	−0.5	−0.3

Low-rise buildings (up to three storeys)

Length to width ratio: 2:1
Shielding condition: Exposed
Wind speed reference level: Building height

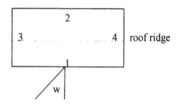

Table 9.D4 Wind pressure coefficient data

Location						Wind angle (°)			
		0	**45**	**90**	**135**	**180**	**225**	**270**	**315**
Face 1		0.5	0.25	−0.5	−0.8	−0.7	−0.8	−0.5	0.25
Face 2		−0.7	−0.8	−0.5	0.25	0.5	0.25	−0.5	−0.8
Face 3		−0.9	0.2	0.6	0.2	−0.9	−0.6	−0.35	−0.6
Face 4		−0.9	−0.6	−0.35	−0.6	−0.9	0.2	0.6	0.2
Roof	Front	−0.7	−0.7	−0.8	−0.7	−0.7	−0.7	−0.8	−0.7
(< 10° pitch)	Rear	−0.7	−0.7	−0.8	−0.7	−0.7	−0.7	−0.8	−0.7
	Average	−0.7	−0.7	−0.8	−0.7	−0.7	−0.7	−0.8	−0.7
Roof	Front	−0.7	−0.7	−0.7	−0.6	−0.5	−0.6	−0.7	−0.7
(11–30° pitch)	Rear	−0.5	−0.6	−0.7	−0.7	−0.7	−0.7	−0.7	−0.6
	Average	−0.6	−0.65	−0.7	−0.65	−0.6	−0.65	−0.7	−0.65
Roof	Front	0.25	0	−0.6	−0.9	−0.8	−0.9	−0.6	0
(> 30° pitch)	Rear	−0.8	−0.9	−0.6	0	0.25	0	−0.6	−0.9
	Average	−0.18	−0.45	−0.6	−0.45	−0.18	−0.45	−0.6	−0.45

Low-rise buildings (up to three storeys)

Length to width ratio: 2:1
Shielding condition: Surrounded by
                                obstructions
                                equivalent to half
                                the height of the
                                building
Wind speed reference level: Building height

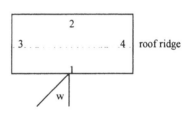

Table 9.D5 Wind pressure coefficient data

Location		0	45	90	135	180	225	270	315
					Wind angle (°)				
Face 1		0.25	0.06	−0.35	−0.6	−0.5	−0.6	−0.35	0.06
Face 2		−0.5	−0.6	−0.35	0.06	0.25	0.06	−0.35	−0.6
Face 3		−0.6	0.2	0.4	0.2	−0.6	−0.5	−0.3	−0.5
Face 4		−0.6	−0.5	−0.3	−0.5	−0.6	0.2	0.4	0.2
Roof	Front	−0.6	−0.6	−0.6	−0.6	−0.6	−0.6	−0.6	−0.6
(< 10° pitch)	Rear	−0.6	−0.6	−0.6	−0.6	−0.6	−0.6	−0.6	−0.6
	Average		−0.6	−0.6	−0.6	−0.6	−0.6	−0.6	−0.6
Roof	Front	−0.6	−0.6	−0.55	−0.55	−0.45	−0.55	−0.55	−0.6
(11–30° pitch)	Rear	−0.45	−0.55	−0.55	−0.6	−0.6	−0.6	−0.55	−0.55
	Average	−0.5	−0.6	−0.55	−0.6	−0.5	−0.6	−0.55	−0.6
Roof	Front	0.15	−0.08	−0.40	−0.75	−0.6	−0.75	−0.40	−0.08
(> 30° pitch)	Rear	−0.6	−0.75	−0.40	−0.08	−0.15	−0.08	−0.4	−0.75
	Average	−0.2	−0.40	−0.40	−0.40	−0.20	−0.40	−0.40	−0.40

Low-rise buildings (up to three storeys)

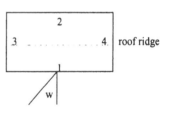

Length to width ratio: 2:1
Shielding condition: Surrounded by
obstructions
equivalent to the
height of the building
Wind speed reference level: Building height

Table 9.D6 Wind pressure coefficient data

Location		Wind angle (°)							
		0	**45**	**90**	**135**	**180**	**225**	**270**	**315**
Face 1		0.06	−0.12	−0.2	−0.38	−0.3	−0.38	−0.2	−0.12
Face 2		−0.3	−0.38	−0.2	−0.12	0.06	−0.12	−0.2	−0.38
Face 3		−0.3	0.15	0.18	0.15	−0.3	−0.32	−0.2	−0.32
Face 4		−0.3	−0.32	−0.2	−0.32	−0.3	0.15	0.18	0.15
Roof	Front	−0.49	−0.46	−0.41	−0.46	−0.49	−0.46	−0.41	−0.46
(< 10° pitch)	Rear	−0.49	−0.46	−0.41	−0.46	−0.49	−0.46	−0.41	−0.46
	Average	−0.49	−0.46	−0.41	−0.46	−0.49	−0.46	−0.41	−0.46
Roof	Front	−0.49	−0.46	−0.41	−0.46	−0.4	−0.46	−0.41	−0.46
(11–30° pitch)	Rear	−0.4	−0.46	−0.41	−0.46	−0.49	−0.46	−0.41	−0.46
	Average	−0.45	−0.46	−0.41	−0.46	−0.45	−0.46	−0.41	−0.46
Roof	Front	0.06	−0.15	−0.23	−0.6	−0.42	−0.6	−0.23	−0.15
(> 30° pitch)	Rear	−0.42	−0.6	−0.23	−0.15	−0.06	−0.15	−0.23	−0.6
	Average	−0.18	−0.4	−0.23	−0.4	−0.18	−0.4	−0.23	−0.4

10

Solar control

DESCRIPTION OF SOLAR GEOMETRY

The earth orbits around the sun once a year and it revolves on its polar axis once every 24 hours. The earth's movement around the sun follows an elliptical path, with its polar axis always inclined at the same direction pointing towards the distant North Star, at an angle of 23.45° with respect to the plane of the orbit around the sun (Figure 10.1). In the Northern hemisphere, on two days of the year, 21 March (or 22 March in a leap year, but for simplicity the 21st is usually referred to), *Spring or Vernal Equinox*, and 21 September, *Fall or Autumnal Equinox*, the earth's axis is perpendicular to the sun's rays and the sun rises and sets due east and west respectively. During these days, the sun stands directly over the Equator and everywhere on the earth, except at the poles, day and night are exactly 12 hours long. The Northern hemisphere is used as reference for the following descriptions but because of the symmetrical movement of the sun relative to the Equator, a reproduction of the same phenomena occurs in the Southern hemisphere six months later. At noon on 21 December, *Winter Solstice*, and on 21 June, *Summer Solstice*, the sun stands directly overhead the Southern (Tropic of Capricon) and Northern (Tropic of Cancer) Tropics respectively. The sun gradually reaches its highest position before the summer solstice, it stops on the solstice and then it begins to fall, reaching its lowest position just before the winter solstice. Thus, the summer solstice is associated with the longest day of the year and the winter solstice with the shortest day.

The angular position of the sun, at solar noon, relative to the plane passing through the earth's equator is called the *solar declination* (Figure 10.1). The declination angle ranges from 23.45° on 21 June to –23.45° on 21 December. At the Spring and Autumnal Equinoxes it is 0°, as the sun is over the Equator at solar noon on these two days.

It is easier to understand the angles made by the sun if the earth's orbit is considered as circular with the sun at its centre and it is assumed that the sun's rays are parallel everywhere on the earth. The second assumption is justified by the (much bigger) size of the sun compared to the earth and the immense distance between sun and earth (1.495×10^{11} m ± 1.7%).

The eccentricity of the earth's orbit around the sun and the earth's inclination results in differences between the *solar time* (ST), measured by the sun's position in the sky, and local clock time, measured by conventional clocks. Solar noon is defined as the instant when the sun is exactly over the south–north axis (to the south in the Northern hemisphere and to the north in the Southern hemisphere) and it has its

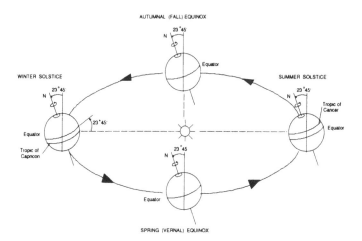

Figure 10.1 Sun–earth movement

highest position in the sky for the day. In most places, solar time differs from local clock time but it can be converted from one to the other through a simple expression, which is described later. The time of the day expressed in terms of an angle of the sun's motion relative to solar noon, is referred as *solar hour angle* (ω). This is measured in degrees, on a plane perpendicular to the earth's axis, and, by convention, it is taken as negative before solar noon and positive after solar noon. As it takes 24 hours for the earth to revolve on its axis (360°), it revolves 15° (360/24) every hour.

At any place on the earth, the angular position of the sun changes with time during the day and from one day to another (Figure 10.2). For a specific place, the angular location of the sun can be specified at any instant by two angles:

- the solar altitude (α_s) and
- the solar azimuth (γ_s).

The solar altitude (α_s) is the angle measured above the horizon in a vertical plane passing through the sun, and it ranges from 0° to 90° (Figure 10.2). Its supplementary angle in the vertical plane is known as the zenith angle (θ_z). The solar azimuth (γ_s) is measured in the horizontal plane and is the angle between the vertical plane passing through the sun and the north axis, ranging from 0 to 360° clockwise from the north [1]; alternatively [2, 3] it can be measured in the Northern hemisphere from the south, ranging from 0° to 180°, positive in the clockwise direction (west) and negative in the anti-clockwise one (east). If the latitude (φ) of a place is known, the sun's altitude and azimuth for any day and time of the year can be defined using simple trigonometric formulae.

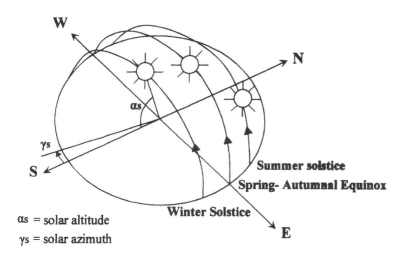

W

N

αs

γs

S

Summer solstice

Spring- Autumnal Equinox

Winter Solstice

E

αs = solar altitude

γs = solar azimuth

Figure 10.2 Movement of the sun and solar angles

PREDICTION OF SUN'S POSITION

The trigonometric expressions used to describe the position of the sun in the sky are based on the solar time (ST) and not on the local time. The local time is defined by reference to the Standard Longitude (or Meridian) of Greenwich (0°) and the world is divided into time zones based on standard longitudes. The European Community is divided into three time zones:

- 0°, including Ireland, UK and Portugal, which have Greenwich Mean Time (GMT)
- 15°, including Denmark, The Netherlands, Germany, Belgium, France, Luxembourg, Italy, Spain, which are one hour ahead of GMT
- 30°, including Greece, which are two hours ahead of GMT.

During summer, all the European countries change into daylight saving time mode, which is one hour ahead of the standard time. The transformation from the local time (LT) to the *solar time* is made through the expression:

$$ST = LT + \frac{L_{st} - L_{loc}}{15°/h} + \frac{E}{60 \text{ min/h}} - D \quad \text{(in hours)}$$

where
LT = Local Time (in hours)

L_{st} = the standard meridian for the reference of the local time, taken positive east of the Standard Longitude (Greenwich) and negative west of it (in degrees);

L_{loc} = the longitude (meridian) of the place, measured in the same way as described previously (in degrees);

D = the difference in time due to summer daylight saving mode (in hours);

E = a value given from the equation of time taken from Figure 10.3 (in minutes) [3]; it can alternatively be estimated from the following expression (in minutes) (from Spencer [4], as cited by Iqbal [5]):

$$E = (0.000075 + 0.001868 \times \cos B - 0.032077 \times \sin B$$

$$-0.014615 \times \cos 2B - 0.04089 \times \sin 2B) \times 229.18$$

where $B = 360 \times (n - 1)/365$ (in degrees) and n is the day of the year.

The *solar declination* (δ), in degrees, can be calculated from the expression [6]:

$$\delta = 23.45° \times \sin[280.1 + ((360°/365) \times n)]$$

either by the expression cited in[7]:

$$\delta = \sin^{-1}(0.3978 \times \sin(y - 80.2 + 1.92 \times \sin(y - 2.89)))$$

where $y = 360 \times n/365.25$ (in degrees).

The approximate values of the declination angles for the 21st of each month are given in Table 10.1 [2].

The *solar hour angle* (ω), is given as:

$$\omega = 15° \times (ST - 12)$$

which is negative before solar noon and positive after solar noon.

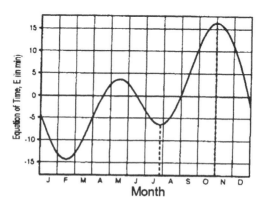

Figure 10.3 Graph of the equation of time, E, in minutes [3]

Table 10.1 Approximate declination angles on the 21st day of each month

Spring	Summer	Declination angle		Autumn	Winter
21 March		0	0	21 Sept.	
21 April	21 Aug.	11° 20'	–11° 20'	21 Oct.	21 Febr.
21 May	21 July	20° 10'	–20° 10'	21 Nov.	21 Jan.
	21 June	23° 27'	–23° 27'		21 Dec.

The angular characteristics of a surface relative to the sun geometry are presented in Figure 10.2. The *solar altitude* (α_s), in degrees, is calculated from the formula (as cited in [3] and [8]):

$$\sin \alpha_s = \cos \theta_z = \cos \varphi \cos \delta \cos \omega + \sin \varphi \sin \delta.$$

At solar noon the altitude of the sun can be expressed as:

$$\alpha_s = 90 - \varphi + \delta.$$

The *solar azimuth* (γ_s), in the case that it is measured from the South – positive clockwise and negative anti-clockwise – is calculated by the expression given by Braun and Mitchel [9]. The azimuth angle is calculated in terms of a pseudo surface azimuth angle γ_s' in the first or the fourth quadrant and by the relationship of the hour angle ω to the hour angle ω_{ew}, when the sun is due west or east.

$$\gamma_s = C_1 C_2 \gamma_s' + C_3 ((1 - C_1 C_2)/2) \times 180$$

where

$$\sin \gamma_s' = (\cos \delta \sin \omega)/\cos \alpha_s$$

or

$$\tan \gamma_s' = \frac{\sin \omega}{\sin \delta \cos \omega - \cos \varphi \tan \delta}$$

$$C_1 = \begin{cases} 1 & \text{if } abs(\omega) \le \omega_{ew} \\ -1 & \text{if } abs(\omega) > \omega_{ew} \end{cases}$$

$$C_2 = \begin{cases} 1 & \text{if } (\varphi - \delta) \ge 0 \\ -1 & \text{if } (\varphi - \delta) < 0 \end{cases}$$

$$C_3 = \begin{cases} 1 & \text{if } \omega \ge 0 \\ -1 & \text{if } \omega < 0 \end{cases}$$

$$\cos \omega_{ew} = (\tan \delta / \tan \varphi).$$

The *surface solar azimuth* (γ) is specified by the horizontal angle between the normal to the surface and a vertical plane through the sun. This angle is zero due south, positive west and negative east ($-180° \leq \gamma \leq 180°$) (Figure10.4)

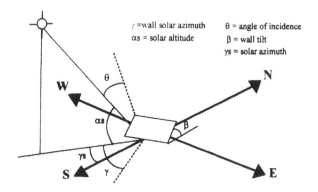

Figure 10.4 Solar angles related to tilted surfaces

The *angle of incidence* (θ) of the direct solar radiation on an inclined plane is taken as the angle between the normal to the surface and the sun's beam on that surface (Figure 10.4) The angle of incidence (θ) of the sun's beam to a plane of tilt (β) from the horizontal plane, is given by the expressions as cited in [3]:

$$\cos\theta = \cos\beta\sin\alpha_s + \sin\beta\cos\alpha_s\cos(\gamma_s - \gamma).$$

and by:

$$\cos\theta = \sin\delta\sin\varphi\cos\beta - \sin\delta\cos\varphi\sin\beta\cos\gamma + \cos\delta\cos\varphi\cos\beta\cos\omega$$

$$+ \cos\delta\sin\varphi\sin\beta\cos\gamma\cos\omega + \cos\delta\sin\beta\sin\gamma\sin\omega.$$

Only the positive values are considered in the calculations, as they indicate that the surface is facing the sun. The optimum tilt for the maximum noontime irradiance on a south facing surface is given by the angle ($\varphi - \delta$) [2].

GRAPHICAL DESIGN TOOLS

The information obtained for the sun's geometry by applying the trigonometric expressions can be also derived by using graphical charts. The graphical tools are quite useful for designers, who are more familiar with drawings than calculations. Following a simple procedure, they can draw important information at the design stage of buildings. These tools can assist in the design of openings and shading devices and in the assessment of their performance; they can also be used to derive the magnitude of shading imposed by or on neighbouring buildings and by other landscape elements.

Sunpath charts

Sunpath diagrams are obtained individually for each latitude and they give the solar altitude and azimuth angle for any day and time.

When the sun's orbit in the sky is projected onto a horizontal plane and after graphical corrections have been made, if the 'stereographic' projection method is used, we obtain *stereographic sunpath charts*. Figures 10.5, 10.6 and 10.7 give the stereographic sunpath charts for different latitudes. Concentric circles and radial lines form a coordinate grid with the zenith represented at the centre (as 90°), where the observer is assumed to be, and the horizon at the circumference of the circle (as 0°). The altitude of the sun is indicated by equally spaced concentric circles and the solar azimuth by radial lines. Curved lines running from east to west represent the daily path of the sun for every month (usually the 21st of each month) and these are crossed by the hour lines which represent solar time. The altitude and the azimuth of the sun at any time can be found by the intersection between the curve for a certain time and an hour line.

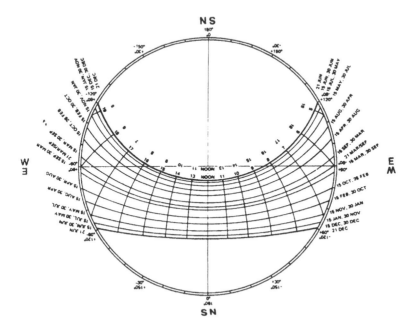

Figure 10.5 Stereographic sunpath chart for latitude 37° [10]

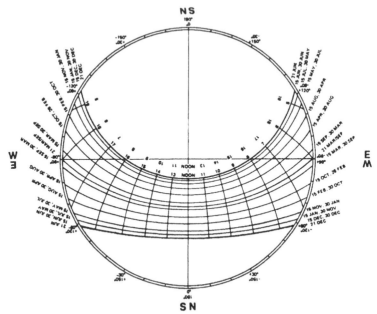

Figure 10.6 Stereographic sunpath chart for latitude 41° [10]

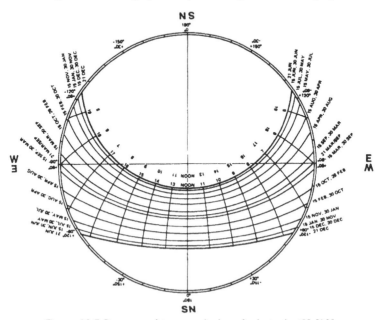

Figure 10.7 Stereographic sunpath chart for latitude 45° [10]

Another form of a sunpath chart, more commonly used in overshadowing assessment, is the vertical *cylindrical sunpath diagram*, where the sky vault is projected onto a vertical cylindrical surface, which is cut along the north solar azimuth and laid flat. An orthogonal coordinate system is used where the solar azimuth is plotted against the solar altitude (Figures 10.8, 10.9 and 10.10) with the south represented in the centre of the azimuth axis. The sunpath for each month is represented by curved lines which are crossed by the solar hour lines. At their intersection, the altitude and azimuth angles can be read for a specific day and time. The intersection of the sunpath lines with the horizontal axis gives the time of the sunrise (east of the south) and sunset (west of the south) for each month. For representation convenience, the charts cover only a sector between 120° east and 120° west, which means that the sunpath lines around mid-summer sunrise and sunset for some high latitudes are not illustrated on the chart.

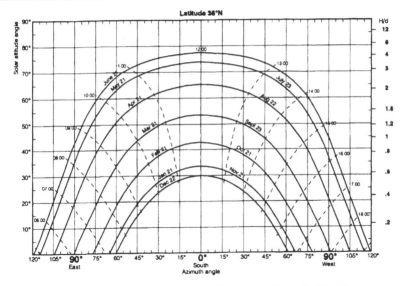

Figure 10.8 Cylindrical sunpath diagram for latitude 36° [7]

The period when shading is provided by distant obstructions can also be identified with the aid of these charts by plotting the altitude angle of the obstruction and the surface solar azimuth (Figure 10.11).

Shadow mask chart

The shadow mask chart (Figure 10.12) which is used in conjunction with the vertical cylindrical sunpath chart of a location is employed either to determine the period when shade is provided by obstructions or by horizontal or vertical projections on the building facade (shading devices) or, inversely, to identify the size of a shading device for a specified opening.

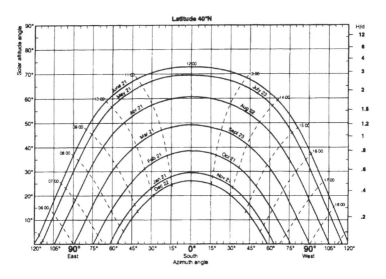

Figure 10.9 Cylindrical sunpath diagram for latitude 40° [7]

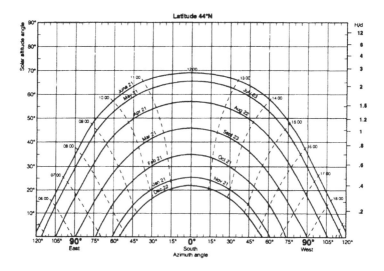

Figure 10.10 Cylindrical sunpath diagram for latitude 44° [7]

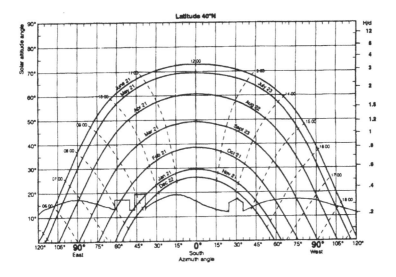

Figure 10.11 Determination of shading by distant obstacles

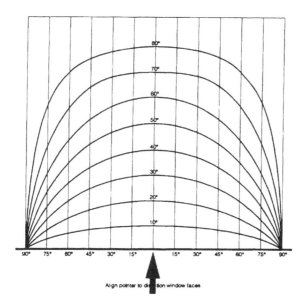

Figure 10.12 Shadow mask [7]

The curves on the chart, ranging from 10° to 80°, correspond to the vertical angle of horizontal obstructions or projections and the horizontal axis corresponds to the horizontal angles of the vertical projections or obstructions. Both charts, the shadow mask and the sunchart, have to be plotted on the same scale and overlaid one on the top of the other (Figure 10.13). Having defined the angle of the shading elements, as it is described in the following section, we transfer it on the curves of the shadow mask chart. The wall azimuth angle has to be identified (e.g. wall azimuth + 45° for a south-west wall) and the 0° of the horizontal axis of the shadow chart should be positioned on the corresponding value of the azimuth axis (e.g. 45° south west) on the sunpath chart for the specified latitude. For any specific day and time, the shading effect can be identified by taking the intersection of the shading curve of the shadow mask with the date and time curve on the sunpath chart.

Following the inverse procedure, the size of any shading element can be identified for the desired period of shading.

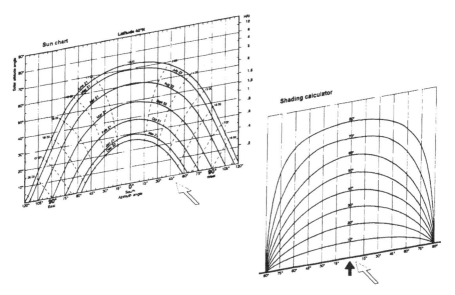

Figure 10.13 Procedure for aligning overlay on the solar chart [7]

PERFORMANCE OF SHADING ELEMENTS

Horizontal shading devices. For complete shading of the opening (100% shade), the vertical angle between the line connecting the edge of the shading device to the base of the opening and the horizontal plane is measured (Figure10.14). The corresponding angle is taken on the curves of the shadow mask chart. Having determined the wall azimuth of the opening, we position the 0 of the mask chart on the corre-

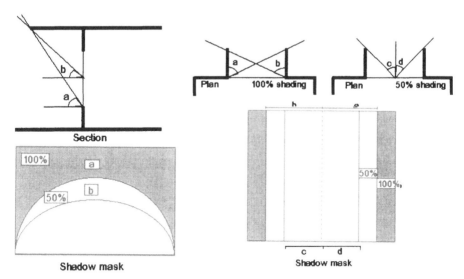

Figure 10.14. Shadow mask of   *Figure 10.15 Shadow mask of vertical shading*
horizontal overhangs   *devices*

sponding azimuth angle of the sunpath chart for the latitude of the place. In this way, we can identify on the sunpath chart the days and the hours when we have complete shade. The same procedure can be followed for any desired shading, say for 50% shading, by defining the vertical angle line to the centre of the opening.

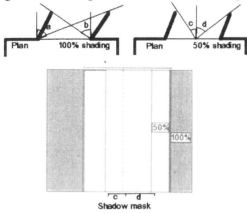

Figure 10.16 Shadow mask of oblique shading device

Vertical shading devices. For a vertical projection (either vertical or oblique to the opening plane), we define (Figure 10.15, Figure 10.16) the horizontal angles from the element edge to the opposite side of the opening (for 100% shading) or to the centre of the opening (for 50% shading) or to any other chosen position. We transfer these angles to the horizontal axis of the shadow mask and we draw the verticals to these points. In this way we obtain the shadow mask of the shading device with which, used in conjunction with the sunpath chart and, following the procedure described above, we can find the period of shadowing.

Combination of vertical and horizontal shading devices (egg-crate type shading devices). The same procedure described above is used separately for each type of shading device and the two resultant shading masks are combined to obtain the desired shadowing period (Figure 10.17).

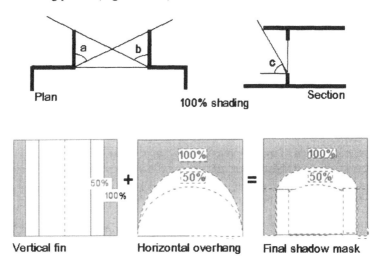

Figure 10.17 Shadow mask of an egg-crate type shadow device

Shadow tables

It is quite important for the designer to identify in the early stages of the design the solar access to the building. It can be found by drawing the shadow produced by surrounding obstacles. By tracing the shadow pattern for each building and any other obstacle (e.g. tree), it is possible to identify the solar access to a specified space.

   The simplest way to understand shadow patterns is by examining the shadow produced by a vertical pole (Figure 10.18). When the altitude of the sun is known, by employing simple trigonometrical expressions, we can estimate the shadow on the horizontal plane of a vertical pole ($s = H/\tan\alpha_s$) or inversely, for a specified shadow length, the height of the object that produces the shadow ($H = s\tan\alpha_s$) can be determined. The shadow pattern for a building is constructed by assuming that the building consists of a series of poles (Figure 10.19). In this way, the shadow

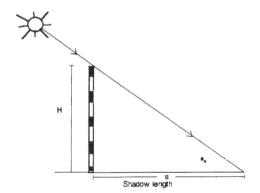

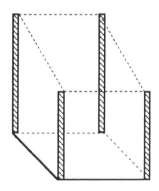

Figure 10.18 Shadow of a vertical pole   *Figure 10.19 Representation of a building as a series of poles*

produced from neighbouring buildings on the one studied or the inverse can be examined early in the design stage.

Table 10.2 [11] gives the values of the shadow length of a vertical pole of height ranging from 0.1 m to 10.0 m, for sun altitude ranging from 5° to 80°.

The procedure to work out the shadow pattern on the horizontal plane of a building or any other obstacle is as follows (Figure 10.20):

1 Determine the sun altitude and azimuth for the specified latitude, day and time (from a solar chart, trigonometric expressions or published tables).
2 On the plan of the building, draw the shadow trace from each edge parallel to the direction of the azimuth angle.
3 Specify from Table 10.2 the shadow length corresponding to the height of the edge, sun altitude and azimuth.
4 On the shadow trace, take the length of the shadow for the specified height as it is defined in the previous step.
5 Connecting the shadow length of each edge, find the shadow pattern of the specified building, on the horizontal plane.

The shadow produced on the vertical plane as it is received on the facade of neighbouring buildings can also be found (Figure 10.20):

Table10.2 Shadow length of vertical obstacles for different solar altitudes [11]

		Vertical obstacle's height (m)								
		0.10	0.20	0.30	0.40	0.50	0.60	0.70	0.80	0.90
	5	1.14	2.29	3.43	4.57	5.72	6.86	8.00	9.14	10.29
	6	0.95	1.90	2.85	3.81	4.76	5.71	6.66	7.61	8.56
	7	0.81	1.63	2.44	3.26	4.07	4.89	5.70	6.52	7.32
	8	0.71	1.42	2.13	2.85	3.56	4.27	4.98	5.69	6.40
	9	0.63	1.26	1.89	2.53	3.16	3.79	4.42	5.05	5.68
	10	0.57	1.13	1.70	2.27	2.84	3.40	3.97	4.54	5.10
	11	0.51	1.03	1.54	2.06	2.57	3.09	3.60	4.12	4.63
	12	0.47	0.94	1.41	1.88	2.35	2.82	3.29	3.76	4.23
	13	0.43	0.87	1.30	1.73	2.17	2.60	3.03	3.47	3.90
	14	0.40	0.80	1.20	1.60	2.01	2.41	2.81	3.21	3.61
S	15	0.37	0.75	1.12	1.49	1.87	2.24	2.61	2.99	3.36
o	16	0.35	0.70	1.05	1.39	1.74	2.09	2.44	2.79	3.14
l	17	0.33	0.65	0.98	1.31	1.64	1.96	2.29	2.62	2.94
a	18	0.31	0.62	0.92	1.23	1.54	1.85	2.15	2.46	2.77
r	19	0.29	0.58	0.87	1.16	1.45	1.74	2.03	2.32	2.61
	20	0.27	0.55	0.82	1.10	1.37	1.65	1.92	2.20	2.47
A	21	0.26	0.52	0.78	1.04	1.30	1.56	1.82	2.08	2.34
l	22	0.25	0.50	0.74	0.99	1.24	1.49	1.73	1.98	2.23
t	23	0.24	0.47	0.71	0.94	1.18	1.41	1.65	1.88	2.12
i	24	0.22	0.45	0.67	0.90	1.12	1.35	1.57	1.80	2.02
t	25	0.21	0.43	0.64	0.86	1.07	1.29	1.50	1.72	1.93
u	26	0.21	0.41	0.62	0.82	1.03	1.23	1.44	1.64	1.85
d	27	0.20	0.39	0.59	0.79	0.98	1.18	1.37	1.57	1.77
e	28	0.19	0.38	0.56	0.75	0.94	1.13	1.32	1.50	1.69
	29	0.18	0.36	0.54	0.72	0.90	1.08	1.26	1.44	1.62
	30	0.17	0.35	0.52	0.69	0.87	1.04	1.21	1.39	1.56
(°)	31	0.17	0.33	0.50	0.67	0.83	1.00	1.16	1.33	1.50
	32	0.16	0.32	0.48	0.64	0.80	0.96	1.12	1.28	1.44
	33	0.15	0.31	0.46	0.62	0.77	0.92	1.08	1.23	1.39
	34	0.15	0.30	0.44	0.59	0.74	0.89	1.04	1.19	1.33
	35	0.14	0.29	0.43	0.57	0.71	0.86	1.00	1.14	1.29
	36	0.14	0.28	0.41	0.55	0.69	0.83	0.96	1.10	1.24
	37	0.13	0.27	0.40	0.53	0.66	0.80	0.93	1.06	1.19
	38	0.13	0.26	0.38	0.51	0.64	0.77	0.90	1.02	1.15
	39	0.12	0.25	0.37	0.49	0.62	0.74	0.86	0.99	1.11
	40	0.12	0.24	0.36	0.48	0.60	0.72	0.83	0.95	1.07
	41	0.12	0.23	0.35	0.46	0.58	0.69	0.81	0.92	1.04
	42	0.11	0.22	0.33	0.44	0.56	0.67	0.78	0.89	1.00
	43	0.11	0.21	0.32	0.43	0.54	0.64	0.75	0.86	0.97
	44	0.10	0.21	0.31	0.41	0.52	0.62	0.72	0.83	0.93
	45	0.10	0.20	0.30	0.40	0.50	0.60	0.70	0.80	0.90
	46	0.10	0.19	0.29	0.39	0.48	0.58	0.68	0.77	0.87
	47	0.09	0.19	0.28	0.37	0.47	0.56	0.65	0.75	0.84
	48	0.09	0.18	0.27	0.36	0.45	0.54	0.63	0.72	0.81
	49	0.09	0.17	0.26	0.35	0.43	0.52	0.61	0.70	0.78
	50	0.08	0.17	0.25	0.34	0.42	0.50	0.59	0.67	0.76
	51	0.08	0.16	0.24	0.32	0.40	0.49	0.57	0.65	0.73
	52	0.08	0.16	0.23	0.31	0.39	0.47	0.55	0.63	0.70
	53	0.07	0.15	0.23	0.30	0.38	0.45	0.53	0.60	0.68

Table10.2 (continued)

Vertical obstacle's height (m)									
1.00	2.00	3.00	4.00	5.00	6.00	7.00	8.00	9.00	10.00
11.43	22.86	34.29	45.72	57.15	68.58	80.01	91.44	100.9	114.3
9.51	19.03	28.54	38.06	47.57	57.09	66.60	76.11	85.63	95.14
8.14	16.29	24.43	32.58	40.72	48.87	57.01	65.15	73.30	81.44
7.12	14.23	21.35	28.46	35.50	42.69	49.81	56.92	64.04	71.15
6.31	12.63	18.94	25.26	31.57	37.88	44.20	50.51	56.82	63.14
5.67	11.34	17.01	22.69	28.36	34.03	39.70	45.37	51.04	56.71
5.14	10.29	15.43	20.58	25.72	30.87	36.01	41.16	46.30	51.45
4.70	9.41	14.11	18.82	23.52	26.23	32.93	37.64	42.34	47.05
4.33	8.66	12.99	17.33	21.66	25.99	30.32	34.56	38.98	43.31
4.01	8.02	12.03	16.04	20.05	24.06	28.08	32.09	36.10	40.11
3.73	7.46	11.20	14.93	18.66	22.39	26.12	29.86	33.59	37.32
3.49	6.97	10.46	13.95	17.44	20.92	24.41	27.90	31.39	34.87
3.27	6.54	9.81	13.08	16.35	19.63	22.90	26.17	29.44	32.71
3.08	6.16	9.23	12.31	15.39	18.47	21.54	24.62	27.70	30.78
2.90	5.81	8.71	11.62	14.52	17.43	20.33	23.23	26.14	29.04
2.75	5.49	8.24	10.99	13.74	16.48	19.23	21.98	24.73	27.47
2.61	5.21	7.82	10.42	13.03	15.63	18.24	20.84	23.45	26.05
2.48	4.95	7.43	9.90	12.38	14.85	17.33	19.80	22.28	24.75
2.36	4.71	7.07	9.42	11.78	14.14	16.49	18.85	21.20	23.56
2.25	4.49	6.74	8.98	11.23	13.48	15.72	17.97	20.21	22.46
2.14	4.29	6.43	8.58	10.72	12.87	15.01	17.16	19.30	21.45
2.05	4.10	6.15	8.20	10.25	12.38	14.35	16.40	18.45	20.50
1.96	3.93	5.89	7.85	9.81	11.78	13.74	15.70	17.66	19.63
1.88	3.76	5.64	7.52	9.40	11.28	13.17	15.05	16.93	18.81
1.80	3.61	5.41	7.22	9.02	10.82	12.63	14.43	16.24	18.04
1.73	3.46	5.20	6.93	8.66	10.39	12.12	13.86	15.59	17.32
1.66	3.33	4.99	6.66	8.32	9.99	11.65	13.31	14.98	16.64
1.60	3.20	4.80	6.40	8.00	9.60	11.20	12.80	14.40	16.00
1.54	3.08	4.62	6.16	7.70	9.24	10.78	12.32	13.86	15.40
1.48	2.97	4.45	5.93	7.41	8.90	10.38	11.86	13.34	14.83
1.43	2.86	4.28	5.71	7.14	8.57	10.00	11.43	12.85	14.28
1.38	2.75	4.13	5.51	6.88	8.26	9.63	11.01	12.39	13.76
1.33	2.65	3.98	5.31	6.64	7.96	9.29	10.62	11.94	13.27
1.28	2.56	3.84	5.12	6.40	7.68	8.96	10.24	11.52	12.80
1.23	2.47	3.70	4.94	6.17	7.41	8.64	9.88	11.11	12.35
1.19	2.38	3.58	4.77	5.96	7.15	8.34	9.53	10.73	11.92
1.15	2.30	3.45	4.60	5.75	6.90	8.05	9.20	10.35	11.50
1.11	2.22	3.33	4.44	5.55	6.66	7.77	8.88	10.00	11.11
1.07	2.14	3.22	4.29	5.36	6.43	7.51	8.58	9.65	10.72
1.04	2.07	3.11	4.14	5.18	6.21	7.25	8.28	9.32	10.36
1.00	2.00	3.00	4.00	5.00	6.00	7.00	8.00	9.00	10.00
0.97	1.93	2.90	3.86	4.83	5.79	6.76	7.73	8.69	9.66
0.93	1.87	2.80	3.73	4.66	5.60	6.53	7.46	8.39	9.33
0.90	1.80	2.70	3.60	4.50	5.40	6.30	7.20	8.10	9.00
0.87	1.74	2.61	3.48	4.35	5.22	6.09	6.95	7.82	8.69
0.84	1.68	2.52	3.36	4.20	5.03	5.87	6.71	7.55	8.39
0.81	1.62	2.43	3.24	4.05	4.86	5.67	6.48	7.29	8.10
0.78	1.56	2.34	3.13	3.91	4.69	5.47	6.25	7.03	7.81
0.75	1.51	2.26	3.01	3.77	4.52	5.27	6.03	6.78	7.54

Table10.2 (continued)

		Vertical obstacle's height (m)								
		0.10	0.20	0.30	0.40	0.50	0.60	0.70	0.80	0.90
	54	0.07	0.15	0.22	0.29	0.36	0.44	0.51	0.58	0.65
	55	0.07	0.14	0.21	0.28	0.35	0.42	0.49	0.56	0.63
S	56	0.07	0.13	0.20	0.27	0.34	0.40	0.47	0.54	0.61
o	57	0.06	0.13	0.19	0.26	0.32	0.39	0.45	0.52	0.58
l	58	0.06	0.12	0.19	0.25	0.31	0.37	0.44	0.50	0.56
a	59	0.06	0.12	0.18	0.24	0.30	0.36	0.42	0.48	0.54
r	60	0.06	0.12	0.17	0.23	0.29	0.35	0.40	0.46	0.52
	61	0.06	0.11	0.17	0.22	0.28	0.33	0.39	0.44	0.50
A	62	0.05	0.11	0.16	0.21	0.27	0.32	0.37	0.43	0.48
l	63	0.05	0.10	0.15	0.20	0.25	0.31	0.36	0.41	0.46
t	64	0.05	0.10	0.15	0.20	0.24	0.29	0.34	0.39	0.44
i	65	0.05	0.09	0.14	0.19	0.23	0.28	0.33	0.37	0.42
t	66	0.04	0.09	0.13	0.18	0.22	0.27	0.31	0.36	0.40
u	67	0.04	0.08	0.13	0.17	0.21	0.25	0.30	0.34	0.38
d	68	0.04	0.08	0.12	0.16	0.20	0.24	0.28	0.32	0.36
e	69	0.04	0.08	0.12	0.15	0.19	0.23	0.27	0.31	0.35
	70	0.04	0.07	0.11	0.15	0.18	0.22	0.25	0.29	0.33
	71	0.03	0.07	0.10	0.14	0.17	0.21	0.24	0.28	0.31
(°)	72	0.03	0.06	0.10	0.13	0.16	0.19	0.23	0.26	0.29
	73	0.03	0.06	0.09	0.12	0.15	0.18	0.21	0.24	0.28
	74	0.03	0.06	0.09	0.11	0.14	0.17	0.20	0.23	0.26
	75	0.03	0.05	0.08	0.11	0.13	0.16	0.19	0.21	0.24
	76	0.02	0.05	0.07	0.10	0.12	0.15	0.17	0.20	0.22
	77	0.02	0.05	0.07	0.09	0.12	0.14	0.16	0.18	0.21
	78	0.02	0.04	0.06	0.09	0.11	0.13	0.15	0.17	0.19
	79	0.02	0.04	0.05	0.08	0.10	0.12	0.14	0.16	0.17
	80	0.02	0.04	0.05	0.07	0.09	0.11	0.12	0.14	0.16

6   The end points of the shadow trace (*c* and *d*), as they were specified on the horizontal plane (plan drawing), are projected to the section drawing. Working on the section drawing, we connect the projected end points (*c'* and *d'*) with the top edges of the building that gives shade (*A*) and in this way we identify the direction of the sun's beam. The intersection of the sun's beam with the facade of the shaded building (*B*) (points *a'* and *b'*) gives the length of the shadow on the facade. Moving to the facade drawing, we project the intersection of the shadow trace with the shaded building (*B*) (points *a* and *b*) on the facade of the latter building. Transferring the heights of points *a'* and *b'* on the facade, we define its shaded area.

Table10.2 (continued)

Vertical obstacle's height (m)									
1.00	2.00	3.00	4.00	5.00	6.00	7.00	8.00	9.00	10.00
0.73	1.45	2.18	2.91	3.63	4.36	5.09	5.81	6.54	7.27
0.70	1.40	2.10	2.80	3.50	4.20	4.90	5.60	6.30	7.00
0.67	1.35	2.02	2.70	3.37	4.05	4.72	5.40	6.07	6.75
0.65	1.30	1.95	2.60	3.25	3.90	4.55	5.20	5.84	6.49
0.62	1.25	1.87	2.50	3.12	3.75	4.37	5.00	5.62	6.25
0.60	1.20	1.80	2.40	3.00	3.61	4.21	4.81	5.41	6.01
0.58	1.15	1.73	2.31	2.89	3.46	4.04	4.62	5.20	5.77
0.55	1.11	1.66	2.22	2.77	3.33	3.88	4.43	4.99	5.54
0.53	1.06	1.60	2.13	2.66	3.19	3.72	4.25	4.79	5.32
0.51	1.02	1.53	2.04	2.55	3.06	3.57	4.08	4.59	5.10
0.49	0.98	1.46	1.95	2.44	2.93	3.41	3.90	4.39	4.88
0.47	0.93	1.40	1.87	2.33	2.80	3.26	3.73	4.20	4.66
0.45	0.89	1.34	1.78	2.23	2.67	3.12	3.56	4.01	4.45
0.42	0.85	1.27	1.70	2.12	2.55	2.97	3.40	3.82	4.24
0.40	0.81	1.21	1.62	2.02	2.42	2.83	3.23	3.64	4.04
0.38	0.77	1.15	1.54	1.92	2.30	2.69	3.07	3.45	3.84
0.36	0.73	1.09	1.46	1.82	2.18	2.55	2.91	3.28	3.64
0.34	0.69	1.03	1.38	1.72	2.07	2.41	2.75	3.10	3.44
0.32	0.65	0.97	1.30	1.62	1.95	2.27	2.60	2.92	3.25
0.31	0.61	0.92	1.22	1.53	1.83	2.14	2.45	2.75	3.06
0.29	0.57	0.86	1.15	1.43	1.72	2.01	2.29	2.58	2.87
0.27	0.54	0.80	1.07	1.34	1.61	1.88	2.14	2.41	2.68
0.25	0.50	0.75	1.00	1.25	1.50	1.75	1.99	2.24	2.49
0.23	0.46	0.69	0.92	1.05	1.39	1.62	1.85	2.08	2.31
0.21	0.43	0.64	0.85	1.06	1.28	1.49	1.70	1.91	2.13
0.19	0.39	0.58	0.78	0.97	1.17	1.36	1.56	1.75	1.94
0.18	0.35	0.53	0.71	0.88	1.06	1.23	1.41	1.59	1.76

Following a similar procedure, the solar access to the interior of a space can be found (Figure10.21):

1 On the plan drawing, lines parallel to the azimuth direction are drawn from the two ends of the opening.
2 From Table 10.2 the shadow length is found for the height of the sill (h_1) and the top of the opening (h_2).
3 By taking the length of the shadow lines on the plan, the interior sunlight area can be identified for the specified time of the day.

In a similar way, the length of any shading device can be investigated for the period that shading is desirable.

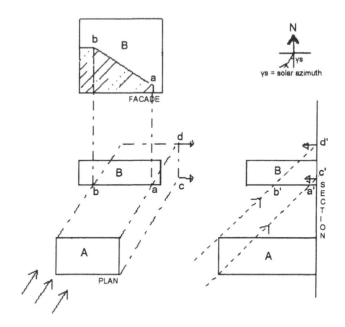

Figure 10.20 Graphical procedure for identifying the shading pattern
of neighbouring buildings

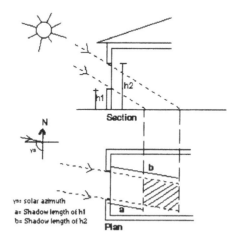

Figure 10.21 Graphical procedure for
identifying the solar access to the interior
of a space

Shadow pattern plot

A simple graphical design tool for the designer to visualize the solar access and shading pattern is to draw the plot of the shadow pattern boundary. The *shadow pattern plot* produces the shadow cast of vertical obstructions during the day and this can be used with the drawings.

By drawing the shadow pattern for each building and any other obstacle (tree, etc.), it is possible to explore quickly the solar access and shading of buildings. The pattern is drawn on the same scale as the drawings and the procedure to produce them is the following (Figure 10.22):

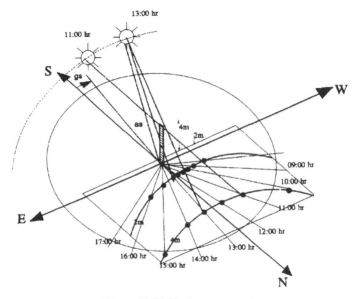

Figure 10.22 Shadow pattern plot

1 For the specified latitude and day, find the sun altitude and azimuth for each hour.
2 Draw lines in the direction of the sun azimuth starting from the base of an imaginary pole.
3 Find, from Table 10.2 for the specified sun altitude and azimuth, the shadow length for different heights, e.g. from 1.0 to 5.0 m, every 1.0 m interval.
4 Take this length on the line corresponding to the specified sun azimuth.
5 Connecting the trace of the shadow for the same pole height, define the shadow pattern curve for the related height on the specified day.
6 Following the same procedure, draw the shadow pattern for different heights on the same graph.

For each latitude and each day the shadow pattern graph is different. An acceptable simplification is to work them out for the 21st day of each month. Using these graphs, the direction and length of the shadow produced from any obstruction can be determined, along with the period that a specified building is shaded.

SOLAR CONTROL REVIEW

The selective exclusion of solar radiation by shading buildings and outdoor spaces (Figure 10.23) is of great value in hot climates. Shading elements add a new element to the architectural form of the building and become a regulatory measure to protect adjacent buildings from the solar radiation. Special elements, like arcades, are also used to achieve an architectural style and to enhance the shading effect. In hot and humid climates, which are characterized by sparsely spread buildings with large openings in order to enhance circulation of air, shading is essential to control solar gains.

Shading elements on the facade of a building can be multi-functional and add to the architectural style of the building, ranging from a simple overhang or a fin to a balcony that shades the space below it (Figure 10.24). The balconies of modern multi-storey buildings and the long overhangs of contemporary buildings offer the same shading effect and could additionally achieve interesting aesthetic effects. Although shading of the building's envelope is essential during the hot period of the year, solar protection of the openings is the most crucial point. As application of big overglazed areas in buildings is widespread in most countries around the world, overheating of specific buildings, like offices, becomes a problem even in countries

Figure 10.23 Shading of open spaces

Figure 10.24 Use of long overhangs for shading

with a cold climate. Thus, manipulation of shading elements and of the thermal properties of glazing should become an effective tool to improve the thermal and visual performance of buildings and to control daylighting and glare effects. Their forms can also add a play of shadows and light in the protected spaces. Their application throughout the years has become part of local cultures and a characteristic element of traditional architectures.

Ancient Greek and Roman architecture enriched buildings with colonnades, which have the dual function of shading the building and of offering sun-protected outdoor spaces. Anonymous architecture in hot climates can show examples of using the plot of the settlement to create dense forms which offer shaded outdoor spaces.

SOLAR CONTROL ISSUES

Provision of shading in buildings is one of the initial priorities that have to be considered at the early stages of the design. Blocking the sun before it reaches the envelope, and especially its interior, can have a considerable effect on the thermal performance of the building, its energy requirements and the thermal and visual comfort of its occupants. Although shading is mainly associated with exclusion of the solar heat from the building during the hot months, its application is an interplay of several objectives:

- obstruction of the solar heat gains, mainly the direct ones, from reaching the envelope and the interior of the building;
- non-interference with winter solar gains;
- control of the intense daylight, especially during the summer months, by diffusing it in a uniform way in most of the space;

- unobstructed view from the windows;
- admission and even regulation of the ventilation of adjacent spaces.

These design objectives and the applied shading techniques differ according to the latitude, the location, the type of building, its operational schedule, the specific use of the various spaces (occupants' activities, internal gains) and the expected comfort conditions. While exclusion or admission of the solar gains varies according to climatic variables, a controlled manipulation of daylighting becomes essential all year round and is an aspect that should be examined before the design of any shading system.

Residences in northern latitudes are unlikely to face intense overheating problems, while solar heat gains, daylighting and view are desirable. Residences in southern latitudes need manipulation of the solar gains and daylighting while ensuring at the same time an unobstructed view to the exterior. Control of solar gains is advisable in most offices, conservatories and atria throughout the world and provision of diffuse natural light is also essential. School buildings have similar design requirements, especially in southern latitudes, and prevention of overheating becomes crucial. Buildings for special use, like museums, art galleries and sports halls, exclude the penetration of solar gains, while admission of controllable diffuse light is desirable.

SOLAR CONTROL TECHNIQUES

Modulation of the solar heat gains entering a building can be achieved through [12]:

- orientation and aperture geometry
- shading devices
- control of solar-optical properties of opaque and transparent surfaces.

Orientation and aperture geometry

Orientation of the openings, combined with their size and tilt can modulate the solar gains passing through them. South-orientated openings accept high solar heat gains during winter and can easily be shaded during summer, owing to the high position of the sun in the sky. Shading of west and east windows presents difficulties because of the low position of the sun in the sky and, additionally, west-orientated openings are associated with external conditions of high solar radiation and ambient temperature during summer. Thus, it is advisable that windows in this orientation should be minimized or replaced with other design solutions, such as south- or north-facing openings in these facades (Figure 10.25). North-facing windows accept very limited solar heat gains, restricted in the summer early morning and late afternoon hours while they have the advantage of permitting diffuse daylight to enter the adjoining

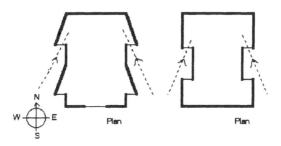

Figure 10.25 Simplified solutions for south- and north-facing windows in east- and west-orientated walls

spaces. Although north facing windows are widely considered as energy wasteful, in the southern latitudes, the gross energy balance of north facing windows in mid-winter is not worse than the one of the south facing windows in the colder climate of northern latitudes [12].

The size of the openings in each orientation should be defined according to the annual energy requirements (heating, cooling, lighting) of the specific building. This cannot be globally defined, but depends on the latitude of the place, the location, the functions and architecture of the building. Tilt of the openings can also contribute to the shading effect, since outwards tilt, facing the ground, restricts direct solar gains. A big variation in the tilt, especially of large glazed areas, is limited because of associated structural support problems. Skylights impose difficulties in shading, as they face the overhead noon sun directly. Clerestory windows perform better throughout the year, both in shading and daylighting, as their vertical glazed area can easily be shaded by simple means (Figure 10.26).

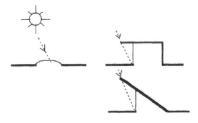

Figure 10.26 Clerestory windows are better shaded than skylights

Shading devices

The principal role of shading devices is to protect openings from direct solar radiation, while their second is to protect openings from diffuse and reflected radiation. Shading devices can be classified into:

- **Fixed shading elements**: These are mainly *external* elements, including horizontal overhangs, vertical fins, combination of horizontal and vertical elements closely spaced (egg-crate type) (Figure 10.27), balconies or *internal* elements like louvres and light-shelves (Figure 10.28). Louvres can be strategically positioned in order to obstruct the summer sun's rays from entering the space, while they permit the entrance of the winter sun. During the summer, they operate as light diffusing elements, provided they are painted white, by scattering the light uniformly into the space below.
- **Adjustable (or retractable) shading elements**: These can be external shading elements in the form of tents, awnings, blinds, pergolas, or internal elements like curtains, rollers, venetian blinds; they can also be positioned between the panes of the window. Window shutters are also included in this category. Adjustable shading devices can be lifted, rolled or drawn back from the window either manually or automatically by responding to radiation and daylighting levels.

The use of traditional slanted windows (Figure 10.29) is another shading approach, which for a given aperture, maximizes the view to the sky and provides shade to the opening through its own configuration.

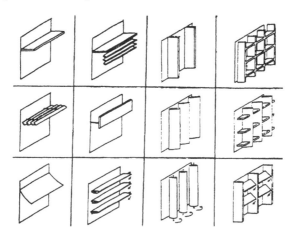

Figure 10.27 Options of external shading devices [13]

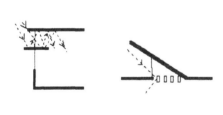

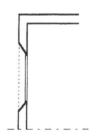

Figure 10.28 Example of light-shelves and louvres at top windows

Figure 10.29 Sketch of a slanted window

As the direct component of the solar radiation is of predetermined direction, it can be effectively obstructed by external shading elements. The diffuse component, due to its wider angle of incidence, is more difficult to control and thus internal shading elements or shading elements between the glazing panes are more effective. The reflected part can be controlled by changing the reflectivity properties of the surfaces at the opposite sides [14].

PERFORMANCE EVALUATION

Two methods are usually employed for the evaluation of the efficiency of the different solar protection approaches. The one that is widely used is the calculation of the *shading coefficient*. The shading coefficient is taken as the ratio of the total solar heat gains entering through the combination of glass-shading element to that entering a single unshaded glass window [13]. For fixed shading devices, the average daily solar transmission is usually considered for the calculation of the coefficient. For evaluation of the performance of shading screens, the average of the coefficients in all directions is usually derived, while for blinds, the value quoted is the average of all the orientations where the slats are adjusted to prevent direct rays. The total shading coefficient for a glass is the sum of the short-wave and long-wave shading coefficients, calculated for radiation at normal incidence. For any other angle of incidence, the shading coefficient is compared with that for clear glass in the same situation, which results in deriving shading coefficients that are almost constant at all incident angles of solar radiation [15]. Thus, the coefficient should be considered as an approximate value, as the position of the sun changes during the day, together with the proportion of the incident direct and diffuse solar radiation on the shading system. However, work carried out by the IEA [16] shows considerable deviations in the total transmission with the change of the incidence angle, the orientation and the period of the year and, thus, the shading coefficient can be regarded as a misleading and inadequate index for assessment [12].

 The other approach used to evaluate the thermal effect of a shading system is to compare the indoor air temperatures obtained with the shading system to those achieved with the same window unshaded.

Another index, which is used by some North American glazing manufacturers but only for evaluation of the glazing performance, is the so-called 'coolness index' [15]. This is defined as the daylight transmittance divided by the shading coefficient.

EVALUATION OF SHADING DEVICES

Choice of the appropriate shading device from the wide range of fixed and adjustable elements depends on the latitude, sky conditions (the direct-diffuse-reflected solar radiation component), orientation, building type and overall design of the building. External shading devices are more effective as they obstruct the sun radiation before it reaches the interior of the building. Internal shading elements eliminate from the solar radiation which has already penetrated only the portion that can be reflected at their surfaces and then transmitted outside through the glazing, while the remainder of the radiation is absorbed, convected and radiated to the room. The effectiveness of internal shades is thus mainly determined by their reflectivity. Furthermore, internally applied shading devices may conflict with daylighting and ventilation requirements, as in most cases they block the openings. Based on the evaluation of the shading coefficient, it is argued [13] that, on average, external shading elements are about 35% more effective than internal ones. Application of awnings can effectively reduce summer heat gains by up to 65% on south facades and by up to 80% on east and west facades [17]. Table 10.3 [17] shows the effectiveness of various shading devices.

Table 10.3 Solar gain factors for various shading elements (strictly for UK only, but approximately correct world-wide) [17]

Shading element		Solar gain factor (*) Glazing Type	
Position	Type	Single	Double
Internal	Dark-green open-weave plastic blind	0.62	0.56
	White venetian blind	0.46	0.46
	White cotton curtain	0.41	0.40
	Cream holland-linen blind	0.30	0.33
Mid-pane	White venetian blind	–	0.28
External	Dark-green open-weave plastic blind	0.22	0.17
	Canvas roller blind	0.14	0.11
	White louvred sunbreaker, blades at 45°	0.14	0.11

* The solar gain factor of a transparent material is the fraction of incident solar energy passing through the material.

The colour and the material of the shading element also determine its shading effectiveness. The difference in the effectiveness of external shading elements compared to internal ones increases with the darkness of the colour. A study [13] has shown that off-white venetian blinds give 20% more shade protection than dark ones, while for roller blinds the effect can reach to 40%. An aluminium blind can add more 10% protection than a coloured one. For internal curtains the differences are less, as light coloured ones are only 18% more effective than dark ones.

The increased efficiency of dark-coloured external shading elements, compared to internal ones of the same colour, is valid only for the case of closed windows. For open windows, the orientation with regard to the wind direction is more critical than the colour and the position of the shading element. Thus, a dark shading element on the windward side will heat the air entering the space, while the same element on the leeward side will have a weaker effect [2].

In many cases fixed devices are preferred because of their simplicity, low maintenance cost and sometimes low construction cost. However, movable shading devices are more flexible as they respond better to the dynamic nature of the sun's movement, allow better control of the diffuse radiation and glare and, in most cases, cause less or negligible sun obstruction during winter. Nontheless, in some spaces, provision of shade during summer is more important than unobstructed solar access during winter.

The same overall shading performance as that produced by a single shading element can be achieved by using several small shading units that effectively block the sun (Figure 10.30).

Figure 10.30 Different shading devices with simi-
lar shading effect

ORIENTATION

Horizontal overhangs are effective for south-facing windows because they can effectively obstruct the direct sun's rays during summer, as the sun is in a high position in the sky. Long verandas and roof overhangs work satisfactorily in hot climates and in many cases are accomplished with canvas tents or pergolas. In east and west orientations, a combination of horizontal and vertical elements (egg-crate type) can increase the effectiveness if the vertical elements are inclined at 45° to the south [2]. If horizontal or vertical elements are used at this orientation, they should be quite long in order to be effective, but caution should be exercised not to restrict the view and eliminate the solar gains during winter. Horizontal shading elements are more effective than vertical ones in the south-east and south-west orientations, but the egg-crate combination is the most effective. The egg-crate-type shading elements are considered effective for east and west orientations in hot climates and for south-east and south-west facades in extremely hot climates, although it has been found that horizontal overhangs are preferable in these latter orientations.

VEGETATION

Rational planting of vegetation around the building can offer significant shade. This can be in the form of trees (deciduous or evergreen), bushes, pergolas, trellis, etc., and its performance has been more thoroughly examined in Chapter 5. The shade offered by vegetation is better than any artificial shading device as vegetation also modifies the thermal properties of the surrounding air and improves the microclimate around the building. Its contribution can be evaluated by taking into account other environmental aspects, such as to purify the air by supplying oxygen (O_2), filtering the dust and also contributing to the reduction of the ambient temperature in cities.

The position and orientation of vegetation follows the basic principles of shading design, while the choice of the type of plant and the shape of shade each offers during the year should be carefully examined. Horizontal overhangs, like pergolas, are preferable at the south orientation, while trees should be preferred on the east-south-east and west-south-west orientations. Bushes can also be used to shade east- and west-facing windows, while vertical trellises covered with climbing plants are effective on east and west facades. Creepers are also useful in shading the envelope of the building.

Solar-optical properties of glazing

The penetration of solar energy into the interior of buildings depends on the thermal properties of the glazed surface, which for solar control purposes are defined as (Figure 10.31):

- *reflectivity*, the fraction of solar radiation at normal incidence that is reflected by the glass;
- *solar transmittance*, the fraction of the normal incident solar radiation that is directly transmitted through the glass;
- *absorptance*, the fraction of the solar radiation at normal incidence that is absorbed by the glass.

The total solar transmittance is the fraction of the solar radiation at normal incidence that is transferred through the glass. It is composed of the direct component of the solar radiation (short-wave radiation) and the part of the solar radiation absorbed in the glass and dissipated inwards (long-wave radiation). It should be distinguished

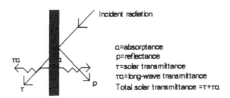

Figure 10.31 Thermal characteristics of glass

from the thermal transmittance (or total transmission, as it is sometimes referred to) through the glass, which is made up of the transmission of the long-wave radiation which results from the radiation emitted from objects, the diffuse solar radiation from the sky and the diffuse component reflected from the ground. The light transmittance – the fraction of the visible part of the spectrum of the solar radiation passing through the glass – is also of importance, as it defines the amount of daylight entering the building.

The thermal properties of transparent materials depend on the angle of incidence of the radiation. The direct solar transmittance remains quite steady until about 50° and drops sharply above 60°. The reflectivity also sharply decreases above an angle of about 60°. For simplicity, in most calculations a mean value is taken (Table 10.4) [15].

Table 10.4 Solar optical properties of typical blind materials and glass for louvres closed and at 45° to the glazing [15]

Blind material	Reflectivity		Absorptance		Transmittance	
	0°	45°	0°	45°	0°	45°
Opaque material						
high-performance blind	0.70	0.50	0.30	0.39	0	0.11
medium-performance blind	0.55	0.40	0.45	0.53	0	0.07
low-performance blind	0.40	0.30	0.60	0.65	0	0.05
Translucent material						
high-performance blind	0.50	0.50	0.10	0.14	0.40	0.36
medium-performance blind	0.40	0.40	0.20	0.27	0.40	0.33
low-performance blind	0.30	0.30	0.30	0.40	0.40	0.30
Clear glass	0.07	0.09	0.13	0.13	0.80	0.78
Body-tinted glass	0.05	0.07	0.51	0.53	0.44	0.40

Control of the heat passing through the glass in the interior of buildings can be achieved with use of special treated glasses, such as:

- body-tinted glasses, with high absorptance
- surface coated glasses, with increased reflectivity
- variable transmission glasses
- translucent glazing materials
- special sun-control membranes
- temporary glazing coatings
- single- and double-glazed units with laminated glass incorporating blinds and louvres.

Reduction of the solar transmittance is usually associated with reduction of the visible part of the solar spectrum, although there are some body tints and coatings which preferentially attenuate the non-visible part of the solar spectrum, leaving unchanged the greater part of the visible radiation.

GLAZING TYPES

Body-tinted glasses (or absorptive glasses). These are characterized by high absorptivity, especially of the long-wave radiation (35–75% of the incident radiation). The high absorptivity is also extended to the visible part of the solar radiation, which results in change of the characteristics of the indoor perceived daylight. Body-tinted glasses are produced by addition of small quantities of metal oxides (iron, cobalt, selenium), and, depending on the proportion of each one, a range of colours – bronze, grey, blue and green – can be produced. Their solar absorption, transmission and colour shade vary with the glass thickness, while their reflectivity is slightly less than that of clear glass. Because of their lower light transmission they are used in some applications in order to reduce intense sunlight. Their optical–thermal properties are shown in Table 10.5 [15].

Surface-coated glasses. Application of special coatings on the glass surface can modify its long-wave and solar radiation transmittance to desired levels. The different glazing types are:

- **Reflective glasses**, which are used for solar control by increasing the reflection of the direct solar radiation with special reflective coatings. These glasses are characterized by an increased solar absorptance compared to a clear glass. They are produced in a wide range of colours. Their properties, compared to other glass types, are shown in Table 10.5 [15].

  Coatings of silver, copper and gold are effective for solar control owing to their increased reflectivity for the near-infrared radiation, which corresponds to about 50% of the solar radiation. The mirror silvering of some glass surfaces is achieved by a coating of metal, mostly silver, onto the surface of clear glass, protected by a layer of copper and a paint back. Its application in most cases is objectionable because of its glare disruption and

Table 10.5 Thermal–optical properties of different glass types (T= Transmittance, R = Reflectivity, A = Absorptance) [15]

Glass type		Light		Solar radiation Direct			Total	Shading coefficient		
Glass thickness	Light/ heat ratio	T	R	T	R	A	T	Short wave	Long wave	Total
Clear Float										
4 mm		0.89	0.08	0.82	0.07	0.11	0.86	0.94	0.04	0.98
6 mm		0.87	0.08	0.78	0.07	0.15	0.83	0.90	0.05	0.95
10 mm		0.84	0.07	0.70	0.07	0.23	0.78	0.80	0.09	0.89
12 mm		0.82	0.07	0.67	0.06	0.27	0.76	0.77	0.10	0.87
Reflective										
Silver										
6 mm	10/23	0.10	0.38	0.08	0.32	0.60	0.23	0.09	0.17	0.26
10 mm	10/23	0.10	0.37	0.08	0.30	0.62	0.23	0.09	0.18	0.27
6 mm	20/34	0.20	0.23	0.16	0.18	0.66	0.34	0.18	0.21	0.39
10 mm	20/34	0.20	0.22	0.15	0.17	0.69	0.34	0.17	0.22	0.39
Bronze										
6 mm	10/24	0.10	0.19	0.06	0.21	0.73	0.24	0.07	0.20	0.27
10 mm	10/24	0.10	0.18	0.05	0.19	0.76	0.24	0.06	0.21	0.27
Blue										
6 mm	20/33	0.20	0.20	0.15	0.21	0.64	0.33	0.17	0.21	0.38
10 mm	20/33	0.20	0.20	0.15	0.19	0.66	0.33	0.17	0.21	0.38
6 mm	30/39	0.30	0.16	0.21	0.18	0.18	0.39	0.24	0.21	0.45
10 mm	30/38	0.30	0.15	0.20	0.17	0.63	0.38	0.23	0.21	0.44
6 mm	40/50	0.40	0.10	0.32	0.10	0.58	0.50	0.38	0.19	0.57
10 mm	40/49	0.40	0.09	0.31	0.09	0.60	0.49	0.36	0.20	0.56
Body-tinted										
Green										
6 mm	76/62	0.72	0.06	0.46	0.05	0.49	0.62	0.53	0.19	0.72
Blue										
6 mm	54/62	0.54	0.05	0.46	0.05	0.49	0.62	0.53	0.19	0.72
Bronze										
4 mm	61/70	0.61	0.06	0.58	0.06	0.36	0.70	0.67	0.13	0.80
6 mm	50/62	0.50	0.05	0.46	0.05	0.49	0.62	0.53	0.19	0.72
10 mm	33/51	0.33	0.04	0.29	0.04	0.67	0.51	0.33	0.26	0.59
12 mm	27/47	0.27	0.04	0.23	0.04	0.73	0.47	0.26	0.28	0.54
Grey										
4 mm	55/68	0.55	0.05	0.55	0.05	0.40	0.68	0.63	0.16	0.79
6 mm	42/60	0.42	0.05	0.42	0.05	0.53	0.60	0.48	0.21	0.69
10 mm	25/49	0.25	0.04	0.25	0.04	0.71	0.49	0.29	0.27	0.56
12 mm	19/45	0.19	0.04	0.19	0.04	0.77	0.45	0.22	0.29	0.51

the reflected part of the solar radiation that is diverted to the surrounding buildings and outdoor spaces.

The combination of a special coating with body-tinted glasses can be produced for a selective performance and specific applications.

- **Low emissivity (low E) glasses**, which are used for reduction of heat losses and improved thermal insulation performance. Special coatings on the surface of the glass reduce its long-wave radiation emissivity, from a value of 0.9 for a clear glass to less than 0.1, and consequently increase their reflectivity – the infrared reflectivity of an appropriate coated glass in the 3–30 μm wavelength range increases to over 80%. The high insulating effect (low U value) of the low-emissivity coatings in a double-glazed window is due to reduced long-wave radiation exchange between the panes, which for an air-filled gap amounts to about 60% of the total heat exchange across the space.

Special coatings can allow maximum daylight and short-wavelength infrared radiation transmittance and reduce heat losses by reflecting the long-wave heat back to the room. Typical coatings of this kind are the three-layer lower E coatings, with the main component a thin metal layer, usually of gold, silver or copper, placed between layers of tin oxide (dielectric layer). The thickness of this layer is chosen to give maximum visible light transmission.

Water droplets on the surface of the coated glass modify its performance because of the high emissivity of the water. During winter, the inner surface of a double glazed unit with low E coating presents higher temperatures than an untreated one, reducing in this way the radiant asymmetry and thermal discomfort close to the window.

In the past, the coatings were 'soft' and, in order to protect them, they were produced in sealed double-glazed units or as two panes bonded together with a plastic material or a resin (lamination). The recent development of hard low-emissivity coatings allows the retrofit of existing windows by installing individually coated glazing in secondary frames. The hard coatings have a slightly higher solar heat transmission than the soft ones.

The comparative performance in light and total solar transmission performance for the different glass types are presented in Figure 10.32 [15].

Variable transmission glasses (VT glass). New products are under development that permit the building envelope to be used dynamically, responding to the outdoor climate and interior thermal needs by modifying the transmission properties of the glass. Among the different ways of achieving variable transmittance are the use of photochromic, thermochromic and electrochromic materials. Thermochromic glasses are considered to lose transparency when activated.

Electrochromic glasses appear to offer the best potential for use. They have an electrochromic coating, which is a transparent multi-layer coating, more complex than other ordinary coatings. The coating is activated by a small electrical voltage – generated from the building services control or manually by the occupants – which changes the tint of the coating and thus its solar and light transmittance. The original transmittance is restored by reversing the voltage.

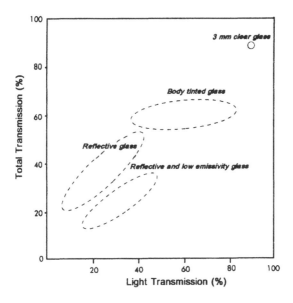

*Figure 10.32 Light and total solar transmission of
different glass types [15]*

The ideal properties of variable transmission glasses have been suggested as [18]:

- solar transmittance from 10–20% for coloured glass to 50–70% for bleached glasses;
- 50–70% for visible transmittance;
- 10–20% for near-infrared reflectivity, which reaches 70% for coloured glasses [18].
- the switching voltage is 1–5 volts, with switching speed 1–60 s and a lifetime of 20–50 years.

As there is no universal single VT glass to meet a wide range of cooling and electrical energy-saving requirements, a variety of material options may be developed to meet specific uses and climates. Although preliminary estimates show savings, which are smaller in energy terms but bigger in capital and running of lighting and cooling equipment (Figure 10.33) [16], the final manufacturing costs are not yet known and thus final comparisons of cost-effectiveness cannot be made.

Recent research has been carried out for development of a special type of glass which will also perform as a communications medium in parallel to its light and thermal response to the environment [15]. With use of electroluminescent display technology, moving decoration is integrated with the information technology to

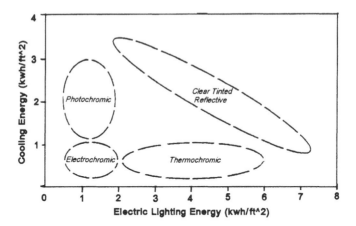

Figure 10.33 Comparative performance of different glass types [16]

provide electronic displays of information or simply colour patterns in response to transmission changes, thus instantly changing the character of the envelope of the building.

Translucent glazing materials. These are characterized by low thermal and solar transmittance. They diffuse daylight and they can be effectively used in spaces where direct optical communication with the external environment is not crucial, such as skylights or an atrium roof. A uniform light quality can be achieved with highly diffusing materials, although they are characterized by lower light transmittance, which under overcast conditions reduces considerably the light levels indoors.

Transparent membranes. These are transparent polymer films with a very thin metallic coating, controlled by a computer, sandwiched between two other polymer sheets [19]. They were introduced in America in 1978 and laboratory measurements have shown an indoor air-temperature reduction of 5–7°C during summer. They reduce the glare and the ultraviolet disruptive effects by about 80%.

Temporary glazing coatings. These come in a form of a translucent gel that is spread onto the internal surface of common glazing. They last for some months and can easily be removed. They are used as a secondary sun-control medium, as they can significantly reduce the level of internal light.

GLAZING TYPE AND CLIMATE
The fundamental requirement to avoid overheating in hot climates can be promoted by minimizing the admission of solar heat in the building's interior. Although rational design of the number and size of openings will give the best performance,

contemporary buildings with extensively glazed facades demand different handling. The building has to be insulated from the external heat gains during the day and its interior has to be protected from the glare which is produced by the intense direct and reflected sunlight without diminishing the level of natural light and without restricting the view. For these requirements, glass with low solar thermal and light transmission is necessary. For improved solar control and higher insulation, a double-glazing unit with a low-emissivity coating on the inner pane provides solar control by reflecting back the heat absorbed by the glass. In the same manner, the external pane is kept warm, avoiding condensation on its surface. Especially in humid climates, the water droplets of the hot ambient humid air can condense on the external surface of the cool windows of air-conditioned buildings. The improved U-value of the glazing also insulates cool air-conditioned overglazed buildings from the external hot environment during the day and reduces their cooling load. In very hot climates, when buildings can experience overheating for a long period of the year, use of glass with very low thermal-transmission properties can be beneficial, despite the reduction of solar gains during winter.

In cold climates with low radiation, use of glass must maximize solar gains and light transmission during the day and provide high insulation, near to that provided by a wall, during the night. Clear glass can perform satisfactorily for collection of solar heat, while addition of heavily insulated shutters for night use can achieve the required thermal performance.

In temperate climates, there should be a balance between the maximum solar collection and night insulation during winter and the control of the solar heat transmission during summer, while adequate levels of natural light inside the buildings should be ensured. Both solar and light transmission can be higher than those required in a hot climate.

The variety of thermal requirements imposed for different climates and building types cannot be met with a unique glazing type. A careful study of the properties of the different glass types needs to be made according to the specific building and the surrounding environment. The proposed applicability of the different glass types for different orientations is shown in Figure 10.34 [19].

REFERENCES

1 *CIBSE Guide, Weather and Solar Data* (1986). Section A2. Chartered Institution of Building Services Engineers, London.
2 Givoni, B. (1976). *Man, Climate and Architecture*, 2nd ed. Applied Science Publishers, London.
3 Duffie, J.A. and W.A. Beckman (1994). *Solar Engineering of Thermal Processes*, 2nd ed. Wiley-Interscience, New York.

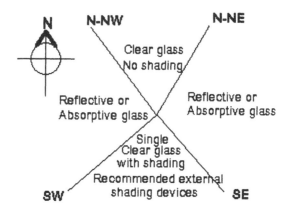

Figure 10.34 Recommended glass types for different orientations

4   Spencer, J.N. (1971). 'Fourier series representation of its position of the sun', *Search*, Vol. 2, No. 5, p. 172.

5   Iqbal, M. (1983). *An Introduction to Solar Radiation*. Academic Press, Toronto.

6   Cooper, P.I. (1969). 'The absorption of solar radiation in solar stills', *Solar Energy*, Vol. 12, p. 3.

7   Page, J.K. (1993). 'Climate and Design', *Energy in Architecture; The European Passive Solar Handbook*', eds J. Goulding, J. Owen Lewis and T. Steemers, Ch 2. Publication No EUR 13446, Batsford for the Commision of the European Community, London,

8   Kreider, J.F. and A. Rabl (1994). *Heating and Cooling of Buildings; Design for Efficiency*. Mechanical Engineering Series, McGraw-Hill International Editions, New York.

9   Braun, J.E. and J.C. Mitchell (1983). 'Solar geometry for fixed and tracking surfaces', *Solar Energy*, Vol. 31, p. 439.

10 Markus, T. A. and E. N. Morris (1980). *Buildings, Climate and Energy*. Pitman Publishing, London.

11 Tzonos, P. (1985) *Solar Control*. Thessaloniki (in Greek).

12 Yannas, S. (1990). 'Solar control techniques', Workshop on Passive Cooling, Commission of the European Communities, EC Joint Research Centre, Ispra (Italy), 2–4 April.

13 Olgyay, V. (1963). *Design with Climate; Bioclimatic Approach to Architectural Regionalism*. Princeton University Press, Princeton, NJ.

14 Lechner, N. (1991). *Heating, Cooling, Lighting: Design Methods for Architects*. J. Wiley & Sons, New York.

15 Button, D. and B. Pye (eds) (1993). *Glass in Building: A Guide to Modern Architectural Glass Performance*. Pilkington Glass Ltd, Flachglas AG, Libby Owens Ford, Pilkington Floatglas AB, Pilkington (Australia) Ltd with Butterworth Architecture, Oxford, London.

16 Brunello, Z. (1987). *Solar Properties of Fenestration, Solar and Thermal Properties of Windows; Expert Guide*, Annex XII. International Energy Agency.

17 Goulding, J., J. Owen Lewis and T. Steemers (eds) (1993).'Passive cooling', *Energy in Architecture; The European Passive Solar Handbook*, Ch. 6. Publication No EUR 13446, Batsford for the Commission of the European Community, London.

18 Selkowitz S. and C. Lambert (1989). 'Application of large area chromogenics to architectural glazings', in *Large Area Chromogenics: Materials and Devices for Transmittance Control*, eds C.M. Lambert and C.G.Grangvis. Optical Engineering Press, Washington, DC.

19 Papadopoulos M. and C. Axarli (1982). *Passive Solar Design and Passive Solar Systems of Buildings*. Thessaloniki, (in Greek).

20 Molina, J.L., S. Alvarez and E.A. Rodriguez (1991). *A Methodology for Calculating Monthly Based Shading Factors*, University of Seville and Commission of the European Communities.

FURTHER READING

1 *Cooling and Heating. Load Calculation Manual.* ASHRAE GRP158.
2 Santamouris, M. *Manual of PASSPORT.* PASSYS II Programme, Commission of the European Communities.

APPENDIX A
SIMPLIFIED METHOD FOR CALCULATION OF THE SHADING COEFFICIENT

A simplified method for calculation of the mean monthly shading coefficients of selected shading devices had been developed, based on a methodology presented in [20]. It is a very simple method to be followed and can be helpful during the initial stage of the design of a building. It is applicable in the following cases:

- **Shading caused by facade shading devices,** including recessed windows, side fins and overhangs.
- **Shading caused by shading devices which are very close to the window,** such as internal venetian blinds, internal roller shades, external sun screens. This category also includes shading caused by the use of multi-layer fenestration with shading features, such as anti-sun and dark coloured glazing.

This method calculates the mean monthly shading coefficient, SC, which is necessary for the calculation of the solar radiation passing through openings in the interior of the building. The method is manual and the user has to complete specific forms, which are presented in the following pages. The whole process is divided into groups of calculations, each one concerning a specific part of the complete procedure. The following magnitudes should be calculated for each case:

- Facade obstacle direct radiation shading factor, SF2
- Diffuse radiation shading factor, SD
- Reflected radiation shading factor, SR.

For application of shading devices, such as sun screens, coloured glazing, curtains, blinds, roller shades, after the calculation of the previously mentioned shading factors, the values for the internal and external shading coefficients, SCin and SCex, should be defined. These values can be found in Tables 10.A5 to 10.A9.

After the calculation of the above factors, the radiation transmitted through the opening total solar radiation per unit surface (H) is given as:

$$H = H_b \, SF2 + H_d \, SD + H_g \, SR$$

where H is the total transmitted solar radiation per unit surface, H_b the mean monthly beam solar radiation on the window surface, H_d the mean monthly diffuse solar radiation on the window surface and H_g the mean monthly ground-reflected solar radiation on the window surface.

The total absolute solar radiation on a surface is calculated as:

$$Q_s = H \times A \times SCin \times SCex$$

where Q_s is the solar radiation transmitted to the space, A the window surface area (m^2), SCin the shading coefficient for internal shading devices and SCex the shading coefficient for external shading devices.

Several assumptions are made for the development of the method:

- The method is valid for North European latitudes (i.e. 35° to 66°).
- For the case of side fins, it is assumed that both right and left fins are normal to the window plane and located at equal distances from the window sides.
- For the case of an overhang, it is assumed that it is normal to the window plane.
- The method is valid mainly for vertical planes, but it can also be used without significant error for tilted surfaces in the range of tilt that is met in buildings.

THE FORMS

Form 1:   Steps 1 to 5:General information data (necessary for all groups of calculations)

1   Determine the site latitude (F) (in deg), (35° to 66°)   _____
2   Determine the surface azimuth (G) (in deg)
    (−90° for east, 0° for south, 90° for west, 180° for north)   _____
3   Determine the surface tilt (B) (in deg)   _____
4   Select from Table 10.A1 the solar declination (D) (in deg)   _____
5   Calculate AR3 = $F - D$ [AR3 = Step 1 − Step 4]   _____

Form 2:   Calculations for remote obstacles

In absence of any remote obstacles, proceed to Step [101]   _____

20   Determine the angle between the centre of the window and the
      top of the remote obstacle (ROA) (in deg)
      In absence of any remote obstacles, set ROA = 0   _____

21   Calculate FWR = (ROA + B – 90) [FWR = Step 20 + Step 3 – 90]   _____

22   Calculate SFW = sin(FWR) [SFW = sin(Step 20)]   _____

23   Calculate FRW = 1 – SFW [FRW = 1 – Step 22]   _____

24   Calculate RFW = FRW/2 [RFW = (Step 23)/2]   _____

25   Calculate view factor Fw-r, as Fw-r = 1 – RFW

      [Fw-r = 1 – (Step 24)]   _____

Form 3:   Calculation of the direct radiation shading factor SF2 in case of recessed window

101   Determine the depth of the recessed window (R) (in m)   _____

102   Determine the height of the window (H) (in m)   _____

103   Determine the width of the window (W), (in m)   _____

104   Calculate the ratio R/H as P1 [= Step 101/Step 102]   _____

105   Calculate the ratio R/W as P2 [= Step 101/Step 103]   _____

106   Select from Table 10.A2 the factor A1 corresponding to the
      window orientation   _____

107   Select from Table 10.A2 the factor B1 corresponding to the
      window orientation   _____

108   Select from Table 10.A2 the factor A2 corresponding to the
      window orientation   _____

109   Select from Table 10.A2 the factor B2 corresponding to the
      window orientation.   _____

110   Calculate BBG = P1+P2 [BBG = Step 104 + Step 105]   _____

111   Calculate GBB = –0.8632 × BBG [GBB = –0.8632 × (Step 110)]   _____

112   Calculate BGB = exp(GBB) [BGB = exp(Step111)]   _____

113   Calculate Fw-f=1 – BGB [Fw-f = 1 – Step112]   _____

114   Calculate MT1=B1 × AR3 [MT1 = Step 107 × Step 5]   _____

115   Calculate MT2 = B2 × AR3 [MT2 = Step109 × Step 5]   _____

116   Calculate MT3 = A1 + MT1 [MT3 = Step 106 + Step 114]   _____

117   Calculate MT4 = MT3 × P1 [MT4 = Step 116 × Step 104]   _____

118   Calculate MT5 = A2 + MT2 [MT5 = Step 108 + Step 115]   _____

119   Calculate MT6 = P2 × MT5 [MT6 = Step 105 × Step 118]   _____

120   Calculate MT7 = MT4 + MT6 [MT7 = Step 117 + Step 119]   _____

121   Calculate the DIRECT RADIATION SHADING FACTOR
      (SF2 = exp (MT7))   _____

Form 4: Calculation of the direct radiation shading factor SF2 in case of side fins

201   Determine the width of the window (W) (in m)   _____

202   Give the length of the side fins (L) (in m)   _____

203   Give the distance between the window and the side fins (D) (in m)   _____

204   Calculate $P1 = L/W$ [P1 = Step202/Step 201]   _____

205   Calculate $P2 = D/W$ [P2 = Step 203/Step 201]   _____

206   Select from Table 10.A3 the factor A1 corresponding to the window orientation   _____

207   Select from Table 10.A3 the factor B1 corresponding to the window orientation   _____

208   Select from Table 10.A3 the factor A2 corresponding to the window orientation   _____

209   Select from Table 10.A3 the factor B2 corresponding to the window orientation   _____

210   Calculate $MY1 = P1 \times P1$ [MY1 = Step 204 × Step 204]   _____

211   Calculate $MY2 = P2 \times P2$ [MY1 = Step 205 × Step 205]   _____

212   Calculate $MY3 = MY1 + MY2$ [MY3 = Step 210 + Step 211]   _____

213   Calculate $MY4 = (MY3)^{0.5}$ [MY4 = (Step 212)$^{0.5}$]   _____

214   Calculate $MY5 = P2/MY4$ [MY5 = Step 205/Step 213]   _____

215   Calculate $MY6 = 1 - MY5$ [MY6 = 1 – Step 214]   _____

216   Calculate $Fw\text{-}f = 0.3282 \times MY6$ [Fw-f = 0.3282 × Step 215]   _____

217   Calculate $MY7 = AR3 \times B1$ [MY7 = Step 5 × Step 207]   _____

218   Calculate $MY8 = A1 + MY7$ [MY8 = Step 206 + Step 217]   _____

219   Calculate $MY9 = MY8 \times P1$ [MY9 = Step 218 × Step 204]   _____

220   Calculate $MY10 = B2 \times AR3$ [MY10 = Step 209 × Step 5]   _____

221   Calculate $MY11 = A2 + MY10$ [MY11 = Step 208 + Step 220]   _____

222   Calculate $MY12 = MY11 \times P1 \times P2$

[MY12 = Step 221 × Step 204 × Step 205]

223   Calculate the DIRECT RADIATION SHADING FACTOR ($SF2 = 1 + MY9 + MY12$) [SF2 = 1 + Step 219 + Step 222]   _____

Form 5:   Calculation of the direct radiation shading factor SF2 in case of an overhang

301   Give the window height (H) (in m)   _____

302   Give the overhang length (L) (in m)   _____

303   Give the distance between the window and the overhang (D) (in m)   _____

304   Calculate P1 = L/H [P1 = Step 302/Step 301]   _____

305   Calculate P2 = D/H [P2 = Step 303/Step 301]   _____

306   Select from Table 10.A4 the factor A1 corresponding to the window orientation   _____

307   Select from Table 10.A4 the factor B1 corresponding to the window orientation   _____

308   Select from Table 10.A4 the factor A2 corresponding to the window orientation   _____

309   Select from Table 10.A4 the factor B2 corresponding to the window orientation   _____

310   Calculate NY1 = P1 × P1 [NY1 = Step 304 × Step 304]   _____

311   Calculate NY2 = P2 × P2 [NY2 = Step 305 × Step 305]   _____

312   Calculate NY3 = NY1+NY2 [NY3 = Step 310 + Step 311]   _____

313   Calculate NY4 = (NY3)$^{0.5}$ [NY4 = (Step 312)$^{0.5}$]   _____

314   Calculate NY5 = P2/NY4 [NY5 = Step 305/Step 313]   _____

315   Calculate NY6 = 1 − NY5 [NY6 = 1 − Step 314]   _____

316   Calculate the factor Fw-f = 0.3282 × NY6 [Fw-f = 0.3282 × Step 315]   _____

317   Calculate NY7 = AR3 × B1 [NY7 = Step 5 × Step 307]   _____

318   Calculate NY8 = A1 + NY7 [NY8 = Step 306 + Step 317]   _____

319   Calculate NY9 = NY8 × P1 [NY9 = Step 318 × Step 304]   _____

320   Calculate NY10 = AR3 × B2 [NY10 = Step 5 × Step 309]   _____

321   Calculate NY11 = NY10 + A2 [NY11 = Step 320 + Step 308]   _____

322   Calculate NY12 = NY11 × P1 × P2

[NY12 = Step 312 × Step 304 × Step 305]   _____

323   Calculate the DIRECT RADIATION SHADING FACTOR (SF2 = 1 + NY9 + NY12)   _____

Form 6:   Calculation of diffuse and reflected radiation shading factors SD and SR

401   Set the value of the factor Fw-f, as previously calculated.
       In absence of any facade obstacles, set Fw-f = 0   _____

402   Define the factor F1w-f = Fw-f.
       In the case of an overhang, F1w-f = 0   _____

403   Calculate IY1 = 1 − Fw-f [IY1 = 1 − Step 401]   _____

404   Calculate IY2 = 1 − Fw-r [IY2 = 1 − Step 25]   _____

405   Calculate IY3 = 2 × IY1 × IY2 [IY3 = 2 × (Step 403) × (Step 404)]   _____

406   Calculate IY4 = cos(B) [IY4 = cos(Step 3)]   _____

407   Calculate IY5 = 1 + IY4 [IY5 = 1 + Step 406]   _____

408   Calculate the DIFFUSE RADIATION SHADING FACTOR
       (SD = IY3/IY5) [SD = Step 405/Step 407]   _____

409   Calculate IY6 = 1 − F1w-f [IY6 = 1 − Step 402]   _____

410   Calculate IY7= 2 × IY6 × Fw-r [IY7 = 2 × Step 409 × Step 25]   _____

411   Calculate IY8=1 − IY4 [IY8 = 1 − Step 406]   _____

412   Calculate the REFLECTED RADIATION SHADING
       FACTOR (SR = IY7/IY8) [SR = Step 410/Step 411]   _____

Form 7:   Calculation of the mean monthly shading coefficient SC

501   Define the mean monthly beam solar radiation for the opening,
       without the shading device (H_b)   _____

502   Define the mean monthly diffuse solar radiation for the open-
       ing, without the shading device (H_d)   _____

503   Define the mean monthly ground reflected solar radiation for
       the opening, without the shading device (H_g)   _____

50   Calculate H1 = H_b+H_d+H_g [H1 = Step 501 + Step 502 + Step 503]   _____

505   Calculate IB1 = SF2 × H_b _____

506   Calculate IB2 = SD × H_d _____

507   Calculate IB3 = SR × H_g _____

508   Calculate IB4 = IB1 + IB2 +I B3 [IB4 = Step 505 + Step 506 + Step 507]   _____

509   Calculate SC = IB4/H1 [SC = Step 508/Step 504]   _____

When any complementary shading devices, such as sun screens, coloured glazing, curtains, blinds or roller shades, are used, the user has to specify the values of SCex and SCin from Tables 10.A5 to 10.A9. An example for each case of shading is given at the end of the section.

TABLES

Table 10.A1 Solar declination angle (in deg)

Month	5th Day	15th Day	25th Day
January	−22.6	−21.3	−19.3
February	−16.4	−13.6	−9.8
March	−6.8	−2.8	1.2
April	5.6	9.4	13.0
May	16.1	18.8	20.9
June	22.5	23.3	23.4
July	22.8	21.5	19.6
August	16.8	13.8	10.3
September	6.2	2.2	−1.8
October	−5.8	−9.6	−13.1
November	−6.5	−19.1	−21.2
December	−22.6	−23.3	−23.4

For monthly based calculations, the 15th day of each month is usually chosen as the reference day.

Table 10.A2 Factors for recessed windows

Orientation	A1	B1	A2	B2
S	−5.695	0.081	−1.342	0.009
SE–SW	−2.418	0.032	−1.479	0.017
E–W	−0.868	0.009	0.232	−0.022
NE–NW	0.336	−0.013	−0.320	−0.074
N	−1.193	0.036	−1.825	−0.163

Table 10.A3 Factors for side fins

Orientation	A1	B1	A2	B2
S	−1.175	0.012	0.860	−0.008
SE–SW	−0.799	0.009	0.684	−0.006
E–W	0.118	−0.014	0.005	0.010
NE–NW	0.155	−0.041	0.680	0.009
N	0.275	−0.133	0.641	0.039

Table 10.A4 Factors for overhangs

Orientation	A1	B1	A2	B2
S	−3.023	0.045	1.285	−0.006
SE–SW	−1.255	0.015	0.905	−0.008
E–W	−0.684	0.005	0.610	−0.004
NE–NW	−0.654	0.006	0.616	−0.006
N	−0.726	0.007	0.609	−0.007

Table 10.A5 Shading coefficients for transparent sheeting

	Transmittance		
Type of plastic	Visible	Solar	SC
Acrylic			
Clear	0.92	0.85	0.98
Grey tint	0.16	0.27	0.52
	0.33	0.41	0.63
	0.45	0.55	0.74
	0.59	0.62	0.80
	0.76	0.74	0.89
Bronze tint	0.10	0.20	0.46
	0.27	0.35	0.58
	0.61	0.62	0.80
	0.75	0.75	0.90
Reflective			
Aluminium metallized polyester film on plastic	0.14	0.12	0.21
Polycarbonate			
Clear (3.2mm)	0.88	0.82	0.98
Grey (3.2mm)	0.50	0.57	0.74
Bronze (3.2mm)	0.50	0.57	0.74

Table 10.A6 Shading coefficients for external louvred sun screens

| Profile angle | Group 1 | | Group 2 | |
(in deg)	Transmittance	SC	Transmittance	SC
10	0.23	0.35	0.25	0.33
20	0.06	0.17	0.14	0.23
30	0.04	0.15	0.12	0.21
≥40	0.04	0.15	0.11	0.20

| Profile angle | Group 3 | | Group 4 | |
(in deg)	Transmittance	SC	Transmittance	SC
10	0.40	0.51	0.48	0.59
20	0.32	0.42	0.39	0.50
30	0.21	0.31	0.28	0.38
≥40	0.07	0.18	0.20	0.30

| Profile angle | Group 5 | | Group 6 | |
(in deg)	Transmittance	SC	Transmittance	SC
10	0.15	0.27	0.26	0.45
20	0.04	0.11	0.20	0.35
30	0.03	0.10	0.13	0.26
≥40	0.03	0.10	0.04	0.13

Profile angle = tan (Sh/P) where Sh/P=Shadow length per unit length of projection.
Group 1:   Black, 9 louvres per cm, width over spacing ratio 1.15.
Group 2:   Light colour, high reflectance, otherwise same as Group 1.
Group 3:   Black or dark colour, 6.7 louvres per cm, width over spacing ratio 0.85.
Group 4:   Light colour or unpainted aluminium, high reflectance, otherwise same as Group 3.
Groups 5 & 6: Same as Groups 1 & 3 respectively except two lights 0.65 cm clear glass with 1.30 cm
         air space.

Table 10.A7 Shading coefficients for double glazing with an intermediate shading element

| Type of glass | Nominal each pane* (cm) | Solar transmittance | | Venetian blinds | | Louvred Sun-screen |
		Outer	Inner	Light	Medium	
Clear out, in	7.6/81.3, 2.5/20.3	0.87	0.87	0.33	0.36	0.43
Clear out, in	2.5/10.2	0.80	0.80	–	–	0.49
Heat absorbing[†] out				0.28	0.30	0.37
Clear in	2.5/10.2	0.46	0.80	–	–	0.41

*Refer to manufacturer's literature for values
[†]Grey, bronze and green tinted heat absorbing glass

Table 10.A8 Shading coefficients for glass without or with interior shading

Type of glass	Nominal thickness of each light,* (cm)	Solar transmittance†	Without interior shading		Venetian blinds		Roller shades		
					Medium	Light	Opaque		Translucent Light
			$h=22.7$§	$h=17.3$			Dark	Light	
Single									
Clear	7.6/81.3 to 2.5/10.2	0.87–0.80	1.00	1.00	0.64	0.55	0.59	0.25	0.39
Clear	2.5/10.2 to 2.5/5.1	0.80–0.71	0.94	0.95	"	"	"	"	"
Clear	7.6/20.4	0.72	0.90	0.92	"	"	"	"	"
Clear	2.5/5.1	0.67	0.87	0.88	"	"	"	"	"
Clear pattern	2.5/20.3 to 22.9/81.3	0.87–0.79	0.83	0.85	"	"	"	"	"
Heat absorbing pattern	2.5/20.3		0.83	0.85	"	"	"	"	"
Heat absorbing pattern	7.6/40.7 to 2.5/10.2		0.69	0.73	0.57	0.53	0.45	0.30	0.36
Tinted	2.5/20.3 to 17.8/81.3	0.59–0.45	0.69	0.73	"	"	"	"	"
Heat absorbing‡	7.6/40.7 to 2.5/10.2	0.46	0.69	0.73	"	"	"	"	"
Heat absorbing or pattern		0.44-0.30	0.60	0.64	0.54	0.52	0.40	0.28	0.32
Heat absorbing‡	7.6/20.3	0.34	0.60	0.64	"	"	"	"	"
Heat absorbing or pattern	2.5/5.1	0.44–0.30 0.24	0.53	0.58	0.42	0.40	0.36	0.28	0.31
Reflective glass			0.30		0.25	0.23			
			0.40		0.33	0.29			
			0.50		0.42	0.38			
			0.60		0.50	0.44			
Double									
Clear out	7.6/81.3, 2.5/20.4	0.71							
Clear in			0.88	0.88	0.57	0.51	0.60	0.25	0.37
Clear out	2.5/10.2	0.61	0.81	0.82					
Clear in									
Heat absorbing out	2.5/10.4	0.36	0.55	0.58					
Clear in					0.39	0.36	0.40	0.22	0.30
Reflective glass			0.20		0.19	0.18			
			0.30		0.27	0.26			
			0.40		0.34	0.33			
Triple									
Clear	2.5/10.2		0.71						
Clear	2.5/20.3		0.80						

* Refer to manufacturer's literature for values

† For vertical blinds with opaque white and beige louvres in the tightly closed positon, Sc is 0.25 and 0.29 when used with glass of 0.71 to 0.80 transmittance

‡ Gray, bronze and green tinted heat absorbing glass

§ The external heat transfer coefficient h of 22.7 and 17.0 W m^{-2} K^{-1} is taken to correspond to wind velocities of 3.35 and 2.24 m s^{-1}

Table 10.A9 Shading coefficients for curtains

Type of curtain	Transmittance	Inside	Outside
White jalousies	5%	0.25	0.10
	10%	0.30	0.15
	30%	0.45	0.35
White curtains	50%	0.65	0.55
	70%	0.80	0.75
	90%	0.95	0.95
Coloured textiles	10%	0.42	0.17
	30%	0.57	0.37
	50%	0.77	0.57
Aluminium-coated textiles	5%	0.20	0.08

APPENDIX B

EXAMPLE

Form 1:   Steps 1 to 5:General information data (necessary for all groups of calculations)

1 Determine the site latitude (F) (in deg), ($35°$ to $66°$)   40°

2 Determine the surface azimuth (G) (in deg)
   ($-90°$ for east, $0°$ for south, $90°$ for west, $180°$ for north)   20°

3 Determine the surface tilt (B) (in deg)   40°

4 Select from Table 10.A1 the solar declination (D) (in deg),   21.5°

5 Calculate AR3 $= F - D$ [AR3 = Step 1 – Step 4]   18.5

Form 2:   Calculations for remote obstacles

In absence of any remote obstacles, proceed to Step [101]   _____

20 Determine the angle between the centre of the window and the
    top of the remote obstacle (ROA) (in deg).
    In absence of any remote obstacles, set ROA $= 0$ 20°

21 Calculate FWR $= (ROA + B - 90)$ [FWR = Step 20 + Step 3 – 90]   −30°

22 Calculate SFW $= \sin(FWR)$ [SFW = sin(Step 20)]   −0.5

23 Calculate FRW $= 1 - SFW$ [FRW = 1 – Step 22]   1.5

24 Calculate RFW $= FRW/2$ [RFW = (Step 23)/2]   0.75

25 Calculate view factor Fw-r, as Fw-r $= 1 - RFW$

   [Fw-r = 1 – (Step 24)]   0.25

Form 3:   Calculation of the direct radiation shading factor SF2 in case of recessed window

101	Determine the depth of the recessed window (R) (in m)	0.3
102	Determine the height of the window (H) (in m)	1.5
103	Determine the width of the window (W), (in m)	2.0
104	Calculate the ratio R/H as P1 [= Step 101/Step 102]	0.2
105	Calculate the ratio R/W as P2 [= Step 101/Step 103]	0.15
106	Select from Table 10.A2 the factor A1 corresponding to the window orientation	-5.695
107	Select from Table 10.A2 the factor B1 corresponding to the window orientation	0.081
108	Select from Table 10.A2 the factor A2 corresponding to the window orientation	-1.342
109	Select from Table 10.A2 the factor B2 corresponding to the window orientation.	0.009
110	Calculate BBG = P1+P2 [BBG = Step 104 + Step 105]	0.26
111	Calculate GBB = –0.8632 × BBG [GBB = –0.8632 × (Step 110)]	-0.224
112	Calculate BGB = exp(GBB) [BGB = exp(Step111)]	0.799
113	Calculate Fw-f=1 – BGB [Fw-f = 1 – Step112]	0.201
114	Calculate MT1=B1 × AR3 [MT1 = Step 107 × Step 5]	1.50
115	Calculate MT2 = B2 × AR3 [MT2 = Step109 × Step 5]	0.166
116	Calculate MT3 = A1 + MT1 [MT3 = Step 106 + Step 114]	-4.195
117	Calculate MT4 = MT3 × P1 [MT4 = Step 116 × Step 104]	-0.839
118	Calculate MT5 = A2 + MT2 [MT5 = Step 108 + Step 115]	-1.176
119	Calculate MT6 = P2 × MT5 [MT6 = Step 105 × Step 118]	-0.176
120	Calculate MT7 = MT4 + MT6 [MT7 = Step 117 + Step 119]	-1.015
121	Calculate the DIRECT RADIATION SHADING FACTOR (SF2 = exp (MT7))	**0.36**

Form 4:   Calculation of the direct radiation shading factor SF2 in case of side fins

201	Determine the width of the window (W) (in m)	2.0
202	Give the length of the side fins (L) (in m)	1.0
203	Give the distance between the window and the side fins (D) (in m)	0.5
204	Calculate P1 = L/W [P1 = Step202/Step 201]	0.5
205	Calculate P2 = D/W [P2 = Step 203/Step 201]	0.25
206	Select from Table 10.A3 the factor A1 corresponding to the window orientation	-0.799
207	Select from Table 10.A3 the factor B1 corresponding to the window orientation	0.009
208	Select from Table 10.A3 the factor A2 corresponding to the window orientation	0.684
209	Select from Table 10.A3 the factor B2 corresponding to the window orientation	-0.006
210	Calculate MY1 = P1 × P1 [MY1 = Step 204 × Step 204]	0.25
211	Calculate MY2 = P2 × P2 [MY1 = Step 205 × Step 205]	0.0625
212	Calculate MY3 = MY1 + MY2 [MY3 = Step 210 + Step 211]	0.3125
213	Calculate MY4 = $(MY3)^{0.5}$ [MY4 = $(Step 212)^{0.5}$]	0.559
214	Calculate MY5 = P2/MY4 [MY5 = Step 205/Step 213]	0.447
215	Calculate MY6 = 1− MY5 [MY6 = 1 – Step 214]	0.552
216	Calculate Fw-f = 0.3282 × MY6 [Fw-f = 0.3282 × Step 215]	0.181
217	Calculate MY7 = AR3 × B1 [MY7 = Step 5 × Step 207]	0.166
218	Calculate MY8 = A1 + MY7 [MY8 = Step 206 + Step 217]	-0.633
219	Calculate MY9 = MY8 × P1 [MY9 = Step 218 × Step 204]	-0.316
220	Calculate MY10 = B2 × AR3 [MY10 = Step 209 × Step 5]	-0.111
221	Calculate MY11 = A2+MY10 [MY11 = Step 208 + Step 220]	-0.573
222	Calculate MY12 = MY11 × P1 × P2 [MY12 = Step 221 × Step 204 × Step 205]	0.0716
223	Calculate the DIRECT RADIATION SHADING FACTOR (SF2 = 1 + MY9 +MY12) [SF2 = 1 + Step 219 + Step 222]	**0.755**

Form 5:   Calculation of the direct radiation shading factor SF2 in case of an
overhang

301	Give the window height (H) (in m)	2.0
302	Give the overhang length (L) (in m)	1.2
303	Give the distance between the window and the overhang (D) (in m)	1.0
304	Calculate P1 = L/H [P1 = Step 302/Step 301]	0.6
305	Calculate P2 = D/H [P2 = Step 303/Step 301]	0.5
306	Select from Table 10.A4 the factor A1 corresponding to the window orientation	-3.023
307	Select from Table 10.A4 the factor B1 corresponding to the window orientation	0.045
308	Select from Table 10.A4 the factor A2 corresponding to the window orientation	1.285
309	Select from Table 10.A4 the factor B2 corresponding to the window orientation	-0.006
310	Calculate NY1 = P1 × P1 [NY1 = Step 304 × Step 304]	0.36
311	Calculate NY2 = P2 × P2 [NY2 = Step 305 × Step 305]	0.25
312	Calculate NY3 = NY1+NY2 [NY3 = Step 310 + Step 311]	0.61
313	Calculate NY4 = $(NY3)^{0.5}$ [NY4 = (Step 312)$^{0.5}$]	0.78
314	Calculate NY5 = P2/NY4 [NY5 = Step 305/Step 313]	0.64
315	Calculate NY6 = 1 − NY5 [NY6 = 1 − Step 314]	0.36
316	Calculate the factor Fw-f = 0.3282 × NY6 [Fw-f = 0.3282 × Step 315]	0.118
317	Calculate NY7 = AR3 × B1 [NY7 = Step 5 × Step 307]	0.83
318	Calculate NY8 = A1 + NY7 [NY8 = Step 306 + Step 317]	-2.19
319	Calculate NY9 = NY8 × P1 [NY9 = Step 318 × Step 304]	-1.31
320	Calculate NY10 = AR3 × B2 [NY10 = Step 5 × Step 309]	-0.111
321	Calculate NY11 = NY10 + A2 [NY11 = Step 320 + Step 308]	1.174
322	Calculate NY12 = NY11 × P1 × P2 [NY12 = Step 312 × Step 304 × Step 305]	0.35
323	Calculate the DIRECT RADIATION SHADING FACTOR (SF2 = 1 + NY9 + NY12)	**0.04**

Form 6:   Calculation of diffuse and reflected radiation shading factors SD and SR

401	Set the value of the factor Fw-f, as previously calculated.	0.26
	In absence of any facade obstacles, set Fw-f = 0	
402	Define the factor F1w-f = Fw-f.	
	In the case of an overhang, F1w-f = 0	0.26
403	Calculate IY1 = 1 − Fw-f [IY1 = 1 − Step 401]	0.74
404	Calculate IY2 = 1 − Fw-r [IY2 = 1 − Step 25]	0.75
405	Calculate IY3 = 2 × IY1 × IY2 [IY3 = 2 × (Step 403) × (Step 404)]	1.11
406	Calculate IY4 = cos(B) [IY4 = cos(Step 3)]	0.766
407	Calculate IY5 = 1 + IY4 [IY5 = 1 + Step 406]	1.766
408	Calculate the DIFFUSE RADIATION SHADING FACTOR	
	(SD = IY3/IY5) [SD = Step 405/Step 407]	**0.628**
409	Calculate IY6 = 1 − F1w-f [IY6 = 1 − Step 402]	0.74
410	Calculate IY7= 2 × IY6 × Fw-r [IY7 = 2 × Step 409 × Step 25]	0.37
411	Calculate IY8=1 − IY4 [IY8 = 1 − Step 406]	0.234
412	Calculate the REFLECTED RADIATION SHADING	
	FACTOR (SR = IY7/IY8) [SR = Step 410/Step 411]	**1.0**

Form 7:   Calculation of the mean monthly shading coefficient SC

501	Define the mean monthly beam solar radiation for the opening,	
	without the shading device (H_b)	12 MJ
502	Define the mean monthly diffuse solar radiation for the open-	
	ing, without the shading device (H_d)	6 MJ
503	Define the mean monthly ground reflected solar radiation for	
	the opening, without the shading device (H_g)	2 MJ
50	Calculate H1 = H_b+H_d+H_g [H1 = Step 501 + Step 502 + Step 503]	20 MJ
505	Calculate IB1 = SF2 × H_b	4.8
506	Calculate IB2 = SD × H_d	1.8
507	Calculate IB3 = SR × H_g	1.6
508	Calculate IB4 = IB1 + IB2 +I B3 [IB4 = Step 505 + Step 506 + Step 507]	8.2
509	Calculate SC = IB4/H1 [SC = Step 508/Step 504]	**0.41**

11

Ground cooling

The concept of ground cooling is based on the heat-loss dissipation from a building to the ground which, as will be explained in the following, during the cooling period has a temperature lower than the ambient. This dissipation can be achieved either by direct contact of an important part of the building envelope with the ground, or by injecting air into the building that has been previously circulated underground, by means of earth-to-air heat exchangers.

GROUND COOLING BY DIRECT CONTACT

A building exchanges heat with the environment by conduction, convection and radiation. For most buildings, being in contact only with the ground, the main mechanism is convection, since the biggest part of the building envelope is in contact with ambient air. Then comes radiation and finally conduction, since the part of the building envelope in contact with the ground is small. The idea of ground cooling by direct contact is to increase the conductive heat exchange. The building is constructed in such a way that an important part of the building envelope is in contact with the ground. An example of such a construction is shown in Figure 11.1. A house is built on a sloping site, facing south. The north side is buried. Therefore conductive heat exchange is increased and the building temperature drops, since the ground is at a lower temperature during the cooling period. The inverse phenomenon, i.e. conductive heat gains, occurs during the heating season. The important thermal contact also has another major advantage regarding the thermal behaviour of the building. This is the increase of the thermal inertia of the building, since the building boundary becomes the ground, which has an important thermal inertia. What is the advantage of large thermal inertia? It is that the indoor temperature swings, induced by the important variation of the outdoor temperature, are decreased; therefore this results in lower average indoor temperatures during the cooling season and higher ones during the heating season. An approximation of the indoor conditions in such a dwelling is the feeling of the indoor environment that one might have inside a cave.

The example of Figure 11.1 is the Llavaneres semi-buried house, located close to Barcelona, Spain, on the Costa Brava. On the island of Santorini, in the Aegean Sea in Greece there are many semi-buried dwellings built during various historical periods. Construction of these still continues today.

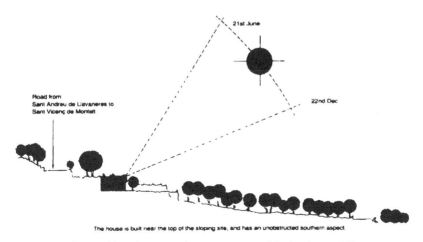

Figure 11.1 Llavaneres house – section of the landscape [1]

As illustrated in Figure 11.1, the house is built near the top of a sloping site, the slope descending towards the south. This gives the building the advantage of contact with the ground at the north facade, thus increasing the heating losses during summer and giving the building an important thermal inertia. Figures 11.2 and 11.3 show the south facade and the roof of the building, which is planted for better integration with the surrounding landscape. Figure 11.4 shows the floor plan section. It should be noted that in order to avoid daylight problems all principal spaces are located in the southern part of the building, which has external openings, while the northern part is reserved for the auxiliary spaces.

Figure 11.2 South facade of the Llavaneres building [1]

*Figure 11.3 Roof of the Llavaneres building, integrated
with the surrounding landscape [1]*

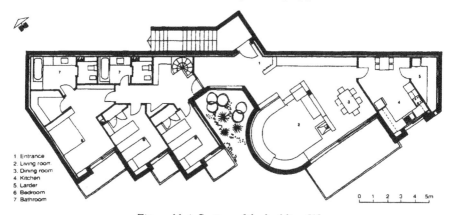

1 Entrance
2. Living room
3. Dining room
4 Kitchen
5. Larder
6 Bedroom
7 Bathroom

0 1 2 3 4 5m

Figure 11.4. Section of the building [1]

Semi-buried buildings are very common on the island of Santorini, Greece. Figure 11.5 shows photographs of existing buildings in the village of Oia in Santorini. Semi-buried buildings under construction in the same village are shown in Figure 11.6.

Other than the thermal inertia of earth contact buildings, there are also reasons why these buildings present a reduced energy consumption. The earth around the envelope of these buildings makes them more air-tight and therefore infiltration during both winter and summer is reduced, leading to lower heating and cooling loads respectively. If the exposed part of the building is on the leeward side of the

Figure 11.5 Examples of semi-buried houses in Santorini, Greece
(A. Argiriou, personal archive)

prevailing winter winds, then infiltration through this part of the building is reduced. Regarding cooling, this is achieved not only because of the increased conductive losses through the ground but also because of the modulation of the solar gains through the building envelope. The massive ground around the building absorbs solar radiation during daytime. This excess heat is dissipated convectively at night by the ambient air and, in places with low relative humidity and clear sky, dissipation is also achieved by long-wave radiation towards the sky vault. If, furthermore, the ground in contact with the building is planted, the plants will absorb the most of the solar radiation and only a small amount will be absorbed by the earth.

Advantages of earth-contact buildings

Earth-contact buildings have advantages that are not only related to their energy performance [2]. These are described in the following subsections.

VISUAL IMPACT – AESTHETICS

This is the reason for the existence of many earth-contact buildings, built before the energy crises. Since these buildings are below ground level and their envelope is covered by earth, they are better integrated into the landscape than most other buildings. The examples given above illustrate this fact. This is the best solution that a designer can choose if he has to design a building in locations of historical interest where modern buildings would be disruptive, for example museums or other buildings in archaeological sites. Some non-residential buildings such as industrial buildings, storage areas or garages are also often built underground, since they are rather massive and often unattractive.

Figure 11.6 Semi-buried houses under construction in Santorini, Greece (A. Argiriou, personal archive)

PRESERVATION OF SURFACE OPEN SPACES

The preservation of open space on the surface of earth-contact buildings is closely related to the aspect of visual impact. Creating an underground building, the roof of which can be used as a garden or other type of open space, is a very interesting solution, particularly in densely built areas, where open space may be limited but, at the same time, necessary.

ENVIRONMENTAL BENEFITS

There are direct and indirect environmental benefits associated with the use of underground buildings. The direct benefits are that the visual impact of the building is minimized and therefore the landscape is only slightly disturbed. Also the roofs of these buildings, if planted, absorb carbon dioxide and generate oxygen. The indirect impact is related to their reduced energy consumption for heating and cooling. The benefits due to energy conservation are not easy to quantify since they depend on the individual design of each building and on the climate. As a rule of thumb, it can be said that the greater the part of the envelope in contact with the ground, the greater the benefits from energy savings.

NOISE AND VIBRATION CONTROL

The heavy mass of this type of buildings absorbs the noise from the ambient and vibrations caused in various ways. Therefore, these dwellings are particularly suitable for locations close to airports, highways, railways, etc. This can also act in a complementary way; the mass of the building absorbs noises produced inside the building, as for example in earth-contact industrial buildings, thus protecting the outside environment.

MAINTENANCE

Earth contact buildings in general require less maintenance than most other build-ings, since the major part of their envelope is sheltered and therefore not degraded by the weather.

OTHER BENEFITS

The above-mentioned benefits are true for all types of earth-contact buildings. For the sake of completeness we will give here some other benefits, the importance of which depends on the type of building use and on the location of the building. These are fire protection and protection against earthquakes, suitability for civil defence, protection against storms and tornados and higher security against outside intrusions.

Limitations of earth-contact buildings

Despite their energy-related benefits, earth-contact buildings also have some poten-tial limitations. The presentation of these limitations is not meant to discourage the reader of this chapter from choosing an earth-contact building as a design solution, but to help him or her to identify the limitations and try to minimize their effect by an appropriate design.

STRUCTURAL AND ECONOMIC LIMITATIONS

Earth-contact buildings require more expensive structures since the roofs must bear the great weight of the soil and the pressure on the lateral buried walls (this pressure increases with depth). These limitations can be overcome by choosing between two construction methods. The first is to use walls surrounded by earth and a conven-tional roof, which is cheaper than a heavily structured earth-covered roof. The sec-ond approach is to use structural elements having the shape of shells or domes, which support higher earth loads efficiently. This results in lower costs and in higher quantities of earth placed over the building structure.

DAYLIGHTING ASPECTS

Energy-efficient design of buildings is not related only to reducing the energy for heating and cooling. Electricity consumed for artificial lighting must also be consid-ered because in many types of buildings this consumption is significant. Therefore, the depth and the percentage of the building envelope that will be sheltered under the earth depend also on the area of the necessary openings that provide natural light inside the building. In the case of the Llavaneres building this problem was solved (see Figure 11.4) by placing as many spaces as possible on the exposed side of the building, while all the auxiliary spaces were placed on the opposite side.

SLOW RESPONSE

Slow response due to the important thermal inertia of the ground is, as has been previously explained, the major advantage of this type of buildings. This advantage could become a disadvantage if some energy conservation strategies, such as night thermostat setback, are to be considered.

CONDENSATION – INDOOR AIR QUALITY

Since during summer the earth temperature is lower than the ambient temperature, condensation may occur on the internal surfaces of the building, if their temperature drops below the dew-point temperature. This problem is directly related to the weather conditions and can be solved by dehumidification or, more commonly, by increased ventilation rates. Ventilation is also the solution to the indoor air-quality problems which can occur in such very well sealed buildings. As well as odours, low infiltration results in an increased concentration of a wide range of detrimental pollutants, including radon. Ventilation, however, will not affect energy savings if the necessary heat-exchange equipment is used.

Modelling the earth-contact buildings

Thermal-performance simulation is necessary in order to determine the comfort conditions in a building and the related heating and cooling energy requirements. In the case of earth-contact building thermal simulation has two additional goals: (i) to evaluate the potential for moisture condensation and (ii) to evaluate the potential for freeze and thaw or frost-heave structural damage.

Various methods for performing earth-contact analysis have been developed. These can be classified into manual and computerized methods. The manual methods are based on the use of various heat sinks. The potential heat sinks from an underground structure are the mean daily ground temperature and the deep ground, the temperature of which is assumed constant, equal to the mean annual ground temperature. In temperate regions the ground temperature is, during winter, the main heat sink for losses from basements and other underground spaces built close to the earth's surface. Therefore, the effect of deep ground can be ignored. This is not the case during the cooling period, when the deep ground is the only heat sink; in this case, however, interactions with surface ground temperature cannot be neglected, since this is a heat source, acting in opposition to losses to the deep ground. The combined influence of these two heat sinks may be represented by a third fictitious heat sink. All methods for calculating heat losses from underground spaces can be classified according to the sink or combination of heat sinks employed in their algorithms. A second classification is related to the estimated parameters. A first category of methods estimates the winter-design heat loss only. These methods are (1) the 'old ASHRAE' method and the methods of (2) Elliot and Baker, (3) Boileau and

Latta and (4) Wang. The second category comprises methods estimating heat losses throughout the year. These are the Mitalas method and the F-Factor method [2]

When computer methods are used, the three-dimensional heat transfer process is modelled. In some cases only the two-dimensional problem is treated, after some assumptions have been made. Such models are the Kusuda and Achenbach model, the Speltz model and the Ceylan and Myers model [2].

Conclusions

Earth-contact buildings present important advantages regarding their energy consumption both during winter and summer, but also a number of other advantages, not always directly related to energy conservation but to more general environmental issues. Their cost is sometimes a drawback to their realization together with a series of other problems, such as the availability of natural light, condensation and indoor air quality. However these problems can be successfully faced if careful studies during the design phase are carried out.

EARTH-TO-AIR HEAT EXCHANGERS

If various constraints do not allow the thermal contact of the building envelope with the ground, then the ground properties can be used for cooling the indoor air of the building indirectly, by means of earth-to-air heat exchangers. An earth-to-air heat exchanger is a pipe buried horizontally at a certain depth, through which the air circulates by means of electric fans. This technique has been developed recently, based on a similar concept employed by the Persians and Greeks in the pre-Christian era. More recent constructions, from the 16th century, have been reported; these use natural cavities ('*covoli*'), found in the hills of Vicenza, Italy [3].

The principle of a modern system is shown in Figure 11.7. Air is sucked by means of a fan from the ambient and enters the building through the buried pipe. During summer, as has previously been explained, since the ground temperature is lower than the ambient air, the air temperature at the outlet of the exchanger is lower than the temperature at the inlet. The opposite phenomenon occurs during winter.

Earth-to-air heat exchangers can be applied in either an open-loop or a closed-loop circuit. The example in Figure 11.7 is an open-loop circuit. In a closed-loop circuit, both inlet and outlet are located inside the building. Plastic, concrete or metallic pipes are used in modern applications. They are used to cool the ambient air before injecting it into the building, or the indoor air if the system is used in a closed loop. The temperature decrease of the air depends upon the inlet air temperature, the ground temperature at the depth of the exchanger, the thermal conductivity of the pipes and the thermal diffusivity of the soil, as well as the air velocity and pipe dimensions. Detailed calculations are needed to optimize such a system. As a

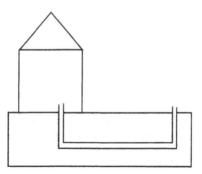

*Figure 11.7 Principle of earth-to-air
heat exchangers, linked to a building*

threshold value for the application of this system, the ground temperature around the
tubes should be at least 5–6 K lower than the air temperature.

Theoretical principles

The principle of ground cooling is directly related to the thermal properties of the
ground. The ground has thermal properties that give it a high thermal inertia. This is
not a new idea. In the island of Delos in Greece during ancient times, as well as in
Morocco, it was usual to cook sheep in a hole in the ground. Heat from a fire was
stored in the ground and used later to cook the meat away from the flames.

The heat transfer mechanisms in soils are, in order of importance: conduction,
convection and radiation. Conduction occurs throughout the soil but the main flow
of heat is through the solid and liquid constituents. Convection is usually negligible,
with the exception of rapid water infiltration after irrigation or heavy rain. Radiation
is important only in very dry soils, with large pores, when the temperature is high.
Therefore, the main parameters influencing the thermal behaviour of the soil are
thermal conductivity and heat capacity. It must be noted here that most soil minerals
have the same density and specific heat. Thermal conductivity and heat capacity can
be jointly expressed under the term of thermal diffusivity:

$$\alpha \equiv \frac{\lambda}{\rho c}$$

where λ is the thermal conductivity, ρ the density and c the specific heat of the soil.
Thermal diffusivity determines the thermal behaviour of the soil. Thermal diffusivity
values for various soils are given in Table 11.1.

From the definition of thermal diffusivity and the values in Table 11.1, it is concluded that heat tends to be stored in the ground, rather than to propagate in it.
Ground can be considered as a semi-infinite medium and therefore the heat transfer
equation becomes

Table 11.1 Thermal diffusivity of various ground types [4]

Soil description	α $(\times 10^{-6}\ m^2.s^{-1})$
Concrete	0.49–0.7
Granite	0.8–1.83
Limestone	0.57–0.59
Marble	1.39
Sandstone	1.06–1.26
Earth coarse	0.139

$$\frac{\partial T}{\partial t} = \alpha \frac{\partial^2 T}{\partial z^2} \tag{11.1}$$

where z is the ground depth. Assuming that the ground surface is subjected to a sinu-soidal variation, then the boundary conditions are

$$T(t,\ 0) = T_a + \Theta_0 \cos \omega t$$

$$T(t,\ \infty) = \text{constant}$$

where T_a is the ambient air temperature, Θ_0 is the amplitude of the assumed sinusoi-dal variation and ω is the angular frequency of the variation (s^{-1}).

The solution of equation (11.1) is

$$T(t, z) = T_a + \Theta_0 e^{-z/d} \cos(\omega t - \frac{z}{d}) \tag{11.2}$$

where $d = (2\alpha/\omega)^{0.5}$ is called the damping depth, because it governs the penetration depth of the 'temperature waves' in the soil. At a depth of $5d$, the temperature varia-tions are almost completely damped out. In order to help understanding of the meaning of the damping depth, the following example is given.

The diurnal angular frequency is $\omega = 2\pi/(24 \times 60 \times 60)$ $s^{-1} = 0.727 \times 10^{-4}$ s^{-1} and therefore the damping depth is 0.148 m for a granite soil (diffusivity given in Table 11.1) and 0.0309 m for coarse earth. Hence the penetration depth of the diurnal variation is 0.740 m for the first ground type and 0.155 m for the second one. For the annual variation, these values are $(365)^{0.5}$ or about 20 times larger [3].

This example illustrates that rapid temperature variations, such as the daily varia-tion of the ambient temperature, have a small damping depth and therefore, below a depth of 1 m, the ground temperature is not influenced. This is the physical principle on which ground cooling is based.

Modelling the ground temperature

The determinant parameter for the evaluation of the ground cooling potential is the ground temperature at various depths. Ideally, this value should be measured. However, only a few weather stations perform measurements of ground surface temperature, while the number of the stations where measurements at various depths are performed is even smaller. This is why algorithms for the calculation of the ground temperature at various depths have been developed. For homogeneous soil of constant thermal diffusivity, the ground temperature at any depth z and time t is [5]:

$$T(z, t) = T_m - A_s \exp\left[-z\left(\frac{\pi}{365\alpha}\right)^{1/2}\right]$$

$$\times \cos\left[\left(\frac{2\pi}{365}\right)\left(t - t_0 - \frac{z}{2}\left(\frac{365}{\pi\alpha}\right)^{1/2}\right)\right] \qquad (11.3)$$

where
T_m = average annual temperature of the soil surface (°C),
A_s = amplitude of surface temperature variation (°C),
z = depth (m),
α = thermal diffusivity of the ground (m² h⁻¹),
t = time elapsed from the beginning of the calendar year (hours).
t_0 = a phase constant (hours) (hours since the beginning of the year of the lowest average ground surface temperature).

This equation shows that the soil temperature at a certain depth depends mainly on the surface temperature and on the thermal characteristics of the soil. Figures 11.8 and 11.9 show the variation of the surface temperature of various soils. Measurements were taken in Athens, by the National Observatory of Athens.

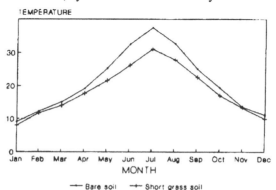

Figure 11.8 Variation of the multi-year mean monthly surface temperature of bare and short-grass-covered soil [6].

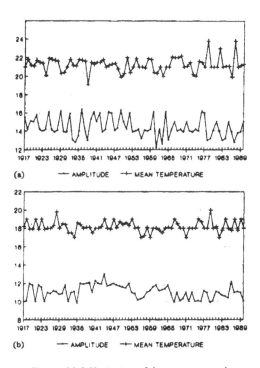

*Figure 11.9 Variation of the mean annual
temperature as well as of the annual amplitude for
(a) bare and (b) short-grass covered soil [6]*

The effect of the thermal inertia of the ground is clearly illustrated in Figure 11.10, which shows the convergence of the ground temperature at a practically constant value as depth increases [2].

Modelling the earth-to-air heat exchangers [7]

SINGLE HEAT EXCHANGER

The transport phenomena that determine the operation of an earth-to-air exchanger are the heat transfer between the air circulating in the exchanger and the ground, and the moisture transfer induced by this heat transfer process. Therefore this process is described by the heat balance differential equation and a mass transfer differential equation. This system of equations is solved, taking into account the initial and boundary conditions.

The heat balance equation, in cylindrical co-ordinates (see Figure 11.11), is

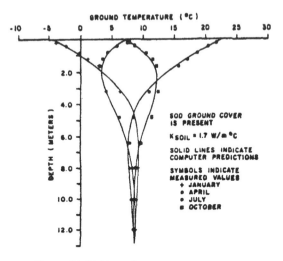

Figure 11.10 Ground temperature variation as
a function of depth [2]

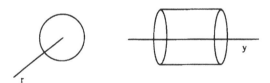

Figure 11.11 Definition of the system of coordinates
for the solution of the heat balance equation

$$\rho c_p \, \frac{\partial T}{\partial t} = \frac{1}{r} \frac{\partial}{\partial r}\left(kr \, \frac{\partial T}{\partial r} \right) + \frac{\partial}{\partial y}\left(k \, \frac{\partial T}{\partial y} \right)$$

$$- l_g \rho_m \frac{1}{r} \frac{\partial}{\partial r}\left(D_{u,\mathrm{vap}} r \, \frac{\partial h}{\partial r} \right) - l_g \rho_m \frac{\partial}{\partial y}\left(D_{u,\mathrm{vap}} r \, \frac{\partial h}{\partial y} \right)$$

with T soil temperature, ρ soil density, c_p specific heat of the soil, k thermal conductivity of the soil, l_g heat of vaporization of the moisture content of the soil, ρ_m moisture density, $D_{u,\mathrm{vap}}$ isothermal diffusivity of moisture in vapour form, and h the moisture content of the soil.

The mass transfer equation is

$$\frac{\partial h}{\partial t} = \frac{1}{r}\frac{\partial}{\partial r}\left(D_T r \frac{\partial T}{\partial r}\right) + \frac{\partial}{\partial y}\left(D_T \frac{\partial T}{\partial y}\right) + \frac{1}{r}\frac{\partial}{\partial r}\left(D_u r \frac{\partial h}{\partial r}\right) + \frac{\partial}{\partial y}\left(D_u \frac{\partial h}{\partial y}\right)$$

where

$$\frac{1}{r}\frac{\partial}{\partial r}\left(D_T r \frac{\partial T}{\partial r}\right) + \frac{\partial}{\partial y}\left(D_T \frac{\partial T}{\partial y}\right)$$

and

$$\frac{1}{r}\frac{\partial}{\partial r}\left(D_u r \frac{\partial h}{\partial r}\right) + \frac{\partial}{\partial y}\left(D_u \frac{\partial h}{\partial y}\right)$$

are the moisture transfer terms due to the temperature and moisture gradients respectively. D_T is the thermal diffusivity of moisture and D_u is the isothermal diffusivity of moisture, given by the following relation:

$$D_u = D_{va} v\alpha a \frac{gM}{RT}\frac{\rho_v}{\rho_l}\left.\frac{\partial \Phi}{\partial \theta}\right|_T$$

where D_{va} is the coefficient of molecular diffusion of water vapour to the air ($L^2 T^{-1}$). This coefficient can be determined from the expression:

$$D_{va} = 4.42 \times 10^{-4}\frac{T^{2.3}}{p}$$

which gives the parameter D_{va} in cm^2 s^{-1} when the temperature T is expressed in K and the total pressure of the gaseous phase in mmHg. $v = p/(p - p_v)$, where p is the total pressure of the gaseous phase and p_v is the partial pressure of the water vapour inside the gaseous phase contained in the soil, α is the tortuosity factor (dimensionless), a is the gas content per unit volume of soil, g is the acceleration of gravity (m s^{-2}), R is the constant of ideal gases (J K^{-1} mol^{-1}), M is the molecular mass of water (kg mol^{-1}), ρ_v is the density of water vapour (M L^{-3}), ρ_l is the density of liquid water (ML^{-3}) and Φ is the hydraulic potential of liquid water (L).

This heat and mass transfer problem has the following initial conditions:

$$T(r, y, 0) = T_0(r)$$

$$h(r, y, 0) = h_0(r).$$

The boundary conditions are:

1 For the r coordinate:

– At a large distance from the pipe (undisturbed soil) ($r = R_a$ = distance at which the ground temperature is not disturbed by the presence of the pipes):

$$T(R_a, y, t) = T(R_a)$$

$$h(R_a, y, t) = h(R_a).$$

– At the outer surface of the pipe the calculated heat flow from the air to the pipe equals the heat flow in the soil:

$$(UA)_{Pipe}(T_a(y, t) - T(R_o, y, t)) = -m'_a c_a dT_a(y) \tag{11.4}$$

where m'_a is the air mass flow rate through the exchanger, c_a the specific heat of the air and T_a the air temperature inside the exchanger. The overall heat transfer coefficient through the pipe is defined as follows:

$$(UA)_{Pipe} = \frac{2\pi l}{(R_{in}h_c)^{-1} + \ln(R_o/R_{in})/k_p} \tag{11.5}$$

with l the length of the earth-to-air heat exchanger, R_o the outer pipe radius, R_{in} the inner pipe radius, k_p the thermal conductivity of the pipe material, h_c the convective heat transfer coefficient between the air in the pipe and the inner pipe surface, and T_a the air temperature inside the pipe.

Another boundary condition is that $T_a(0)$ equals the ambient air temperature.

At $r = R_o$ the migration of moisture is due to the temperature gradient since the pipe walls are not permeable to water. Therefore, the component of the moisture flux due to the moisture gradient is zero:

$$\frac{\partial h(R_o, y, t)}{\partial r} = 0.$$

2   For the y coordinate:
– At a very large distance from both the inlet and outlet of the exchanger, where the temperature field of the soil is undisturbed, y_a and y_b respectively, the temperature equals the temperature of the undisturbed soil:

$$T(r, y_a, t) = T_s(r)$$

$$h(r, y_a, t) = h_s(r)$$

$$T(r, y_b, t) = T_s(r)$$

$$h(r, y_b, t) = h_s(r)$$

where h_s is the moisture content of the undisturbed soil.

The soil temperature close to the exchanger is calculated by super-imposing the temperature field due to the presence of the earth-to-air heat exchanger and the temperature of the undisturbed soil, given in equation (11.1).

N HEAT EXCHANGERS

The thermal behaviour of N earth-to-air heat exchangers is obtained by superimposing the solutions of the system describing the thermal behaviour of a single pipe[8]. Assume that Q_i is the heat through the exchanger i and Q_j is the heat through the exchanger j. These quantities are equal to the right-hand side of equation (11.4), since they are equal to the heat loss of the air circulating in the exchanger. Let $(UA)_{ij}(1 - \delta_{ij})$ be the coupling thermal conductance between the two parallel heat exchangers, $T_{a,i}(y)$ the air temperature in the exchanger i and $T_i(R_o, y, t)$ the temperature at the exchanger–soil interface. Then the difference between the air temperature in the exchanger i and the outer surface of the exchanger is given by

$$T_{a,i} - T_i(R_o, y, t) = \frac{Q_i}{(UA)_i} + \sum_{j=1}^{N} \frac{Q_j}{(UA)_{ij}} (1 - \delta_{i,j})$$

where $\delta_{i,j}$ is the Kronecker delta, defined as follows:

$$\delta_{i,j} = \begin{cases} 1, & i = j \\ 0, & i \neq j \end{cases}.$$

The first term at the right side of this relationship describes the air temperature variation due to the conductance of a single exchanger and the ground, while the second term gives the air temperature variation due to the influence of the other exchangers. The conductance $(UA)_i$ is given by equation (11.5), while the conductance between two pipes i and j is given by

$$(UA)_{ij} = \frac{2\pi l k_p}{\ln\left(\frac{\sqrt{B_{ij}^2 + 4z_i z_j}}{B_{ij}}\right)} (1 - \delta_{i,j})$$

where B_{ij} is the distance between pipes i and j and z_i is the depth of pipe i.

OTHER EARTH-TO-AIR HEAT EXCHANGER MODELS

In addition to the analytical model presented above, there are various other algorithms reported in literature, describing the performance of earth-to-air heat exchangers. Eight of these models have been examined regarding their sensitivity to the inlet air temperature, air velocity, pipe radius, length and depth. The results of these algorithms have also been compared with available experimental data [9]. These algorithms were:

- the Schiller model
- the Santamouris model
- the Rodriguez et al. model
- the Levit et al. model
- the Seroa et al. model
- the Elmer and Schiller model
- the Sodha et al. model
- the Chen et al. model.

The sensitivity analysis and comparison with experimental data performed have shown that all the algorithms have the same behaviour regarding the variation of their input parameters, except the Schiller and Sodha models. The results of these models have the same trend as the results of the other models, but the Schiller model always gives higher temperatures, while the Sodha model always gives lower temperatures.

The difference in the results given by the Schiller model is explained by the fact that this is the only model that calculates the ground temperature at the depth of the exchanger, while all the other models require this temperature as an input.

In order to evaluate and validate these algorithms, their predictions were compared with experimental data. Measurements of the outlet temperature of a 14.8 m long PVC earth-to-air heat exchanger buried at a 1.1 m depth were performed. The pipe diameter was of 0.15 m and the thickness of the pipe wall 0.01 m. Ambient air, heated by a 2 kW electric heater, circulated through the pipe with a velocity of 4.5 m s^{-1}. The inlet and outlet temperatures, the ground surface temperature and the ground temperature at various depths were measured.

A second experiment was also performed, aiming to test the behaviour of the models by applying intermittent air circulation. The fans were operated every day between 08:00 and 20:00, creating an air velocity equal to 10.5 m s^{-1}. Measurements were performed for 15 days during June and July 1983. The temperature measurement points were the same as for the first experiment. The thermal diffusivity of the ground was determined experimentally and used as input for the models.

Figure 11.12 illustrates the variation of the measured temperature at the outlet of the exchanger and the values predicted by the various models. The models can follow the long-term dynamic behaviour of the outlet air temperature, but their predictions of the maximum and minimum values on a daily basis are much less accurate. In order to quantify the accuracy of the various models relative to the experimental results, the root-mean-square error between experimental data and calculated values was determined (Figure 11.13) for the whole experimental period.

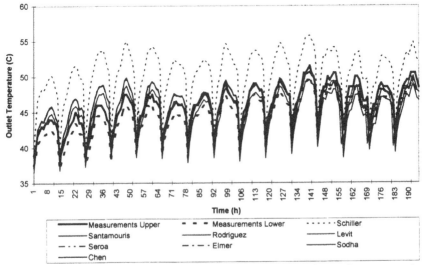

Figure 11.12 Measured and predicted outlet air temperatures during the constant air circulation experiment

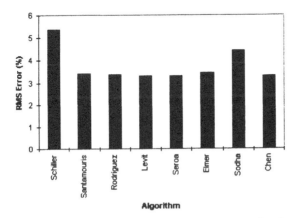

Figure 11.13 The r.m.s. errors between measured and predicted outlet temperatures for the constant air circulation experiment (measurement period 5–13 June 1982)

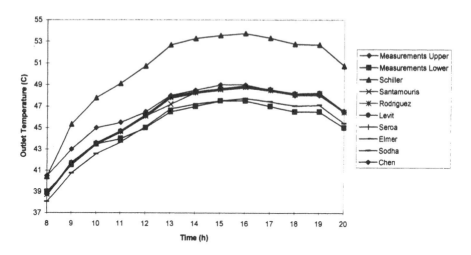

Figure 11.14 Measured and predicted outlet temperatures during the intermittent air-calculation experiment

It is observed that the Schiller and the Sodha algorithms have the highest r.m.s. error. All the other algorithms present the same r.m.s. error of about 3.4%.

The comparison between measurements and predictions for a typical day during the second experiment is illustrated in Figure 11.14. In this figure the upper and lower limits of the error band of the measurements is plotted. All models give results close to the upper limit of the error band, except the Schiller algorithm which gives results 3–4°C higher than the upper limit of the error band, while the Sodha algorithm gives values located in the middle of the error band.

The r.m.s. errors between measurements and predictions for this experiment are given in Figure 11.15. This error was calculated for both upper and lower values of the error band. The Schiller algorithm presents the highest r.m.s. error. The errors of all other models for upper and lower limits of measurement are respectively 2.98% and 4.7%, except the Sodha algorithm, which gives almost the same for both cases, 3.53% and 3.25% respectively.

Cooling potential of earth-to-air heat exchangers

The cooling potential of earth-to-air heat exchangers has been evaluated, using the model presented above. The thermal performance of a plastic tube of 0.125 m in radius and 30 m length, buried at 1.2 m and guiding air at a speed of 5 m.s^{-1} was simulated. Calculations cover the period 1981–1990 for the months June, July and August. Hourly values of the air and ground temperature from 9 a.m. to 9 p.m. were used. The calculated cumulative frequency distributions of the inlet-air temperature are illustrated in Figure 11.16.

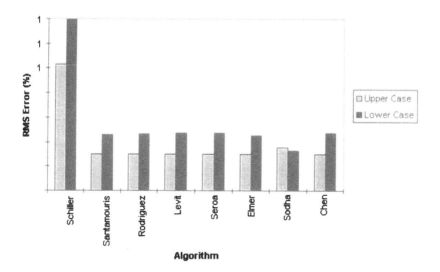

Figure 11.15 R.m.s. errors between measured and predicted outlet temperatures for the intermittent air circulation experiment (measurement period 18 June to 2 July 1983)

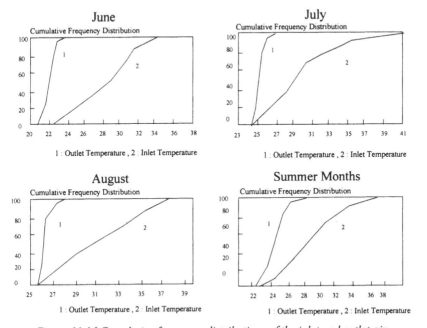

Figure 11.16 Cumulative frequency distributions of the inlet and outlet air temperature [10]

It is estimated that the outlet-air temperature varies in the ranges 20.7 to 23.7°C, 23.2 to 26.8°C and 25.1 to 28.5°C for June, July and August respectively. The corresponding measured inlet-air temperatures vary in the ranges 20.9 to 37.8°C, 23.2 to 40°C and 25.3 to 39.3°C for June, July and August respectively. The overall analysis has shown that the cooling potential of the earth-to-air heat exchangers during the summer period is very important. During June the outlet-air temperature is always lower than 24°C and therefore the heat-exchanger system can be used throughout the month, while for July the outlet-air temperature is lower than 25.5°C for almost 90% of the cases. In August the outlet-air temperature is lower than 26°C for 70% of the period.

In order to evaluate the effect of parameter variations on the performance of the system under real climatic conditions, an extensive sensitivity analysis was performed upon its main design parameters, which are the pipe length and radius, the depth and the air velocity. For each one of these variables a sensitivity analysis has been carried out for a range of values covering existing design practice. Three pipe lengths, of 30, 50 and 70 metres, were tested in the simulations. The cumulative frequency distributions of the outlet-air temperature are shown in Figure 11.17.

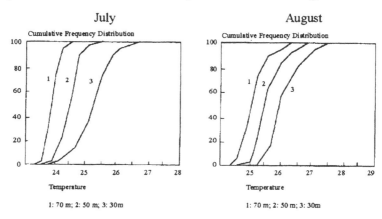

Figure 11.17 Cumulative frequency distributions of outlet-air temperature for the three pipe lengths [10]

It was found that almost 97% of outlet-air temperatures were higher than 26.9°C during July and 28.2°C during August for the 30 m long pipe. For the 50 m long pipe, the temperature was 25.5°C during July and 26.5°C during August. For the 70 m long pipe it was 24.7°C during July and 25.9°C during August.

These calculations show that although the inlet-air temperature is higher in July than in August, the outlet-air temperature is consistently higher during August. This is due to the important thermal inertia of the ground, which has warmed up after a certain time lag.

Pipes with three radii, 0.125 m, 0.180 m and 0.25 m were simulated. Results are reported in Figure 11.18. It was found that a decrease of the pipe radius from 0.250 m to 0.125 m decreases the outlet-air temperature by 1.5 to 2.5°C. An increase of the pipe radius decreases the convective heat transfer coefficient providing a higher air temperature at the outlet of the pipe and therefore a reduction of the cooling capacity of the system.

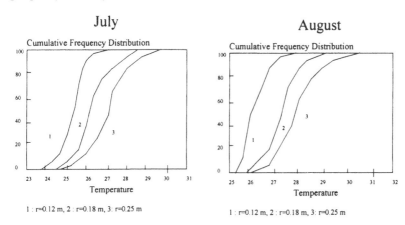

Figure 11.18 Cumulative frequency distributions of outlet-air temperature for the three pipe radii [10]

System performance has been simulated for air velocities of 5, 10 and 20 m s⁻¹. Figure 11.19 shows the calculated cumulative frequency distributions of the outlet-air temperatures for the three velocities. It was found that for the air velocity of 5 m.s⁻¹ the outlet-air temperature is higher during August than during July, varying

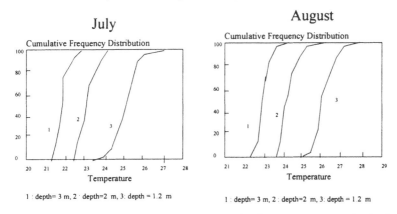

Figure 11.19 Cumulative frequency distributions of outlet-air temperature for the three air velocities [10]

within the range 23.8 to 27.8°C during July and 25.2 to 28.9°C during August. When the air velocity increases to 20 m.s^{-1} the temperature range at the outlet of the exchanger increases slightly and ranges from 23.8 to 28.9°C during July and from 25.3 to 30°C during August. This analysis showed that the air-velocity increase leads to a slight increase of the outlet-air temperature. In this respect the convective heat transfer coefficient, which depends upon air velocity, also increases, contributing thus to a more efficient heat exchange. However, the outlet-air temperature increase is mainly due to the increased mass flow rate.

The performance of any earth-to-air heat exchanger is obviously related to the earth's temperature, which varies with the soil depth. Therefore diurnal, seasonal and annual variations have to be considered in the storage design. Simulations have been carried out for 1.2, 2 and 3 m depths. The results obtained are shown in Figure 11.20. From this figure it can be seen that the cooling capacity of the system increases considerably with depth.

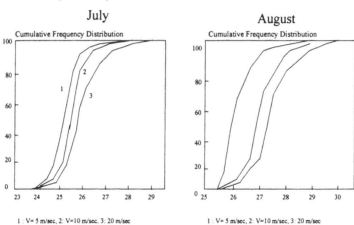

Figure 11.20 Cumulative frequency distributions of outlet-air temperature for the three depths [10]

Assuming a depth of 1.2 m, it is estimated that during July the outlet-air temperature was 25.5°C for 80% of the time and during August was close to 26.5°C. For a 2 m depth, a 2°C reduction of the corresponding temperature is estimated, while for depths of 3 m, the outlet-air temperature was close to 22°C during July and 23.2°C during August.

The effect of the ground surface cover has been investigated in reference [11]. The thermal model describing the performance of multiple parallel earth-to-air heat exchangers was used to assess the performance of four 30 m long PVC tubes of 0.125 radius, buried at a 1.50 m depth. Two series of simulations were performed: one for bare soil and a second for grass-covered soil.

Figure 11.21 illustrates the temperature at the inlet of an internal tube and the distributions of the outlet temperature for bare and grass-covered soil, for July and August. In order to facilitate comparison, the inlet temperature is also plotted on the same graph. This figure shows that the ground cover can influence the performance of earth-to-air heat exchangers. This is because the ground cover affects the amount of solar radiation absorbed by the ground and therefore the surface temperature of the ground. Since bare soil absorbs more solar radiation than grass, this results in higher outlet temperatures of the exchangers operating underneath. Therefore it is concluded that the short-grass soil surface can increase the cooling capacity of earth-to-air heat exchangers.

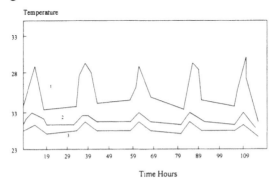

Figure 11.21 Air temperature variation at the outlet of an internal tube buried under bare and grass-covered soil for July and August. The top curve shows the inlet temperature

Parametric analysis of earth-to-air heat exchangers

The models for predicting the performance of the earth-to-air heat exchangers, although very useful for research purposes, are rather difficult for engineers to use in everyday practice. Therefore, simplified parametric models describing the behaviour of earth-to-air heat exchangers are very useful. The model presented here [12] calculates the air temperature at the outlet of an earth-to-air heat exchanger and the cooling energy provided by it. The required input parameters are the pipe length, radius and depth and the air velocity inside the exchanger.

The parametric analysis has been performed in the TRNSYS environment. The proposed numerical model was validated against an extensive set of experimental data. This indicated that the model could accurately predict the temperature and the humidity of the circulating air, the distribution of the temperature and moisture in the soil, as well as the overall thermal performance of the earth-to-air heat exchangers. In order to calculate the overall efficiency of the earth-to-air heat exchangers the following dimensionless coefficient was defined:

$$U = \frac{T_{out} - T_{und}}{T_{in} - T_{und}} \qquad\qquad (11.6)$$

where T_{in} is the inlet-air temperature, T_{out} the outlet-air temperature and T_{und} the undisturbed soil temperature. Taking into account that T_{in} and T_{und} are the performance parameters of the system and their initial values are known, the calculation of U depends on a direct knowledge of the air temperature at the outlet of the pipe.

In order to predict the values of U a systematic analytical process was followed. The analytical numerical code, presented above, was used to calculate the values of U for various sets of input parameters. Then a parametrical approach was followed such that each parameter affecting the variation of a factor is considered to be independent while the others remained unchanged. The parameters upon which U depends are the pipe length (L), the volume air flow rate through the pipe (SV) = $\pi r^2 V$, where r is the radius of the pipe and V the air velocity, and the depth of buried pipe (D). The U-coefficient values that correspond to $L = 30$ m, SV = 0.393 m³ s⁻¹ and D = 2 m, were selected as the U reference profile.

The statistical analysis, based on a regression technique, was extended to create a database of U-coefficient data corresponding to the previously mentioned extensive set of input parameter values. In all, the process consists of developing the reference U-coefficient profiles, the U-coefficient data regression analysis and the calculation of U-coefficient improved predictions.

The U-values corresponding to the reference profiles for SV and D were fitted to a third-degree polynomial function of L. Thus, the equation expressing the U reference profiles is

$$U_{ref}(L) = 0.9952 - 0.0168L + 0.00019L^2 - 9.57 \times 10^{-7} L^3. \qquad (11.7)$$

Furthermore, U-values were normalized for each parameter separately. It should be noted here that the normalization of U-values is a statistical process where a parameter is considered to be the dependent variable for which a regression function has to be determined, while the other variables are considered as independent. The normalization technique implies that the whole set of data should be sorted into relatively small groups, allowing a very high correlation factor in the curve-fitting analysis.

The U-values were sorted into two groups of the system parameters. In the first group the dependent variable is D and the independent parameters are SV and L. In the second group the dependent parameter is SV and D and L are the independent ones. The U-coefficient data in each group was normalized with respect to the reference value of the dependent variable for each value of the two independent variables. Thus, if the first group of $U_{SVi,Lj}(D_t)$ is a numerical predicted value of U-coefficient corresponding to the i, j and t values of the parameters SV, L and D respectively and that $U_{SVi,Lj}$ is the U at the reference value for the dependent variable

(D) and for the given values of the two independent variables (SV) and (L), the relevant normalized U-coefficient value can be written:

$$U_{\text{normSV}i,Lj}(D_t) = \frac{U_{\text{SV}i,Lj}(D_t)}{U_{\text{SV}i,Lj}(D_{\text{ref}})}. \tag{11.8}$$

Similarly, for the second group the normalized U-value is

$$U_{\text{norm}Di,Lj}(\text{SV}_t) = \frac{U_{Di,Lj}(\text{SV}_t)}{U_{Di,Lj}(\text{SV}_{\text{ref}})} \tag{11.9}$$

where i, j and t are values of D, L and SV respectively. These normalized U-values are then expressed as third-degree polynomials, applying a regression analysis. The fitting equations are:

$$U_{\text{normSV}i,Lj}(D_t) = a_0 + a_1 D + a_2 D^2 + a_3 D^3 \tag{11.10}$$

$$U_{\text{norm}Di,Lj}(D_t) = b_0 + b_1 \text{SV} + b_2 \text{SV}^2 + b_3 \text{SV}^3. \tag{11.11}$$

The coefficients of the polynomials in equations (11.10) and (11.11) are given in Appendix A at the end of the chapter (Tables 11.A1 and 11.A2 respectively). The regression analysis was performed using standard regression techniques. The correlation coefficient was very high, between 0.95 to 0.99 in all cases. After the two groups of parameters used in the regression analysis had been considered, a U-correction factor for each group was defined and calculated. For the first group the correction factor is

$$C_{f\text{SV}i,Lj,Dt} = U_{\text{normSV}i,Lj,Dt} = U_{\text{normSV}i,Lj}(D_t) \tag{11.12}$$

where i and j are equal to the values of SV and L considered in the regression analysis. The correction factors for the second group are

$$C_{f Di,Lj,\text{SV}t} = U_{\text{norm}Di,Lj,\text{SV}t} = U_{\text{norm}Di,Lj}(\text{SV}_t). \tag{11.13}$$

The two correction factors were calculated using the curve fitting equations (11.10) and (11.11).

The last step of this parametrical process is the calculation of the actual U-coefficient under given conditions. This is

$$U_k = U_{\text{ref}}(L)\,\text{CF}$$

where U_{ref} is the reference U-coefficient value and CF is the global correction factor which is the product of two specific correction factors, $C_{f\text{SV},L}(D)\,C_{fD,L}(\text{SV})$

On the basis of this parametric analysis, the simplified calculation forms given in the Appendix have been created. This simplified method has been used to calculate

the outlet temperature of a 15 m long and 0.250 diameter earth-to-air heat exchanger, buried at a depth of 1 m. The air velocity in the pipe is 8 m s^{-1}, the air temperature at the inlet is 30°C and the ground temperature at the depth of the exchanger is 17.3°C. Following the steps as indicated in the calculation forms (see
Appendix), the outlet temperature is found to be 29.4°C.

Experimental data from earth-to-air heat exchanger applications

The most recent application of earth-to-air heat exchangers for cooling in Greece is
in the 700 m^2 atrium of the Faculty of Philosophy of the University of Ioannina. The
town of Ioannina is located in the north-western part of Greece. The climate there is
one of the coldest ones in Greece during winter, while summers are hot and humid.
In order to avoid overheating of the atrium during summer, passive-cooling solutions
were applied. These are solar control, stack ventilation and ground cooling. Five
30 m long earth-to-air heat exchangers were used, having a 0.150 m diameter. The
exchangers were placed at a depth of 2 m. The average air velocity in the tubes is
3 m s^{-1}.

Figure 11.22 illustrates the variation of the ambient and indoor temperatures and
also the inlet and outlet temperatures of one of the five earth-to-air heat exchangers,
from 12 August 1993 (day 224) to 29 August 1993 (day 241). Although the ambient
temperature becomes higher than 35°C at noon, the air temperature inside the atrium
is generally lower, showing the effectiveness of the passive cooling features of the
building.

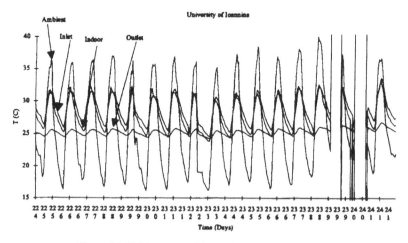

*Figure 11.22 Variation of the ambient, indoor (at 1.5 m
height), inlet- and outlet-air temperatures*

University of Ioannina

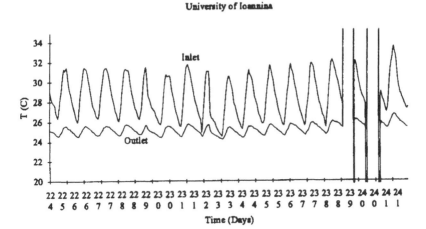

*Figure 11.23 Variation of inlet- and outlet-air temperatures
of an earth-to-air heat exchanger*

The air temperatures at the inlet and outlet of the exchanger have been plotted separately in Figure 11.23. Clearly the air temperature at the outlet is significantly lower. The average inlet temperature is 32°C, while the outlet temperature practically never exceeds 26°C, showing the efficiency of this technique.

Earth-to-air heat exchangers can also be applied to agricultural greenhouses. An example is a 1000 m² greenhouse built in the town of Agrinion in Greece, as one of the Energy Demonstration Projects of CEC, Directorate General for Energy (XVII). There, the exchangers were used to prevent overheating during summer and to manage the greenhouse energy resources better during winter [11]. The indoor and ambient temperature variations, together with the air temperature at the outlet of the exchangers, are shown in Figures 11.24 to 11.26.

From these figures it can be seen that during a winter day the indoor air temperature is usually close to 20°C. Also, during the night the temperature is usually higher than 10°C and close to the required temperature level. It should be noted that the indoor air temperature is always 2–3°C higher than the ambient temperature at night. This is due to the presence of the earth-to-air heat exchangers, as well as to the thermal inertia of a storage wall built on the northern facade of the greenhouse.

A close examination of the daily air temperature profiles at the outlet of the heat exchangers during winter has shown that during night time this temperature is almost 2–3°C higher than the indoor air temperature, thus providing heating energy to the greenhouse. During the daytime the inverse phenomenon occurs: the outlet air temperature is much lower (4–12°C) than the indoor air temperature, which means that the heating energy surplus of the greenhouse is stored, if necessary, in the ground and can be used during the night.

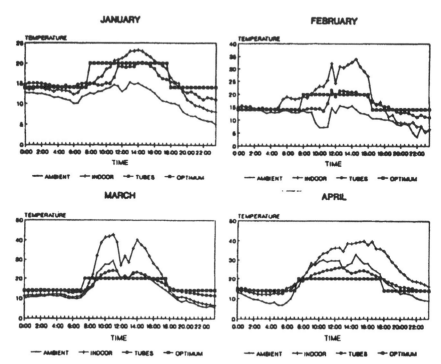

Figure 11.24 Daily measured variation of the ambient and indoor air temperatures and of the air temperature at the outlet of the exchangers for a typical day from January to April

Problems related to earth-to-air heat exchanger applications

Although earth-to-air heat exchangers are rather simple systems and of low cost, attention should be paid to the following, in order to ensure efficient use:

- Possible condensation of water inside the tubes or evaporation of accumulated water, which affects the quality of the air injected into the building. This can be countered by placing appropriate filters at the air outlet. A second option is to link the earth-to-air heat exchangers with the conventional air-conditioning system of the building. In this case the filters of the air-conditioning system are suitable for the exchangers as well. It should be noted however that condensation in such a system is a rather rare phenomenon, which may occur only under extreme conditions.

  Condensation can be avoided by digging the ditch in which the pipe will be placed with a 1% slope towards one side and drilling a small hole in the lower elbow of the pipe. Then the accumulated water will be evacuated though this hole and absorbed by the soil as a result of capillary forces.

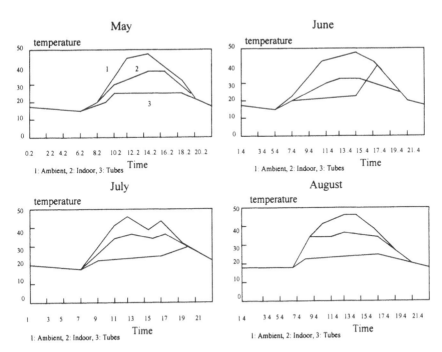

*Figure 11.25 Daily measured variation of the ambient and indoor air
temperatures and of the air temperature at the outlet of the exchangers for a
typical day from May to August*

- In order to maximize the control of such a system it is necessary to provide
  an automatic control algorithm. This algorithm must compare the indoor
  temperature with the temperature at the air outlet and, when the former is
  lower than the latter, the fan must be stopped. Obviously, if the system is to
  be linked with conventional air conditioning, the automatic control has to be
  more sophisticated.

When installing an earth-to-air heat exchanger, attention should also be paid to the
choice of the fan, in order to avoid possible noise problems.

Rules of thumb for ground-cooling applications

Earth-to-air heat exchanger systems should not be installed prior to specific calculations, at least with the simplified method presented here. However some rules of
thumb can be given:

- The length of the exchanger should be at least 10 m.

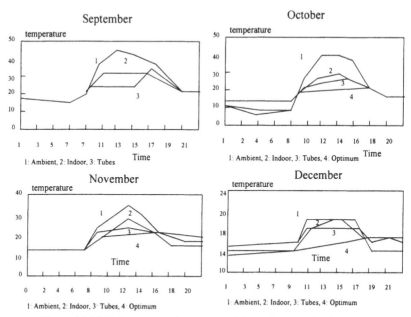

Figure 11.26 Daily measured variation of the ambient and indoor air temperatures and of the air temperature at the outlet of the exchangers for a typical day from September to December

- The diameter of the exchanger should range between 0.2 and 0.3 m.
- The depth of the exchanger should range between 1.5 and 3 m.
- The air velocity through the buried pipe should range between 4 and 8 m s^{-1}.

When placing the earth-to-air heat exchanger, attention should also be paid to obtaining the best possible thermal contact with the surrounding ground. Therefore, it is recommended that the pipe be surrounded with sand 5 cm thick, when it is buried. This is because sand has a good thermal conductivity and, because of its dimensions, it fills the air gaps that would have been formed if raw earth, which would act as an insulating material, had been placed directly over the exchanger.

Coupling and control problems

As previously mentioned, the exchangers can, in some cases, be connected to the conventional air-conditioning system of a building. In this case the air cooled by the exchangers can be used either for direct cooling or for supplying the conventional air-conditioning system with pre-cooled air, thus increasing its CP (coefficient of performance, the parameter characterizing the efficiency of air-conditioning systems). When such a coupling exists, a well-defined control is necessary in order to maximize the overall efficiency of the cooling systems of the building. This control should not only take into account the conventional air conditioning and the earth-to-air heat exchangers, but also all other passive cooling techniques used in the particular building. Below, an example of such a control algorithm will be given, combining three cooling techniques: conventional air conditioning, earth-to-air heat exchangers and ventilation [13].

When the internal temperature, T_{in}, exceeds the desired temperature, the controller checks if the temperature at the outlet of the exchangers T_T is lower than the ambient temperature T_n. If this is so, then the temperature drop between the inlet and the outlet of the exchangers is examined $(T_T - T_n)$. In order to start the exchangers, this difference must be greater than 2 degrees; otherwise the cooling energy produced by the exchangers is of the same order as the electricity consumed by the ventilators circulating the air through the exchangers. In that case, if the temperature of the ambient air is less than the desired temperature inside the building, ventilation is applied. If not, conventional air conditioning is used.

If the temperature drop across the exchangers is satisfactory, i.e. higher than 2 degrees, the controller checks the difference between the internal temperature T_{in} and the temperature at the outlet of the exchangers T_T. If the internal temperature is higher than the outlet temperature and the outlet temperature is higher than the desired temperature of the room, then the air from the exchangers is used to pre-cool the air at the inlet of the air conditioner. If, however, the outlet temperature is lower than the desired temperature in the building, then this air supplies the building directly. In this case the controller checks the air-mass flow rate required to cool the building, in order to activate the appropriate number of exchangers. It is obvious that if the temperature drop across the exchangers is very small, these cannot be used. In this case ventilation or conventional air conditioning will be used.

The operation flow chart of such a controller is shown in Figure 11.27.

CONCLUSIONS

Ground cooling is an interesting passive and hybrid cooling technique that is based on the property of the ground that it maintains, below a certain depth, a practically constant temperature. This temperature is higher than the ambient temperature during winter and lower than the ambient temperature during summer. Ground cooling may be applied to a building by direct thermal contact or by means of an earth-to-air heat exchanger. This is a pipe buried at a depth greater than 1.5 m, through which air is circulated and cooled before being injected into a building. Applications of earth-to-air heat exchangers are of low cost and the systems can be dimensioned using a very simple technique. When applying this technique, attention should be paid to avoiding air-quality problems. Full advantage of such a technique is taken if the operation of the ventilator of the exchanger is subject to automatic control.

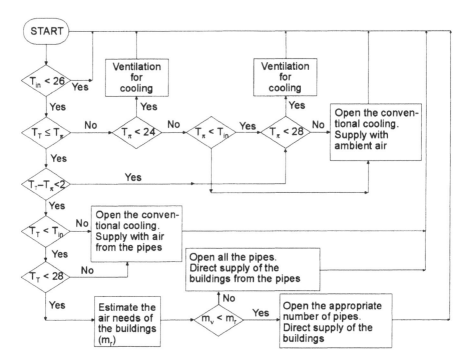

Figure 11.27 Control strategy for coupling earth-to-air heat exchangers with a conventional air-conditioning and ventilation system [13]

REFERENCES

1   *C.E.C. Project Monitor brochures* (1992). ed. J.O. Lewis, Issue 37. University College Dublin.

2   Carmody, J.C., G.D. Meixel, K.B. Labs and L.S. Shen (1985). 'Earth-contact buildings: Applications, thermal analysis and energy benefits', *Advances in Solar Energy*, Vol. 2, Chapter 6, pp. 297–347.

3   Vries, D. A. (1952). 'de Het warmtegeleidingsvermogen van grond', *Med. Landbouwhogeschool,* Wageningen, Vol. 52, pp. 1–73.

4   Pitts, D.R. and L.E. Sissom, (1977). *Heat Transfer.* McGraw-Hill, New York.

5   *Passive Cooling* (1989). ed. J. Cook. MIT Press, Cambridge, MA.

6   Mihalakakou, G., M. Santamouris and D. Asimakopoulos (1992). 'Modelling the earth temperature using multi-year temperatures', *Energy and Buildings*, Vol. 19, pp. 1–9.

7   Mihalakakou, G., M. Santamouris and D. Asimakopoulos (1994). 'Modelling the thermal performance of earth-to-air heat exchangers', *Solar Energy*, Vol. 53, No 2, pp. 301–305.

8   Mihalakakou, G., M. Santamouris and D. Asimakopoulos (1994). 'On the use of ground cooling for heat dissipation', *ENERGY*, Vol. 29, No 1, pp. 17–25.

9 Tzaferis, A., D. Liparakis, M. Santamouris and A. Argiriou (1992). 'Analysis of the accuracy and sensitivity of eight models to predict the performance of earth-to-air heat exchangers', *Energy and Buildings*, Vol. 18, pp. 35–43.

10 Mihalakakou, G., M. Santamouris and D.N. Asimakopoulos (1994). 'On the cooling potential of earth-to-air heat exchangers', *Energy Conversion Management*, Vol. 35, No 5, pp. 395–402.

11 Santamouris, M., A. Argiriou and M. Vallindras (1994). 'Design and operation of a low energy consumption passive solar agricultural greenhouse', *Solar Energy*, Vol. 52, No 3.

12 Mihalakakou, G., M. Santamouris, D. Asimakopoulos and I. Tselepidaki (1995). 'Parametric prediction of the buried pipes cooling potential for passive cooling applications', *Solar Energy*, Vol. 55, No. 5, pp. 163-174.

13 Tombazis, A., A. Argiriou and M. Santamouris (1990). 'Performance evaluation of passive and hybrid cooling components for a hotel complex', *International Journal of Solar Energy*, Vol. 9, pp. 1–12.

APPENDIX A
COEFFICIENTS FOR EQUATIONS (11.10) AND (11.11)

Table 11.A1 Coefficients for equation (11.10)

SV	a_i	$L=10$	$L=15$	$L=20$	$L=25$	$L=30$	$L=35$	$L=40$	$L=45$	$L=50$	$L=55$	$L=60$
0.393	a_0	1.16552	1.489015	1.645340	1.800090	1.951960	2.033872	2.168868	2.220354	2.263630	2.030870	2.35039
	a_1	-0.10921	-0.334799	0.423397	-0.540170	-0.639415	-0.69299	-0.795803	-0.795660	-0.858780	-0.63890	-0.89570
	a_2	0.01641	0.048229	0.057187	0.077390	0.06900868	0.09699	0.116046	0.105440	0.12511190	0.121352	0.120720
	a_3	0.00085	-0.00220	-0.002326	-0.003544	-0.004020	-0.004267	-0.005450	-0.004069	-0.006097	-0.00592	-0.00580
0.425	a_0	1.14940	1.442890	1.598730	1.700350	1.901170	1.967892	2.103820	2.127432	2.137451	2.187293	2.265720
	a_1	-0.101860	-0.31378	-0.398070	-0.501202	-0.611729	-0.612783	-0.767896	-0.772850	-0.794532	-0.824895	-0.852813
	a_2	0.015759	0.04608	0.055895	0.07028	0.088372	0.092743	0.109874	0.101074	0.1050223	0.113278	0.115371
	a_3	-0.0084	-0.002198	-0.002270	-0.003308	-0.003897	-0.004072	-0.004635	-0.004008	-0.005872	-0.001727	-0.001706
0.458	a_0	1.142600	1.418300	1.520700	1.682001	1.863730	1.948295	2.078430	2.1083245	2.109485	2.127295	2.198370
	a_1	-0.098210	-0.309873	-0.376730	-0.491837	-0.598730	-0.632981	-0.739628	-0.740720	-0.743870	-0.792563	-0.843710
	a_2	0.014673	0.045298	0.013793	0.069270	0.076894	0.088731	0.089432	0.096285	0.0099273	0.104496	0.113807
	a_3	-0.00079	-0.002179	-0.002186	-0.003128	-0.003521	-0.003929	-0.003650	-0.003987	-0.005501	-0.005587	-0.005683
0.528	a_0	1.139850	1.407850	1.499300	1.672890	1.721020	1.871430	1.998620	2.076320	2.078360	2.096472	2.178090
	a_1	-0.091420	-0.283985	-0.358780	-0.472893	-0.510870	0.0592743	-0.662782	-0.723214	-0.702890	-0.768270	-0.828930
	a_2	0.013983	0.043987	0.049873	0.068200	0.069785	0.085478	0.08806	0.093821	0.097871	0.095360	0.111708
	a_3	-0.00069	-0.002169	-0.002087	-0.003049	-0.003289	-0.003897	-0.003090	-0.3908	-0.004920	-0.001479	-0.005570
0.565	a_0	1.138740	1.394650	1.482600	1.652890	1.658346	1.850427	1.928548	2.058290	2.08821	2.09627	2.130940
	a_1	0.0889	-0.27298	-0.339670	-0.449880	-0.449326	-0.18783	-0.601979	-0.643972	-0.680667	-0.752912	-0.810804
	a_2	0.013578	0.041962	0.047967	0.064293	0.065225	0.083773	0.077240	0.080521	0.095070	0.096289	0.109953
	a_3	-0.00064	-0.002159	-0.001939	-0.003003	-0.003088	-0.003874	-0.0028	-0.003521	-0.004839	-0.005385	-0.005540
0.604	a_0	1.135090	1.381623	1.4799	1.631498	1.614723	1.850427	1.928548	2.04132	1.96378	2.004350	2.107870
	a_1	-0.0814	-0.264954	-0.324860	-0.450794	-0.419783	-0.587832	-0.601979	-0.693217	-0.677820	-0.738962	-0.792694
	a_2	0.012972	0.040461	0.045879	0.062799	0.062783	0.083773	0.07724	0.092211	0.094872	0.092803	0.109287
	a_3	-0.000621	-0.002114	-0.001891	-0.002960	-0.002981	-0.003874	-0.0028	-0.003887	-0.004737	-0.005227	-0.005508
0.643	a_0	1.132170	1.362930	1.476971	1.592830	1.608790	1.837421	1.908032	2.0017	1.947830	1.992840	2.102380
	a_1	-0.0792	-0.250890	-0.317577	-0.422790	-0.402743	-0.528711	-0.174362	-0.657224	-0.670382	-0.722963	-0.79008
	a_2	0.012235	0.38279	0.0436	0.061897	0.060894	0.081235	0.075189	0.089072	0.094038	0.090225	0.109028
	a_3	-0.000602	-0.002074	-0.001865	-0.002523	-0.002803	-0.003852	-0.002793	0.0003847	-0.004708	-0.005188	-0.005508

Table 11.A1 (continued)

SV	a_i	L = 10	L = 15	L = 20	L = 25	L = 30	L = 35	L = 40	L = 45	L = 50	L = 55	L = 60
0.684	a_0	1.12807	1.359830	1.458030	1.548940	1.597870	1.820071	1.874324	1.97876	1.92135	1.97243	2.099360
	a_1	-0.0773	-0.242711	-0.308975	-0.408294	-0.400873	-0.567213	-0.130893	-0.632743	-0.662893	-0.702814	-0.782694
	a_2	0.01187	0.037003	0.041720	0.058072	0.059713	0.079925	0.070439	0.088072	0.093481	0.089821	0.108987
	a_3	-0.000593	0.0002029	-0.001703	-0.002307	-0.002703	-0.003795	-0.002748	0.0003829	-0.004687	-0.85078	-0.005508
0.726	a_0	1.112890	1.33940	1.43923	1.507980	1.582810	1.787361	1.82976	1.9674	1.89932	1.923763	2.095400
	a_1	-0.0738	-0.230271	-0.29658	-0.392721	-0.399273	0.552984	-0.500945	-0.618732	-0.653894	-0.698714	-0.773845
	a_2	0.011090	0.035923	0.03987	0.0556078	0.058278	0.077251	0.068485	0.086278	0.093087	0.088274	0.108302
	a_3	-0.000582	-0.001989	-0.001678	-0.002289	-0.002508	0.003778	-0.002739	-0.003819	-0.004658	-0.004970	-0.005498
0.770	a_0	1.101332	1.315670	1.42952	1.498730	1.179890	1.75254	1.79863	1.947832	1.890220	1.907893	2.09032
	a_1	-0.070708	-0.248930	-0.28339	-0.377730	-0.396270	-0.548971	-0.49873	-0.608714	-0.642794	-0.672873	-0.762945
	a_2	0.01073	0.035072	0.038432	0.054077	0.056893	0.75544	0.66892	0.85245	0.092994	0.086874	0.108085
	a_3	-0.00542	-0.001917	-0.001551	-0.002147	-0.002498	-0.003758	-0.002720	-0.003807	-0.004647	-0.004729	-0.001398
0.814	a_0	1.100870	1.309270	1.410930	1.48644	1.572060	1.73058	1.7806	1.928539	1.889630	1.897214	2.08627
	a_1	-0.069502	-0.232810	-0.264090	-0.350721	-0.350721	-0.391870	-0.537812	-0.48685	-0.572364	-0.637943	-0.667273
	a_2	0.010128	0.034872	0.036208	0.052272	0.55276	0.073231	0.065754	0.084274	0.092744	0.085273	0.107992
	a_3	-0.0000529	-0.0011920	-0.001527	-0.002098	-0.002430	0.003721	-0.002715	-0.003759	-0.004637	-0.004663	-0.005325
0.860	a_0	1.100425	1.300020	1.390820	1.46653	1.569870	1.699271	1.7659	1.902735	1.884030	1.87764	2.08325
	a_1	-0.06783	-0.227830	-0.258070	-0.322721	-0.390870	-0.522994	-0.48076	-0.552895	-0.624894	-0.652789	-0.752032
	a_2	0.010099	0.032227	0.031879	0.050897	0.014927	0.071892	0.064917	0.081746	0.092675	0.084723	0.107873
	a_3	-0.000513	-0.001898	-0.001486	-0.001927	-0.004111	-0.003708	-0.002712	-0.003769	-0.004602	-0.004527	-0.005298
0.907	a_0	1.1001005	1.298730	1.37760	1.448730	1.569870	1.672720	1.7482	1.897450	1.88403	1.860780	2.06925
	a_1	-0.06593	-0.215631	-0.250912	-0.312783	-0.39087	-0.508741	-0.47076	0.54936	-0.612943	-0.652789	-0.73672
	a_2	0.01004	0.031927	0.034298	0.048874	0.054927	0.069879	0.0640166	0.080694	0.092078	0.083279	0.107089
	a_3	-0.000501	-0.001867	-0.001428	-0.001888	-0.002411	-0.003692	-0.002709	-0.003759	-0.004592	-0.004472	-0.005201
1.005	a_0	1.09238	1.288350	1.36827	1.427830	1.565190	1.652010	1.72058	1.86754	1.879350	1.852704	2.05838
	a_1	-0.061407	-0.209630	-0.242183	-0.302594	-0.389712	-0.488714	-0.46992	-0.539421	-0.609274	-0.648723	-0.724732
	a_2	0.009422	0.030287	0.033781	0.046793	0.054855	0.067823	0.064002	0.079621	0.091989	0.082294	0.106992
	a_3	-0.000471	-0.001768	-0.001408	-0.001809	-0.002300	-0.003677	-0.0027	-0.003728	-0.004578	-0.004227	-0.005098
1.056	a_0	1.089423	1.2799	1.3527841	1.419980	1.548199	1.6321	1.717986	1.83243	1.875908	1.832778	2.045508
	a_1	-0.058381	-0.193694	-0.23484	-0.299862	-00.37674	-0.477213	-0.46835	-0.522743	-0.606392	-0.622794	-0.716440
	a_2	0.009230	0.029781	0.032905	0.0448777	0.053972	0.065471	0.06386	0.079026	0.091265	0.080783	0.106750
	a_3	-0.000436	-0.001730	-0.001390	-0.001789	-0.0021	-0.003633	-0.002699	-0.003708	-0.004558	-0.004128	-0.005173

Table 11.A1 (continued)

SV	a_i	L = 10	L = 15	L = 20	L = 25	L = 30	L = 35	L = 40	L = 45	L = 50	L = 55	L = 60
1.108	a_0	1.087309	1.270730	1.341574	1.410985	1.52073	1.6098	1.707986	1.809273	1.863292	1.8297880	1.98975
	a_1	-0.0117982	-0.189413	-0.227625	-0.290278	-0.35879	-0.452924	-0.46233	-0.503381	-0.594721	-0.60921	-0.69378
	a_2	0.008928	0.028832	0.031069	0.0422794	0.048927	0.0649811	0.06307	0.077843	0.088474	0.080078	0.094387
	a_3	-0.000409	0.001698	-0.001388	-0.001729	-0.00197	-0.003597	-0.002678	-0.003699	-0.004289	-0.00399	-0.004927
1.162	a_0	1.082493	1.26987	1.336128	1.409340	1.50278	1.598991	1.68972	1.799853	1.820721	1.8094	1.95289
	a_1	-0.056327	-0.183020	-0.220809	-0.282764	-0.34894	-0.448742	-0.45878	-0.499763	-0.578932	-0.587223	-0.67843
	a_2	0.008749	0.028287	0.030962	0.0407826	0.045879	0.064722	0.062574	0.076274	0.084271	0.079965	0.093837
	a_3	-0.000402	-0.001659	-0.001370	-0.001654	-0.00186	-0.003567	-0.002653	-0.003683	-0.004099	-0.003827	-0.004680
1.216	a_0	1.080935	1.25921	1.624859	1.399650	1.484762	1.58722	11.65943	1.77213	1.79452	1.8026	1.93012
	a_1	-0.05548	-0.178730	-0.217650	-0.278976	-0.326790	-0.04354	-0.447851	-0.496215	-0.543928	-0.57344	-0.64543
	a_2	0.008431	0.027498	0.030207	0.0382763	0.042479	0.0645273	0.061473	0.075821	0.080721	0.079256	0.090721
	a_3	-0.000399	-0.001638	-0.001360	-0.001594	-0.001736	-0.003549	-0.002609	-0.003654	-0.003927	-0.003792	-0.004280
1.272	a_0	1.079272	1.247010	1.310912	1.39756	1.47496	1.58091	1.63734	1.742183	11.783521	1.799478	1.826372
	a_1	-0.054802	-0.169853	-0.20926	-0.26958	-0.31794	-0.412773	-0.438963	-0.489813	-0.536280	-0.568236	-0.57434
	a_2	0.008280	0.027008	0.029871	0.0367214	0.040896	0.064271	0.0608311	0.074621	0.078271	0.079008	0.079321
	a_3	-0.000395	-0.001601	-0.00134	-0.001505	-0.001582	-0.003508	-0.00254	-0.003633	-0.003876	-0.003703	-0.003902
1.330	a_0	1.078507	1.235614	11.303608	1.39046	1.465173	1.574669	1.62947	1.705842	1.771453	1.796258	1.823225
	a_1	-0.05308	-0.164852	-0.20603	-0.260534	-0.30469	-0.401806	-0.42304	-0.485759	-0.527006	-0.552746	-0.17299
	a_2	0.008077	0.02694	0.029226	0.0354187	0.039477	0.063922	0.059310	0.072673	0.076872	0.078279	0.079625
	a_3	-0.000391	-0.001596	-0.001339	-0.00144	-0.001416	-0.003494	-0.002561	-0.003615	-0.003657	-0.003697	-0.00372

Table 11.A2 Coefficients for equation (11.11)

D	b_i	L = 10	L = 15	L = 20	L = 25	L = 30	L = 35	L = 40	L = 45	L = 50	L = 55	L = 60
0.5	b_0	0.973237	0.984554	0.3268351	0.932432	0.9190869	0.9216535	0.93478	0.9321116	0.896039	0.901643	0.8936885
	b_1	0.0070519	0.017476	2.225822	0.2044293	0.258050	0.2417408	0.1807601	0.1822477	0.339175	0.3139933	0.344522
	b_2	-0.0194166	0.05964	-1.409554	-0.1074301	-0.1675469	-0.1381756	-0.048576	-0.04005	-0.239138	-0.200674	-0.2300972
	b_3	-0.0004534	-0.01989	0.337021	0.0249636	0.046290	0.0328189	-0.005337	-0.0106818	0.0687887	0.0523289	0.06139416
0.75	b_0	0.964251	0.987236	0.9367806	0.9114447	0.900188	0.8787591	0.8902154	0.8804532	0.8673419	0.8602321	0.8403376
	b_1	0.09962	-0.009466	0.188054	0.278246	0.316501	0.4097809	0.3463084	0.389427	0.4507194	0.477553	0.5650315
	b_2	-0.0302	0.121604	-0.0718712	-0.168678	-0.199395	-0.3152629	-0.204793	-0.259733	-0.339269	-0.361111	-0.46734.1
	b_3	-0.0037	-0.051564	0.0063011	0.0404292	0.049742	0.0939591	0.0443726	0.0669399	0.099763	0.104872	0.144468
1	b_0	0.943157	0.942228	0.9330715	0.9141308	0.895504	0.8902966	0.8868784	0.8468693	0.8783546	0.832680	0.70529294
	b_1	0.18709	0.1764782	0.2307171	0.269495	0.333413	0.358894	0.3717613	0.539791	0.414013	0.592354	1.004547
	b_2	-0.12480	-0.0821707	-0.13054	-0.1422083	-0.2056482	-0.23844	-0.248076	-0.4479515	-0.302136	-0.48855	-1.145618
	b_3	0.0308	0.023893	0.0372811	0.03103985	0.052904	0.0675147	0.070307	0.1446922	0.094942	0.154745	0.469766
1.25	b_0	0.934129	0.862693	0.8432887	0.851786	0.824018	0.8483654	0.868781	0.8466222	0.766978	0.83268	0.8698867
	b_1	0.239084	0.3324006	0.420427	0.3893395	0.668460	0.5639846	0.4771419	0.5621656	0.763249	0.592354	0.4367663
	b_2	-0.214145	0.0215894	-0.07337	-0.0254496	-0.607523	-0.465888	-0.3463188	-0.4306529	-0.662467	-0.48855	-0.220565
	b_3	0.07415	-0.0563098	-0.022064	-0.04240454	0.2107695	0.1545722	0.1069714	0.1315264	0.2137057	0.154745	0.037742
1.50	b_0	0.9152	0.751792	0.843289	0.7196968	0.73595	0.775079	0.804484	0.7935002	0.7527	0.829797	0.7274204
	b_1	0.298709	0.7732075	0.420427	0.9176254	0.8591827	0.763326	0.6761029	0.7234378	0.899990	0.6278184	1.019928
	b_2	-0.24388	-0.443959	-0.073366	-0.5957919	-0.50693	-0.475420	-0.404196	-0.4503818	-0.659302	-0.479214	-0.7942481
	b_3	0.07079	0.1012876	-0.22064	0.1535336	0.113296	0.1231506	0.0901246	0.1061139	0.1844449	0.1406238	0.233536
1.75	b_0	0.929652	0.737016	0.843289	0.8704387	0.90007	0.9035841	0.9294046	0.9227829	0.928778	0.763997	0.9200385
	b_1	0.26144	0.837941	0.4204278	0.2158342	0.057796	0.04089	-0.0737209	-0.02418226	-0.031092	0.861373	-0.0034799
	b_2	-0.217096	-0.505118	-0.073366	0.3869876	0.665393	0.7056344	0.88076	0.81105665	0.836506	-0.601421	0.8277673
	b_3	0.06620	0.126232	-0.22064	-0.2006308	-0.32290	-0.340089	-0.4172774	-0.3865548	-0.401961	0.1603747	-0.4003627
2	b_0	0.926914	0.7675465	0.8432887	0.578765	0.4159149	0.3538133	0.3233079	0.3354986	0.3110425	0.939961	0.2790866
	b_1	0.26227	0.748543	0.4204278	1.376202	0.2033441	2.322206	2.459221	2.422317	2.534951	-0.100102	2.690897
	b_2	-0.19433	-0.3895065	-0.0733665	-0.8500351	-1.57369	-1.946369	-2.099215	-2.037891	-2.170177	0.951659	-2.376123
	b_3	0.05065	0.08759	-0.022064	0.1920064	0.4589317	0.606394	0.6585151	0.630029	0.6782876	-0.451305	0.769689
2.25	b_0	0.914591	0.7050645	0.665839	0.4746989	0.3836973	0.31121890	0.292052	0.2992659	0.2832156	0.293936	0.2727471
	b_1	0.27099	1.008623	1.184661	1.984768	2.169122	2.504189	2.598345	2.579707	2.648889	2.621181	2.708974
	b_2	-0.16591	-0.6801089	-0.887789	-1.8881	-1.723887	-2.14481	-2.261188	-2.207337	-2.280849	-2.28992	-2.359529
	b_3	0.033608	0.1884699	0.2641756	0.6765254	0.512640	0.6765856	0.7212318	0.691455	0.7162061	0.735627	0.753427

Table 11.A2 (continued)

D	b_i	$L=10$	$L=15$	$L=20$	$L=25$	$L=30$	$L=35$	$L=40$	$L=45$	$L=50$	$L=55$	$L=60$
2.5	b_0	0.90522	0.667638	0.6014981	0.3764804	0.3447084	0.3202823	0.2968839	0.2330611	0.2511085	0.261737	0.2360517
	b_1	0.30219	1.170535	1.46851	2.398256	2.336022	2.469079	2.57826	2.847635	2.767559	2.752019	2.84505
	b_2	-0.19420	-0.8583918	1.228272	-2.367726	-1.916167	-2.087819	-2.22529	-2.496618	-2.366066	-2.33957	-2.45273
	b_3	0.0457	0.2506312	0.3911678	0.8628598	0.581916	0.6506362	0.708428	0.7903645	0.7306973	0.723340	0.7658028
2.75	b_0	0.924269	0.6740734	0.4725141	0.4081163	0.34289	0.2832801	0.2777359	0.1428270	0.1744601	0.2616875	0.1223982
	b_1	0.255947	0.9987543	1.839883	2.083679	2.357645	2.626279	2.668508	3.202602	3.112508	2.72449	3.325921
	b_2	-0.14585	-0.3810015	-1.367766	-1.62314	-1.935278	-2.26232	-2.322737	-2.87861	-2.708115	-2.275774	-3.011109
	b_3	0.02941	0.0052627	0.3615378	0.4739549	0.5871203	0.7170997	0.7459846	0.9230977	0.8882323	0.6915071	0.9681318
3	b_0	0.886844	0.5259488	0.4639956	0.416583	0.283646	0.233365	0.2469828	0.147332	0.158649	0.1151341	0.1230402
	b_1	0.417362	1.561626	1.804119	2.000029	2.588219	2.811538	2.792266	3.211882	3.199118	3.365295	3.316155
	b_2	-0.34761	-0.969613	-1.208119	-1.434385	-2.190531	-2.459719	-2.43942	-2.88144	-2.876757	-3.059903	-2.986626
	b_3	0.110560	0.233349	0.314976	0.3957886	0.678359	0.7876348	0.780099	0.9267106	0.9267709	0.9926	0.96295
3.25	b_0	0.893665	0.4696789	0.469459	0.3661568	0.27922081	0.2055942	0.1902653	0.075588	0.0732332	0.01898	0.144879
	b_1	0.385441	1.765941	1.76233	2.246018	2.611033	2.943951	3.045912	3.53407	3.543805	3.766664	3.27422
	b_2	-0.29758	-1.143981	-1.096898	-1.764339	-2.198228	-2.611646	-2.752617	-3.272345	-3.267405	-3.519935	-2.943067
	b_3	0.0889856	0.285544	0.2565982	0.5251271	0.6754816	0.8459201	0.9022491	1.074729	1.070178	1.16254	0.95555
3.50	b_0	0.847735	0.417231	0.376006	0.3183943	0.2094475	0.194682	0.1916787	0.0245185	0.0055345	-0.041068	0.0522447
	b_1	0.55266	1.967172	2.108582	2.394992	2.918789	3.004595	3.034765	3.743594	3.823755	4.013946	3.610824
	b_2	-0.45231	-1.349851	-1.449095	-1.835936	-2.56971	-2.658915	-2.6990099	-3.49671	-3.584011	-3.789842	-3.276388
	b_3	0.1305021	0.3633448	0.3796452	0.521886	0.8216636	0.8530601	0.8702025	1.153075	1.183406	1.25467	1.056766
3.75	b_0	0.88303	0.3359354	0.340092	0.232365	0.129263	0.1332628	0.133291	-0.5256	-0.047419	-0.047222	-0.0519329
	b_1	0.4182512	2.266396	2.223635	2.739005	3.22998	3.228278	3.266448	4.048362	4.044835	4.049328	4.07927
	b_2	-0.309536	-1.639372	-1.52558	-2.213976	-2.867527	-2.858651	-2.928204	-3.832799	-3.827527	-3.821276	-3.85663
	b_3	0.085784	0.4551583	0.3944959	0.6569146	0.9131148	0.9086422	0.9422788	1.272034	1.268936	1.267115	1.282988
4	b_0	0.87515	0.3414128	0.2971754	0.214027	0.061206	0.1332628	0.1390147	-0.052571	-0.0567303	-0.0585997	-0.0614299
	b_1	0.45846	2.207632	2.395603	2.784745	3.508352	3.228278	3.236331	4.071914	4.11376	4.111240	4.12-32466
	b_2	-0.3645	-1.481885	-1.7077045	-2.209309	-3.163874	-2.858651	-2.822988	-3.85893	-3.918686	-3.883933	-3.910183
	b_3	0.109153	0.372982	0.4589418	0.6554886	1.028335	0.908642	0.8896344	1.281729	1.307218	1.290864	1.302545
4.25	b_0	0.877611	0.342725	0.3264798	0.166084	0.057348	0.6966724	0.0643832	-0.1793073	-0.007733	-0.048029	-0.1437571
	b_1	+0.378486	2.19454	2.258263	2.983711	3.501942	3.497629	3.552356	3.932336	3.892151	4.065053	4.505578
	b_2	-0.21683	-1.44303	-1.507623	-2.41642	-3.0756658	-3.077707	-3.162026	-3.641212	-3.583629	-3.789436	-4.355048
	b_3	0.042491	0.354063	0.3775601	0.7231219	0.9762707	0.9793283	1.015718	1.201078	10178372	1.255473	1.499909

Table 11.A2 (continued)

D	b_i	$L=10$	$L=15$	$L=20$	$L=25$	$L=30$	$L=35$	$L=40$	$L=45$	$L=50$	$L=55$	$L=60$
4.50	b_0	0.8878090	0.359494	0.3431568	0.1212685	-0.0647950	-0.046087	-0.0481128	-0.087802	-0.096008	-0.133895	-0.124977
	b_1	0.332776	2.116026	2.18007	3.154384	4.025127	4.027384	4.058754	4.2447513	4.292158	4.460814	4.38305
	b_2	-0.156771	-1.322751	-1.389611	-2.615565	-3.731419	-3.7728866	-3.812964	-4.04051	-4.096199	-4.30464	-4.09124
	b_3	0.019882	0.306749	0.328253	0.807617	1.251742	1.277475	1.292591	1.37729	1.399253	1.482066	1.375021
4.75	b_0	0.884379	0.337144	0.335726	0.2201878	-0.0819756	-0.062154	-0.1098455	-0.0956367	-0.0944874	-0.094746	-0.274247
	b_1	0.4016716	2.2040012	2.205328	2.71342	4.029411	4.048858	4.283741	4.23798	4.23792	4.24699	5.014574
	b_2	-0.29393	-1.420383	-1.411832	-2.013251	-3.591674	-3.674356	-3.980989	-3.917023	-3.91274	-3.91968	-4.822285
	b_3	0.090316	0.342389	0.33719	0.5650688	1.161591	1.21104	1.332355	1.307533	1.30562	1.306613	1.650891
5	b_0	0.858521	0.3401984	0.331636	0.0664508	-0.163849	-0.1457857	-0.1773487	-0.1849201	-0.197054	-0.234174	-0.274043
	b_1	0.523549	2.186214	2.220727	3.325306	4.320055	4.384099	4.541629	4.593736	4.652489	4.8356117	4.967864
	b_2	-0.47033	-1.384552	-1.423273	-2.705506	-3.85235	-4.00838	-4.198594	-4.268032	-4.338279	-4.591483	-4.663226
	b_3	0.16939	0.326329	0.341881	0.825806	1.239043	1.3166	1.386763	1.416448	1.443015	1.555015	1.567563
5.25	b_0	0.8148025	0.3366905	0.3295348	0.251416	-0.206809	-0.1894272	-0.206451	-0.2238389	-0.237709	-0.265734	-0.274043
	b_1	0.6596686	2.191744	2.220423	2.586965	4.461898	4.505361	4.65416	4.730609	4.791655	4.926119	4.967864
	b_2	-0.557746	-1.375045	-1.40709	-1.853225	-3.95333	-4.045623	-4.284001	-4.37094	-4.436424	-4.619527	-4.663235
	b_3	0.181006	0.3194259	0.33298	0.5379035	1.27141	1.308498	1.416994	1.448931	1.471484	1.552879	1.5777
5.50	b_0	0.813006	0.3274158	0.3268351	0.1114726	-0.2830512	-0.2778301	-0.277538	-0.295059	-0.305273	-0.339642	-0.3664041
	b_1	0.6385756	2.227647	2.225822	-3.137624	4.694411	4.82404	4.872191	4.94569	4.988927	5.160874	5.292598
	b_2	-0.486005	-1.415437	-1.409554	-2.455989	-4.102561	-4.34069	-4.418575	-4.494796	-4.539485	-4.787701	-4.969428
	b_3	0.139327	0.3361978	0.337021	0.7660176	1.313513	1.421305	1.455344	1.480582	1.496985	1.613237	1.694046
5.75	b_0	0.853445	0.3302605	0.3176731	0.015346	-0.21399	-0.2189623	-0.252132	-0.2489603	-0.269679	-0.306823	-0.327357
	b_1	1.47197	2.208879	2.255272	3.401733	4.343575	4.554145	4.714292	4.700581	4.788508	4.956664	5.05607
	b_2	-0.273699	-1.379269	-1.430856	-2.522707	-3.553716	-3.907322	-4.103426	-4.070438	-4.169687	-4.38992	-4.52238
	b_3	0.0566389	0.3213063	0.3444947	0.7345707	1.087173	1.24039	1.315473	1.298742	1.334454	1.4298873	1.487236
6	b_0	0.8286585	0.2947162	0.2998503	-0.1618487	-0.202211	-0.2030329	-0.243292	-0.290606	-0.3103606	-0.338359	-0.385595
	b_1	0.6164182	2.345598	2.302884	4.089527	4.243664	4.492146	4.688945	4.896543	4.986321	5.111773	5.315577
	b_2	-0.47602	-1.531587	-1.440343	-3.217202	-3.333151	-3.777632	-4.02779	-4.290173	-4.403886	-4.56029	-4.809367
	b_3	0.141876	0.3809683	0.3349724	0.9623275	0.985247	1.186469	1.282913	1.387879	1.434008	1.498927	1.595575

Form 1: Simplified method for ground cooling calculations

Parameters of the earth-to-air heat exchanger

1 Give the length of the tube L (m)

2 Give the radius of the tube r (m)

3 Give the depth of the tube Z (m)

4 Give the air velocity inside the tube u (m s^{-1})

5 Give the inlet air temperature T_{in} (°C)

6 Give the ground temperature at the depth of the exchanger
   T_g (°C)

7 Calculate the air volume flow rate in the tube, $Q = \pi r^2 u$:

   [3.14 × STEP 2 × STEP 2 × STEP 4]

8 Calculate the parameter $P_1 = -0.0161896 \times L$

   [−0.0161896 × STEP 1]

9 Calculate the parameter $P_2 = 0.00019058 \times L \times L$

   [0.00019058 × STEP 1 × STEP1]

10 Calculate the parameter $P_3 = -0.000000957 \times L \times L \times L$

   [−0.000000957 × STEP 1 × STEP 1 × STEP 1]

11 Calculate the dimensionless parameter
   $U = 0.995242 + P_1 + P_2 + P_3$

   [0.995242 + STEP 8 + STEP 9 + STEP 10]

   From Table 11.A1 determine the following parameters (a_0,
   a_1, a_2 and a_3) as a function of the volume flow rate:
   Q (STEP 7) and the tube length (STEP 1)

12 a_0

13 a_1

14 a_2

15 a_3

16 Calculate the parameter $QD_1 = a_1 \times Z$ [STEP 13 × STEP 3]

17 Calculate the parameter $QD_2 = a_2 \times Z \times Z$

   [STEP 14 × STEP 3 × STEP 3]

18 Calculate the parameter $QD_3 = a_3 \times Z \times Z \times Z$

   [STEP 15 × STEP 3 × STEP 3 × STEP 3]

19 Calculate the correction parameter for the depth
   $CV_1 = a_0 + QD_1 + QD_2 + QD_3$

   [STEP 12 + STEP 16 +STEP 17 +STEP 18]

From Table 11.A2 determine the following parameters (b_0,
b_1, b_2 and b_3) as a function of the tube depth Z (STEP 3) and the
tube length L (STEP 1).

20 b_0 _____

21 b_1 _____

22 b_2 _____

23 b_3 _____

24 Calculate the parameter $QV_1 = b_1 \times Q$ [STEP 21 × STEP 7] _____

25 Calculate the parameter $QV_2 = b_2 \times Q \times Q$

    [STEP 22 × STEP 7 × STEP 7] _____

26 Calculate the parameter $QV_3 = b_3 \times Q \times Q \times Q$

    [STEP 23 × STEP 7 × STEP 7 × STEP 7] _____

27 Calculate the correction parameter for the flow rate
    $CV_2 = b_0 + QV_1 + QV_2 + QV_3$

    [STEP 20 + STEP 24 + STEP 25 + STEP 26] _____

28 Calculate the corrected value of $U_{cor} = U \times CV_1 \times CV_2$

    [STEP 11 × STEP 19 × STEP 27] _____

29 Calculate the air temperature at the outlet of the the tube:
    $T_{out} = T_g + U_{cor} \times (T_{in} - T_g)$ [STEP6 + STEP28 × (STEP5 – STEP6)] _____

APPENDIX B
EXAMPLE OF APPLYING THE SIMPLIFIED METHODOLOGY
FOR GROUND COOLING

An earth-to-air heat exchanger is given having the following characteristics:

Tube length:	15 m
Tube radius:	125 mm
Depth at the tube location:	1 m

The following data are also given:

Inlet temperature:	30°C
Ground temperature:	17.3°C

By applying the proposed methodology, the calculated temperature at the outlet of
the exchanger is 29.4°C.

Form 2:   Simplified method for ground cooling calculations

Parameters of the earth-to-air heat exchanger

1 Give the length of the tube L (m) ____15____

2 Give the radius of the tube r (m) ____0.125____

3 Give the depth of the tube Z (m) ____1____

4 Give the air velocity inside the tube u (m s^{-1}) ____8____

5 Give the inlet air temperature T_{in} (°C) ____30____

6 Give the ground temperature at the depth of the exchanger
   T_g (°C) ____17.3____

7 Calculate the air volume flow rate in the tube, $Q = \pi r^2 u$:
   [3.14 × STEP 2 × STEP 2 × STEP 4] ____0.3925____

8 Calculate the parameter $P_1 = -0.0161896 \times L$
   [−0.0161896 × STEP 1] ____-0.242844____

9 Calculate the parameter $P_2 = 0.00019058 \times L \times L$
   [0.00019058 × STEP 1 × STEP1] ____0.0428805____

10 Calculate the parameter $P_3 = -0.000000957 \times L \times L \times L$
   [−0.000000957 × STEP 1 × STEP 1 × STEP 1] ____-0.003229875____

11 Calculate the dimensionless parameter
   $U = 0.995242 + P_1 + P_2 + P_3$
   [0.995242 + STEP 8 + STEP 9 + STEP 10] ____0.792048625____

From Table 11.A1 determine the following parameters (a_0,
a_1, a_2 and a_3) as a function of the volume flow rate:
Q (STEP 7) and the tube length (STEP 1)

12 a_0 ____1.489015____

13 a_1 ____-0.334799____

14 a_2 ____0.048229____

15 a_3 ____-0.0022____

16 Calculate the parameter $QD_1 = a_1 \times Z$ [STEP 13 × STEP 3] ____-0.334799____

17 Calculate the parameter $QD_2 = a_2 \times Z \times Z$
   [STEP 14 × STEP 3 × STEP 3] ____0.048229____

18 Calculate the parameter $QD_3 = a_3 \times Z \times Z \times Z$
   [STEP 15 × STEP 3 × STEP 3 × STEP 3] ____-0.0022____

19 Calculate the correction parameter for the depth
   $CV_1 = a_0 + QD_1 + QD_2 + QD_3$
   [STEP 12 + STEP 16 +STEP 17 +STEP 18] ____1.200245____

From Table 11.A2 determine the following parameters (b_0, b_1, b_2 and b_3) as a function of the tube depth Z (STEP 3) and the tube length L (STEP 1).

20	b_0	0.942228
21	b_1	0.1764782
22	b_2	-0.0821707
23	b_3	0.023893
24	Calculate the parameter $QV_1 = b_1 \times Q$ [STEP 21 × STEP 7]	0.069267694
25	Calculate the parameter $QV_2 = b_2 \times Q \times Q$ [STEP 22 × STEP 7 × STEP 7]	-0.01265891
26	Calculate the parameter $QV_3 = b_3 \times Q \times Q \times Q$ [STEP 23 × STEP 7 × STEP 7 × STEP 7]	0.00144474
27	Calculate the correction parameter for the flow rate $CV_2 = b_0 + QV_1 + QV_2 + QV_3$ [STEP 20 + STEP 24 + STEP 25 + STEP 26]	1.000281523
28	Calculate the corrected value of $U_{cor} = U \times CV_1 \times CV_2$ [STEP 11 × STEP 19 × STEP 27]	0.950920033
29	Calculate the air temperature at the outlet of the the tube: $T_{out} = T_g + U_{cor} \times (T_{in} - T_g)$ [STEP6 + STEP28 × (STEP5 – STEP6)]	29.37668442

12

Evaporative cooling

Evaporative cooling is the process that uses the effect of evaporation as a natural heat sink. Sensible heat from the air is absorbed to be used as latent heat necessary to evaporate water. The amount of sensible heat absorbed depends on the amount of water that can be evaporated.

Evaporative cooling is a very old process, having its origins some thousand years ago, in ancient Egypt and Persia[1]. Modern evaporative coolers are based on the prototypes built in the early 1900s in the USA.

Evaporative cooling can be direct or indirect. During direct evaporative cooling, the water content of the cooled air increases as the air is in contact with the evaporated water. During indirect evaporative cooling, evaporation occurs inside a heat exchanger and the water content of the cooled air remains unchanged.

One might wonder why indirect evaporative cooling is considered since the requirement of a heat exchanger makes this technology more complicated and therefore more expensive. The answer is evident if all the parameters determining thermal comfort in buildings are considered. One must always have in mind that thermal comfort during summertime is not necessarily the same as a low indoor temperature. Other parameters, such as indoor relative humidity, air velocity, etc., can result in thermal discomfort if their values are not within a certain range. Since significant evaporation rates may increase indoor relative humidity and create discomfort, direct evaporative cooling can be applied only in spaces where relative humidity is very low. Otherwise indirect evaporative cooling is more suitable.

When evaporation occurs naturally, this is called passive evaporation. A space can be cooled by passive evaporation, provided that there are surfaces of standing or cooling water, such as basins or fountains. If modern systems are used where evaporation has to be controlled, then evaporation is promoted by means of some mechanical device. These systems are called hybrid evaporative systems. It is obvious that hybrid systems do not have zero energy consumption like passive ones. However, this energy is very low compared to conventional air conditioners.

PHYSICS OF EVAPORATIVE COOLING

Definitions

In this section the physics related to humid air phenomena and processes will be discussed in order to give a better understanding of the evaporative cooling principles. The physics dealing with the thermodynamic conditions of humid air is called psychrometry. The word 'psychrometry' is composite from two Greek words *'psychros'* (= cold) and *'metro'* (= measure).

Atmospheric air consists of a large number of gases, as well as water vapour and other contaminants (e.g. smoke, pollen and other gaseous pollutants) not normally present in the free air, far from sources of pollution.

By definition, *dry air* is obtained when all humidity and contaminants are removed from atmospheric air. The approximate percentage composition of atmospheric air is: nitrogen, 78.084; oxygen, 20.9476; argon, 0.934; carbon dioxide, 0.0314; neon, 0.001818; helium, 0.000524; methane, 0.0002; sulphur dioxide, 0 to 0.0001; hydrogen, 0.00005 and minor constituents such as krypton, xenon and ozone, 0.0002 [2].

Moist air is a mixture of dry air and water vapour. The amount of water vapour in moist air varies from zero (dry air) to a maximum that depends on temperature and pressure. When this maximum is achieved, then this condition is called *saturation*. During saturation there is a neutral equilibrium between the moist air and the condensed water phase (liquid or solid).

The thermodynamic properties of moist air are graphically represented on a chart called a *psychrometric chart*. The skeleton of a psychrometric chart is given in Figure 12.1. It must be noted here that the thermodynamic properties of moist air do not depend only upon the temperature and water content, but also upon atmospheric pressure. Therefore each atmospheric chart is valid for a specific value of the atmospheric pressure. In this chapter, however, normal pressure conditions will be assumed, since the practical consequences of considering a modified temperature are not significant.

The *x*-axis of the psychrometric chart represents the temperature of the humid air, also called *dry bulb temperature*. On the *y*-axis, the *humidity ratio* is reported: this is the ratio of the mass of the water vapour contained in an air sample to the mass of dry air contained in the same sample.

Relative humidity is defined as the ratio of the mole fraction of water vapour x_w in a given moist air sample to the mole fraction x_{ws} in an air sample, saturated at the same temperature and pressure. Practically, this means that the higher the relative

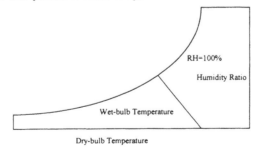

Figure 12.1 Skeleton of the psychrometric chart

humidity is, the closer to saturation conditions the moist air sample is. The curved lines on the psychrometric chart are the relative humidity lines. On this graph only the 100% relative humidity line is shown. The 0% relative humidity line coincides with the dry-bulb temperature line.

The first instrument used for measuring the water content of humid air was the psychrometer. The psychrometer consists of two thermometers, one having a bulb covered with a wick that has been thoroughly wetted with water. When the wet-bulb is placed in an air stream, water may evaporate from the wick, the quantity of which depends on the water content of the air stream. The equilibrium temperature the water eventually reaches is called the *wet-bulb temperature*. The oblique lines on the psychrometric chart are the constant wet-bulb temperature lines. Today electronic instrumentation is available for humidity measurements. Sophisticated psychrometers are still used in laboratories, mainly for calibration purposes.

Evaporative cooling

Evaporative cooling is based on the thermodynamics of the evaporation of water, i.e. the change of the liquid phase of water into water vapour. This phase change requires energy, which is called the *latent heat of evaporation*. This is the energy required to change a substance from liquid phase to the gaseous one, without temperature change. When non-saturated air (i.e. air that does not contain liquid water but only water vapour) comes into contact with water, then evaporation occurs. The necessary latent heat is provided by the air, which is then cooled. It is obvious that during this process the moisture content of the air is increased. Therefore, this is called direct evaporative cooling. This process is represented on the psychrometric chart by a displacement along a constant wet-bulb line, AB (Figure 12.2).

When evaporation occurs in the primary circuit of a heat exchanger, while the air to be cooled circulates in the secondary circuit, the air temperature decreases but its humidity ratio remains constant. It must be noted that, since the air temperature drops, its relative humidity will of course increase, but less than during the direct evaporative cooling process. This is called indirect evaporative cooling. Since the

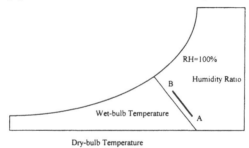

Figure 12.2 Direct evaporative cooling process

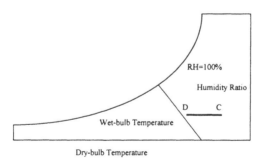

Figure 12.3 Indirect evaporative cooling process

humidity ratio of the air does not change, this process is represented on the psychrometric chart by a displacement along a constant humidity-ratio line, CD (Figure 12.3).

The performance assessment of both direct and indirect evaporative cooling systems is based on the concept of the *saturation efficiency*, defined as

$$\text{Saturation Efficency} = \frac{T_{db,in} - T_{db,out}}{T_{db,in} - T_{wb,in}} \tag{12.1}$$

with $T_{db,in}$ and $T_{db,out}$ the dry-bulb temperatures of the air at the inlet and outlet of the system respectively and $T_{wb,in}$ the wet-bulb temperature of the air at the inlet of the system.

In direct evaporative cooling systems the humidity ratio (or moisture content) of the cooled air increases, raising the relative humidity of the indoor space. If the space is sufficiently ventilated, this increase will most probably not generate discomfort. Otherwise discomfort may occur.

Typically acceptable performance figures for direct evaporative systems, as given in [3], are:

- A saturation efficiency of the cooling process of 70% or better.
- A maximum indoor air velocity of 1 m s^{-1}.
- The air temperature of the indoor space should be around 2 K higher than the discharge air temperature and its relative humidity should be below 70%.
- The resulting temperature of the indoor space should be about 4 K below the outdoor dry-bulb temperature.

Indirect evaporative systems do not create problems related to increased humidity levels. Neither high ventilation rates nor humidity control and de-humidification systems are required. However, as these systems have a more complex technology, their investment, maintenance, operation and repair costs are higher.

EVAPORATIVE COOLING SYSTEMS

Evaporative cooling systems can be classified in two ways. The first classification is related to the contact or not of the cooled air with the evaporated water. As explained in the previous section, according to this classification evaporative cooling systems are divided into direct and indirect types.

The second classification is performed according to the energy required to produce evaporation. If evaporation occurs naturally, then the systems are called passive evaporative systems, but if evaporation is due to a current of air mechanically induced by electric ventilators, the systems are called hybrid evaporative. Air humidification and cooling by evapotranspiration of plants and the use of free water surfaces, like pools and streams, is a passive direct technique. Passive indirect evaporative techniques are roof sprinkling and the use of a roof pond.

Passive direct systems

This category includes the use of vegetation for evapotranspiration, the use of fountains, sprays, pools and ponds as well as the use of volume and tower cooling techniques. Vegetation plays the role of a natural evaporative cooling device. Trees and other plants transpire moisture in order to reject their sensible heat. The heat transferred by evapotranspiration is close to 2320 kJ per kg of evaporated water. The cooling potential due to this evapotranspiration of plants is important. A normal-size deciduous tree evaporates 1460 kg of water during a sunny summer day. The corresponding energy consumption is 870 MJ, which is the cooling effect of five air conditioners. Also, one acre of grass can transfer more than 50 GJ on a sunny day, while evapotranspiration from wet grass can reduce the ground surface temperature by up to 6–8°C below the average surface temperature of bare soil [4].

Current knowledge of the role of vegetation results mainly from observations and theoretical analyses. Reported observations indicate a temperature reduction of 2–3°C due to plant evapotranspiration [5]. It also has been found that the temperatures at the Golden Gate Park of San Francisco were found to be about 8°C cooler than nearby less vegetated areas [6].

Theoretical analysis of the role of plant evapotranspiration has shown that evapotranspiration from one tree can save 250 to 650 kWh of electricity used for air conditioning per year [7]. Fountains, sprays, pools and ponds are particularly effective passive cooling techniques. Very well known applications are Shah Jahan's Taj Mahal in Agra, and the Moorish Palace and the Alhambra in Granada.

The rate of evaporation from a wetted surface depends upon the air velocity and the difference between the water vapour pressure and the air pressure next to the moist surface. Calculations based on mean summer weather give an evaporation rate between 150 and 200 W m^{-2}, which is the cooling potential of this technique.

Evaporative cooling in open spaces is particularly effective in areas that have wet-bulb temperatures below 21°C.

Volume cooling techniques are well known from traditional architecture. The system is based on the use of a tower where water contained in a jar or pads, or water that is sprayed, is precipitated. Ambient air introduced into the tower is cooled by evaporation and then transferred inside the building. On the top of the towers there are low-pressure-drop evaporative cooler pads. The material usually used is cross-corrugated cellulose [8]. In order to save water, the part of the water wetting the pads and not evaporated is recirculated. In most of these applications the wind is directed downwards by means of gravity dampers. The towers operate as a reverse chimney: in a chimney the hot air rises; in a cooling tower the air entering the top of the tower is cooled by evaporation and, since it becomes heavier, it falls to the bottom of the tower to be directed inside the building. Measurements on such a contemporary construction have shown that, with a dry-bulb temperature of incoming air of 35.6°C and a wet-bulb temperature of 22.2°C, the air temperature at the outlet of the tower was close to 24°C.

A variation of the cooling tower application is described in [9]. In this application a 93 m² well-insulated frame building is provided with a cooling tower on one side and a solar chimney on the other side. The cooling tower is 7.6 m high and has a 1.8 × 1.8 m² cross-section. The purpose of the solar chimney is to further enhance ventilation through the building space. Figure 12.4 shows the flow path through the building.

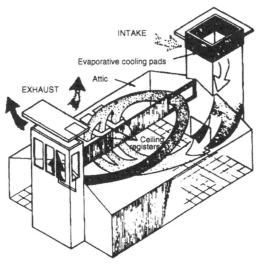

Figure 12.4 Building with attached evaporative cooling tower [9]

Passive indirect systems

These systems mainly include the following techniques:

- roof sprinkling
- roof ponds and
- moving water films.

ROOF SPRINKLING SYSTEMS

Roof sprinkling is a very interesting idea, since in many buildings and mainly in those that have a flat roof, the main part of the external indirect heat gains comes from the roof. Roof sprinkling (or roof spray cooling) is based on evaporation of a water mist layer created by misting sprayheads that cover the roof of the building; when this mist evaporates, it absorbs large amounts of heat.

Monitoring of such a system installed in Chicago has shown that a 5°C average decrease in the indoor temperature is obtained. This spraying system consists of an all-copper network. The sprayheads can be individually adjusted to provide alternative flow rates and directional patterns. Instead of spraying uncontrolled amounts of water on the roof, the system injects water according to the temperature and evaporation rates. A controller measures the roof temperature, determines the temperature variation and calculates the amount of water that can be evaporated at this temperature. Then the necessary amount of water is sprayed, controlled by the electric valves of the system. The goal is to optimize cooling, while using as little water as possible, which in hot and arid areas is in general expensive and not abundant, thus avoiding over-spraying and runoff.

ROOF PONDS

Roof pond systems are much simpler than roof sprinkling. Water ponds are constructed over non-insulated flat roofs. The water surface must be shaded during daytime to avoid excessive water heating. From the definition of the efficiency of evaporative cooling (equation 12.1) the roof temperature must be higher than the wet-bulb temperature of the air. According to Givoni, the necessary condition for applying this technique efficiently, is that the wet-bulb temperature of the air should be lower than 20°C [10].

MOVING WATER FILM

The moving water film technique is based on the flow of a water film over the roof surface. The evaporation process is enhanced by an increase in the relative velocity between the air and the water surface. The cooled water is stored in the basement of the building and then circulated inside the building space, cooling it. The threshold condition for the operation of the system is that the roof temperature should be higher than the wet-bulb temperature of the ambient air.

Direct hybrid air coolers

The principal element of a direct air humidifier is a porous material saturated with water. Air circulates through this porous pad by means of a ventilator and loses part of its heat by evaporating part of the water in the pad. This system is of low capital and operational cost. Its main disadvantage is that it has to be used carefully since it can easily create discomfort, mainly in spaces where latent heat gains are important, because it increases the water content of the indoor air.

This type of equipment has been in production in the USA, Australia and Europe for over 60 years now. Its saturation efficiency is approximately 60 to 90%. The main elements of an evaporative cooling device are the following:

- the wetted surface on which evaporation occurs,
- a water pump,
- a ventilator,
- a water injection system,
- a collection system for the water droplets,
- a box containing the whole device,
- a water tank,
- a water valve maintaining a constant water tank level.

Direct hybrid evaporative coolers are classified according to the type of their wetted surface. There are three main types, called drip-type cooler, spray-type cooler and rotary pad cooler. Generally, the saturation efficiency of these devices ranges between 0.6 and 0.9. All the direct evaporative coolers can be found on the market as room-sized units. The advantage of direct coolers is their low operating cost. Their disadvantage is that humidity control is required and that they cannot perform in locations having a high wet-bulb temperature.

If these systems are used in locations where the increase of relative humidity does not create significant problems, this will have a significant impact on the energy consumption and on the environment. Watt and Lincoln [11] report that if half of the conventional cooling devices in the south-west USA were replaced by simple direct evaporative coolers, this would save 18 million barrels of oil per year.

Indirect evaporative coolers

Hybrid indirect evaporative cooling systems use a heat exchanger. Air circulates through the primary circuit, in which evaporation occurs while the air to be cooled passes through the secondary circuit. This decreases the temperature of the air in the secondary circuit without modifying the humidity ratio. The industrial production of these systems is significant since there are more than ten manufacturers world-wide.

Indirect evaporative coolers can operate only if the indoor wet-bulb temperature is lower than the outdoor dry-bulb temperature. In practice, the indoor wet-bulb temperature should be lower than 21°C [10]. The threshold value for the use of such a system is that the ambient wet-bulb temperature should be lower than 24°C. The performance of this type of system is strongly related to its saturation efficiency. Typical saturation efficiency values for those systems range between 60 and 80%. A sketch of such a cooler is given in Figure 12.5.

Indirect evaporative coolers came on the market after direct evaporative coolers. Their main parts are:

- the heat exchanger in which evaporation occurs,
- two ventilators, usually centrifugal,
- two filters, one at the fresh air inlet and a second at the indoor building air inlet,
- a water pump,
- a water injection system,
- a water tank.

There are three types of indirect evaporative coolers according to the type of the heat exchanger:

- tube type cooler,
- flat plate cooler and
- rotating pad cooler.

Indirect evaporative coolers provide low-cost cooling, like direct evaporative coolers. In addition, they do not require any humidity control, since they do not release any water vapour into the cooled space.

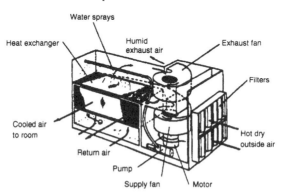

Figure 12.5 Indirect evaporative cooler

The most well known indirect evaporative cooler is the DRICON type cooler [12].

Two-stage evaporative coolers

The two-stage evaporative cooler is a combination of a direct and an indirect evaporative cooler [13]. These systems are used when dry-bulb temperatures lower than those achieved by a single-stage system are required. If a two-stage evaporative system uses as a first-stage unit an indirect evaporative cooler and as a second stage a direct evaporative cooler, the cooling process is explained on the psychrometric chart as follows (Figure 12.6): the air to be cooled, initially at point A, is sensibly cooled by the indirect evaporative cooler until it reaches point B. Since the water content of the air does not change, line AB is parallel to the dry-bulb temperature axis. This air then enters the second stage, where as a result of the direct evaporative cooling process, it reaches point C. This is a constant wet-bulb temperature process and therefore line BC is parallel to the wet-bulb temperature lines.

   The association of a direct with an indirect evaporative cooler results in lower threshold values, reducing thus the possible working time of the equipment and therefore the energy consumed. Also a double-stage cooler offers lower dry-bulb temperatures than the single stage, resulting thus in increased indoor comfort levels. Figure 12.7 shows the percentage of possible working hours of a single- and a double-stage cooler as a function of the period when cooling is required. Performance statistics have been derived after extensive calculation for 14 south European locations.

   Similar data are also presented on a monthly basis in Figure 12.8 for Athens. It is shown that a single direct cooler has to operate 5 to 20% longer than the double stage equipment to achieve the same result.

   As already mentioned, the direct cooler increases the humidity ratio of air and can result in discomfort. In Figure 12.9 the relative humidity range at the outlet of a direct cooler and of a double-stage cooler in Athens is shown. The use of the double-stage cooler results in lower relative humidity values. Consequently the time during which direct coolers can deliver enough cold air to maintain the effective space temperature, as specified by the user, is limited.

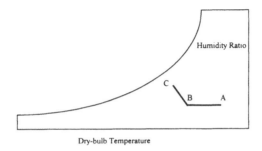

Figure 12.6 Two-stage evaporative cooling process

Evaporative Cooling – Hours of Operation

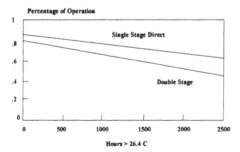

Figure 12.7 Percentage of possible working hours of single-stage and double-stage coolers [14]

Athens, Cooling Needs – Working Hours for Single and Double Stage Coolers

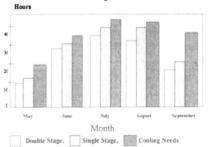

Figure 12.8 Hours of cooling requirements and working hours of single and double-stage evaporative coolers in Athens [14]

Relative Humidity Range – Single and Double Stage Coolers

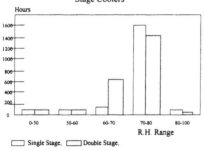
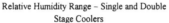

Figure 12.9 Relative humidity at the outlet of single and double-stage coolers [14]

Athens - Comfort Index - Single and Double Stage Coolers

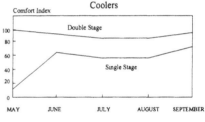

Figure 12.10 Mean monthly comfort index in Athens [14]

Comfort index

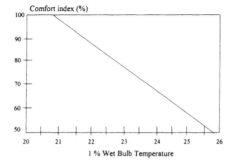

Figure 12.11 Mean monthly comfort index for 14 south European locations [14]

Figure 12.10 shows the mean monthly comfort index in Athens using a single-stage and a double-stage cooler. Clearly for this particular climate, the direct evaporative cooler cannot provide effective air conditioning.

The mean summer comfort index for 14 south European locations when a direct evaporative cooler is used, is given in Figure 12.11 as a function of 1% of the wet-bulb temperature, where it is shown that higher wet-bulb temperatures correspond to lower comfort indexes.

An example of a two-stage evaporative cooler is shown in Figure 12.12.

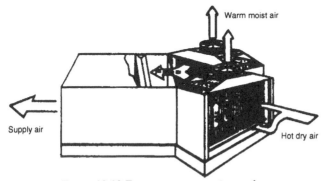

Figure 12.12 Two-stage evaporative cooler.

PERFORMANCE OF EVAPORATIVE COOLING DEVICES

Direct evaporative coolers

The performance of a direct evaporative cooler has been evaluated by simulating the performance of an injection-type direct evaporative cooler.

This cooler has a centrifugal ventilator having a diameter of 0.25 m, which rotates at 1500 r.p.m. The air crosses a battery of 38 porous plates, each having dimensions $1.20 \times 0.60 \times 3.50 \times 10^{-3}$ m. The wet (effective) surface of the battery is 50 m². The water distribution circuit works on a closed loop and has a volume flow rate of 250 m³ h⁻¹.

The software developed for the calculation of the efficiency of this evaporative cooler uses relationships produced by applying an identification technique, based on the Marquard algorithm [15], to experimental data. This software requires as input not only the characteristics of the cooler given in the previous paragraph, but also the hourly values of the ambient air temperature and relative humidity. It must be noted that this method is not universal, but can only be applied to the particular cooler described above.

The following relations were determined for calculating:

• the inlet-outlet temperature difference:

$$T_1 - T_2 = C_1 + C_2 m_f^{C3} X + C_4 m_f^{C3} X^{C5} / N^{C6} \qquad (12.2)$$

with T_1 the inlet air temperature (°C), T_2 the outlet air temperature (°C), m_f the water mass flow rate (kg h^{-1}), X the difference between dry-bulb and wet-bulb air temperature at the inlet of the cooler ($X = T_1 - T_{wb}$) and N the velocity of the centrifugal ventilator (r.p.m.). The constants C_i, $I = 1, \ldots, 6$ take the following values:

$$C_1 = 1.1,\ C_2 = 0.23,\ C_3 = 0.09,\ C_4 = 1.18 \times 10^{-4},\ C_5 = 2.16, C_6 = -0.61.$$

- the air mass flow rate at the outlet of the cooler

$$m_a = C_1' + C_2' N m_f + C_3' N \qquad (12.3)$$

where $C_1' = -39.7$, $C_2' = 1.46 \times 10^{-5}$, $C_3' = 0.2$.

- the mass of water that evaporates during the process:

$$m_e = C_1'' m_f^{C_2''} X + C_3'' m_f^{C_2''} N^{C_4''} X^{C_5''} \qquad (12.4)$$

$$C_1'' = 0.0.7,\ C_2'' = 0.14,\ C_3'' = 0.002 \times 10^{-4},\ C_4'' = 1.4,\ C_5'' = 0.4.$$

Using these relations, the temperature and the relative humidity at the outlet of the cooler and the available cooling energy per day can be calculated.

A simplified methodology for calculating the performance of a direct evaporative cooler is given in the Appendix at the end of this chapter. This methodology has been developed in the form of calculation sheets. An example of completed calculation sheets is also given for the following case:

Cooler characteristics:

- Fan speed = 1500 r.p.m.
- Water mass flow rate in the cooler = 100 kg h^{-1}.

Weather conditions:
- Ambient temperature = 35°C,
- Ambient relative humidity = 40%.

The calculation shows that the temperature at the outlet of the cooler equals 28.4°C.

Indirect evaporative coolers

The performance of indirect evaporative cooling has been evaluated by simulating the performance of the DRICON cooler. This system consists of a heat exchanger made of plastic material, two ventilators for the air circulation in the primary and secondary circuits respectively, a water pump and the injection nozzles.

For these calculations, it was assumed that ambient air is supplied to the primary circuit of the cooler and therefore the saturation efficiency is defined by equation (12.1). Experimentally it has been found that the saturation efficiency of this system is related to the air velocity in the primary circuit of the cooler V_a by the following equation [12]:

$$\text{Saturation efficiency} = 1/(1+0.47V_a^{0.3}). \tag{12.5}$$

The algorithm developed for the calculation of the efficiency of the indirect evaporative cooler requires as input not only the hourly values of the dry-bulb temperature and of the relative humidity of the ambient air, but also the air velocity at the primary circuit of the cooler.

If the air velocity is given, the saturation efficiency is calculated using equation (12.5). Then, also taking into account the other meteorological data, the air temperature at the outlet of the cooler can be calculated by applying equation (12.1).

Simplified calculation forms, given in the Appendix, have also been prepared for indirect evaporative coolers. The following case is given as an example:

Cooler characteristics:
- Air velocity at the inlet of the cooler = 1 m s^{-1}.

Weather conditions:

- Ambient temperature = 35°C,
- Ambient relative humidity = 40%.

The calculated temperature at the outlet of the cooler is found to be 27.5°C.

Case studies

The performances of a direct and an indirect evaporative cooler have been tested by simulating the behaviour of a typical Greek residential building [16], the floor section of which is illustrated in Figure 12.13.

DIRECT EVAPORATIVE COOLING

The direct evaporative cooling device considered in this study is a parallel-plate pad evaporative cooler. This cooler consists of a centrifugal fan (diameter 25 cm). The parallel plate matrix has a 50 m wetted area and is composed of 38 plates. The dimensions of each plate are $1.20 \times 0.60 \times 0.0035$ m and the gap between the plates in the matrix is 4.4×10^{-3} m. The pump of the water distribution system has a capacity of 250 m^3 h^{-1} of water under a head of 3 m.

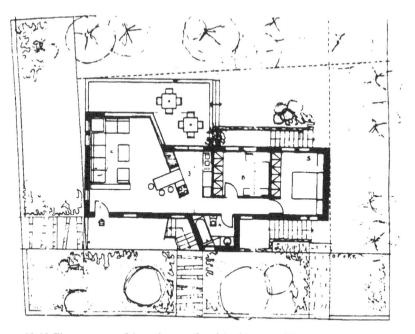

Figure 12.13 Floor section of the reference Greek building used for the evaluation of the efficiency of direct and indirect evaporative coolers

In order to predict the performance of the direct evaporative system, the algorithms presented in [5] were used and linked to the overall calculation program. These algorithms, which were created by using identification procedures, are of sufficient accuracy and have been compared successfully with experimental data.

Using as input data the ambient dry- and wet-bulb temperatures, the fan rate (in r.p.m.) and the water mass flow rate, the temperature and the relative humidity at the outlet of the cooler were calculated. In order to analyse the impact of the parameters regulating the performance of the system for the building, a sensitivity analysis was performed. Therefore, the influence of the fan speed, as well as of the flow rate of the water humidifying the parallel-plate matrix, has been investigated.

The results are presented in Figure 12.14 for a typical day of each month. It is observed that a maximum reduction of the peak indoor air temperature of about 4–6°C is possible. An increase of the water flow rate has an almost negligible effect on the indoor temperature of the reference building. In most of the cases examined, the corresponding temperature variation curves practically coincide. Their difference rises to about 0.5°C when the high fan rate of 3500 r.p.m. is used. It is deduced also that the effect of the fan rate change on the indoor temperature of the reference building is important. The mean temperature decrease is about 1.5°C when the fan rate is increased from 800 to 1500 r.p.m.

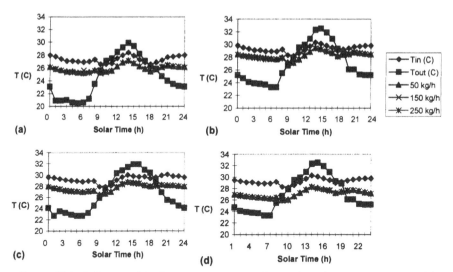

Figure 12.14 Influence of the direct evaporative cooler on the indoor air temperature (a) June, 800 r.p.m., (b) July, 800 r.p.m., (c) August, 800 r.p.m., (d) July, 1500 r.p.m.

INDIRECT EVAPORATIVE COOLING

Indirect evaporative cooling systems can provide efficient cooling of buildings without increasing the moisture content of the indoor air. Various types of indirect evaporative cooling systems have been proposed [6], however plate-type indirect evaporative coolers have given very encouraging results and have already seriously penetrated into the market [7]. The type of cooler considered in the present research consists of a plastic heat exchanger with dimpled sheets of a hydraulic polymer, two fans, a water pump and simple water sprays [17].

In order to calculate the efficiency of the system, the saturation efficiency algorithm for that cooler, proposed in [17], is used. The air temperature at the outlet of the cooler is then calculated, using as inputs the outdoor dry- and wet-bulb temperatures. The impact of the air velocity inside the primary circuit of the cooler on the thermal performance of the building was investigated. Two air velocities were considered, 0.3 m s^{-1} and 0.1 m s^{-1}. The results showing the effect of the air velocity variation on the indoor temperature of the reference building are illustrated in Figure 12.15.

It is deduced that the increase of the air velocity is inversely proportional to the indoor temperature. In all cases, a temperature decrease of at least 1.5°C is obtained, compared with the indoor temperature when no cooling load is provided to the building. During June, acceptable indoor temperature levels are obtained even with the low air velocity. However, in order to create comfortable indoor conditions during daytime in July and August, the high air velocity has to be chosen.

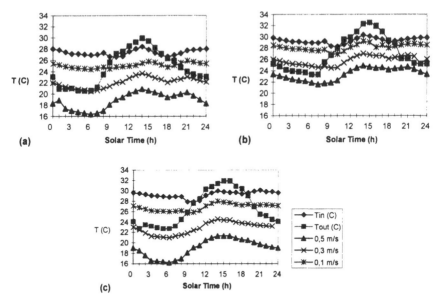

Figure 12.15 Influence of the indirect evaporative cooler on the indoor air temperature (a) June, (b) July (c) August. T_{in} = indoor temperature without coolers, T_{out} = ambient temperature

Impact of evaporative cooling upon comfort

The impact of evaporative cooling on comfort has been assessed by simulating the performance of a direct and an indirect evaporative cooler assuming a 24-hour operation, for a typical day of every month of the cooling season, using the weather data of Athens [17].

This study has shown that the ambient conditions were outside the comfort zone during most hours of the day. When direct evaporative cooling was applied, the condition of the air at the outlet of the cooler was inside the comfort zone, but the air was becoming humid. When indirect evaporative cooling was used instead, with the same inlet values, the relative humidity at the outlet had values inside the comfort zone. This has shown that indirect evaporative coolers are more appropriate for the climate of Athens.

CONCLUSIONS

Evaporative coolers offer an interesting alternative for cooling. They can provide cooling, consuming just the electricity necessary for the operation of air fans and in some cases for small water pumps. This is much less than the electricity required for the operation of conventional air conditioners. Excessive air humidification can be avoided by using indirect coolers instead of direct ones, or by combining the two in the double-stage systems. Research in this field has led to devices that, in consciously designed buildings, can easily compete with conventional air conditioners.

REFERENCES

1 *Passive Cooling* (1989). ed. J. Cook. MIT Press, Cambridge MA.
2 *ASHRAE Fundamentals Handbook* (1977). Ch.5. Amercian Society of Heating, Refrigerating and Air Conditioning Engineers, Atlanta.
3 Goulding, J.R., J. Owen Lewis and T.C. Steemers (eds) (1992). *The European Passive Solar Handbook*. C.E.C.(EUR 13446).
4 Moffat A. and M. Schiller (1981). *Landscape Design that Saves Energy*. William Morrow and Company, New York.
5 Bowen, A. (1980). *Passive Cooling Handbook*. American Section, International Solar Energy Society.
6 Duckworth, E. and J. Sandberg (1954). *Bulletin of the American Meteorological Society*, Vol. 35, p. 198.
7 Akbari, H., J. Huang, P.Martien, L. Rainer, A. Rosenfeld and H. Taha (1988). *ACEEE Summer Study on Energy Efficiency in Buildings*.
8 Kent, K. and T. Lewis Thompson. 'Natural draft evaporative cooling', *ASHRAE Transactions*, pp. 157–160.
9 Cunningham, W.A. and T. Lewis Thompson (1986). 'Passive cooling with natural draft cooling towers in combination with solar chimneys', *Proceedings of the PLEA Conference*, Pecs, Hungary,. pp. S-23–S-24.
10 *Horizontal Study on Passive Cooling* (1990). ed. M. Santamouris. C.E.C. Program BUILDING 2000.
11 Watt, J.R. and A.A. Lincoln. 'Climatic changes and rising markets for evaporative cooling', *ASHRAE Transactions*, pp. 709–717.
12 Pescod, D. (1974). 'An evaporative air cooler using a plate heat exchanger', CSIRO, Mechanical Engineering Division, Technical report No TR 2, Highett, Victoria, Australia.
13 Wu H. (1989). 'Performance monitoring of a two-stage evaporative cooler', *ASHRAE Transactions*, Vol. 95.
14 Santamouris, M. (1990). 'Natural cooling techniques', *Proceedings Workshop on Passive Cooling*, JRC, Ispra, Italy.
15 Achard, P., A.N. Ayoob and L. Elegant (1990). 'Characterisation of an evaporative air cooling system', *Proceedings of the Conference on External Building Components*, Milan, Italy, pp. 221–224.
16 Agas, G., T. Matsaggos, M. Santamouris and A. Argiriou (1991). 'On the use of atmospheric heat sinks for heat dissipation', *Energy and Buildings*, Vol. 17, pp. 321–329.
17 Protechna Ltd & Meletitikh Ltd (1991) *Technical and Economical Evaluation of Passive and Hybrid Techniques Available Worldwide for Cooling of Buildings*, Vol. 3, ed. M. Santamouris. Study financed by the Centre of Renewable Energy Sources, Athens.

APPENDIX
FORMS

Form 1: Simplified method for direct evaporative cooling calculations

1   Give the ambient air temperature (°C)
2   Give the ambient relative humidity (%)
3   Give the fan speed of the cooler (r.p.m.) (If unknown, the
    default value 1500 r.p.m. can be used)
4   Give the water mass flow rate in the cooler (kg h^{-1}) (If unknown,
    the default value 100 kg h^{-1} can be used)
5   Calculate the air flow rate (kg h^{-1}) at the outlet of the cooler, as
    follows: STEP 5 = –39.7 +1.46E–5 × STEP 3 × STEP 4 + 0.2 × STEP 3
6   From the psychrometric chart, calculate the wet-bulb
    temperature at the inlet of the cooler
7   Calculate the following value: STEP 7 = (STEP 4)$^{0.09}$
8   Calculate the following value: STEP 8 = STEP 1 – STEP 6
9   Calculate the temperature at the outlet of the cooler as follows:
    STEP 9 = STEP 1 – 0.23 × STEP 7 × STEP 8
        – 1.18E–4 × STEP 7 × (STEP 8)$^{2.16}$/(STEP 3)$^{-0.61}$

Form 2: Simplified method for indirect evaporative cooling calculations

1   Give the ambient air temperature (°C)
2   Give the ambient relative humidity (%)
3   Give the air velocity at the inlet of the cooler (m s^{-1})
    (If unknown, the default value 1 m s^{-1} can be used)
4   Calculate the efficiency of the cooler: STEP 4 = 1/(1 + 0.47 × (STEP 3)^3)
5   From the psychrometric chart, calculate the wet bulb
    temperature at the inlet of the cooler
6   Calculate the temperature at the outlet of the cooler as follows:
    STEP 6 = STEP 1 – STEP 4 × (STEP 1 – STEP 5)

EXAMPLES

Form 1:   Simplified method for direct evaporative cooling calculations

1	Give the ambient air temperature (°C)	35
2	Give the ambient relative humidity (%)	40
3	Give the fan speed of the cooler (r.p.m.) (If unknown, the default value 1500 r.p.m. can be used)	1500
4	Give the water mass flow rate in the cooler (kg h^{-1}) (If unknown, the default value 100 kg h^{-1} can be used)	100
5	Calculate the air flow rate (kg h^{-1}) at the outlet of the cooler, as follows: STEP 5 = –39.7 +1.46E–5 × STEP 3 × STEP 4 + 0.2 × STEP 3	262.49
6	From the psychrometric chart, calculate the wet-bulb temperature at the inlet of the cooler	24
7	Calculate the following value: STEP 7 = (STEP 4)$^{0.09}$	1.513561248
8	Calculate the following value: STEP 8 = STEP 1 – STEP 6	11
9	Calculate the temperature at the outlet of the cooler as follows: STEP 9 = STEP 1 – 0.23 × STEP 7 × STEP 8 – 1.18E–4 × STEP 7 × (STEP 8)$^{2.16}$/(STEP 3)$^{-0.61}$	28.42465626

Form 2:   Simplified method for indirect evaporative cooling calculations

1	Give the ambient air temperature (°C)	35
2	Give the ambient relative humidity (%)	40
3	Give the air velocity at the inlet of the cooler (m s^{-1}) (If unknown, the default value 1 m s^{-1} can be used)	1
4	Calculate the efficiency of the cooler: STEP 4 = 1/(1 + 0.47 × (STEP 3)^3)	0.680272109
5	From the psychrometric chart, calculate the wet bulb temperature at the inlet of the cooler	24
6	Calculate the temperature at the outlet of the cooler as follows: STEP 6 = STEP 1 – STEP 4 × (STEP 1 – STEP 5)	27.5170068

13

Radiative cooling

Radiative cooling is based on the heat loss by long-wave radiation emission from a body towards another body of lower temperature, which plays the role of the heat sink. In the case of buildings the cooled body is the building and the heat sink is the sky, since the sky temperature is lower than the temperatures of most of the objects upon earth.

There are two methods of applying radiative cooling in buildings. The first method is called direct, or passive, radiative cooling. The building envelope radiates towards the sky and gets cooler, thus enhancing the heat loss from the interior of the building. For physical reasons that will be explained later, the part of the building envelope that radiates the most is a flat roof.

The second method is called hybrid radiative cooling: in this case the radiator is not the building envelope but usually a metal plate. The operation of such a radiator is the opposite of an air flat-plate solar collector. Air is cooled by circulating under the metal plate before being injected into the building. Other systems are combinations of these two configurations.

In this chapter the more important examples of radiative cooling systems reported in the literature will be discussed. Some of them might be considered of reduced application interest, because their development is still at an experimental stage, but their description has been included since they reflect the state of the art in this subject.

PHYSICAL PRINCIPLES OF RADIATIVE COOLING

Understanding the physics of radiative cooling requires a knowledge of the basic definitions and principles of radiative heat transfer, a summary of which is given in this section.

Definitions

ELECTROMAGNETIC RADIATION

Any object at a temperature higher than 0 K emits energy by electromagnetic radiation. This radiation is due to the molecular and atomic agitation associated with the internal energy of the material, which in the equilibrium state is proportional to the temperature of the material. This radiation is called thermal radiation. The major part of it is emitted within a narrow band of the electromagnetic spectrum, between 0.1 μm and 100 μm.

BLACK BODY

All surfaces emit thermal radiation and how the laws of radiation operate depends on the surface itself. It is possible, however, using thermodynamics, to evaluate the maximum thermal energy that can be emitted by radiation at any temperature and at any wavelength. The ideal radiator that would radiate this energy is called a black body and is used as the reference for radiation considerations.

A black body is defined as an ideal body that absorbs all the incident radiation impinging on it, for all wavelengths and all angles of incidence of the radiation. From this definition it can be derived that the black body also emits the maximum radiant energy at every wavelength.

The physical laws describe the emission from the black body. The emission from real bodies is therefore evaluated relative to the emission of the black body under the same conditions, using coefficients called emissivities. Knowing the radiation laws for a real body is a synonym for knowing its emissivities, total (over all wavelengths) or spectral (for each wavelength).

PLANCK'S LAW

Planck's Law gives the spectral emittance of the black body, $M_\lambda^\circ(T)$ (W m^{-2} μm^{-1}), at temperature T (Kelvin) for the wavelength λ and is given by the following relation:

$$M_\lambda^\circ(T) = \frac{C_1 \lambda^{-5}}{e^{C_2/\lambda T} - 1}$$

where

$$C_1 = 3.741 \times 10^8 \text{ W μm}^4 \text{ m}^{-2} \text{ and } C_2 = 14\ 388 \text{ μm K}.$$

STEFAN–BOLTZMANN LAW

This law gives the total hemispherical (i.e. towards all directions) emissive power (W m^{-2}) of a black body:

$$M^\circ = \sigma T^4 \tag{13.1}$$

where $\sigma = 5.67 \times 10^{-8}$ W m^{-2} K^{-4} is the Stefan–Boltzmann constant. The temperature is expressed in K.

The total hemispherical emissive power of a real body can be calculated by the Stefan–Boltzmann law, taking into account its total hemispherical emissivity, ε:

$$M = \varepsilon \sigma T^4. \tag{13.2}$$

Given that the black body is an ideal radiator, which for a given temperature has the maximum total hemispherical emissive power, it is concluded that the total hemispherical emissivity of a real body is always less than one, since one is the black-body emissivity.

INTERACTION BETWEEN RADIATION AND A BODY: REFLECTION, ABSORPTION AND
TRANSMISSION

The radiation incident on a body is partially absorbed, partially reflected and partially transmitted through the body. The fractions of the absorbed, reflected and
transmitted radiation are respectively called absorptivity (α), reflectivity (ρ) and
transmissivity (τ). These parameters are related as follows:

$$\alpha + \rho + \tau = 1. \tag{13.3}$$

These values are total hemispherical values and characterize globally the interaction
between a body and the radiation impinging on it. It must be noted that this expression is accurate for monochromatic radiation, but for the applications examined here
the formalism involving total values can be used.

KIRCHHOFF'S LAW

Kirchhoff's Law describes the relation between the emitting and absorbing properties of a body. According to this law, for every wavelength and for every direction of
propagation of the radiation, the directional spectral emissivity of a body is equal to
its directional spectral absorptivity:

$$\alpha_\lambda(T, \varphi, \theta) = \varepsilon_\lambda(T, \varphi, \theta) \tag{13.4}$$

where λ is the wavelength, φ and θ the angular coordinates defining the direction of
propagation and T is the temperature of the body. This relationship is also valid for
hemispherical properties:

$$\alpha_\lambda(T) = \varepsilon_\lambda(T). \tag{13.5}$$

SIMPLIFICATIONS IN PRACTICE

In building physics and in solar-energy engineering it is principally necessary to
know the optical properties of the materials involved for two zones of the electromagnetic spectrum. Absorptivities are required for the wavelengths of the solar
spectrum to to be able to estimate solar gains. Emissivities are required in the long-
wave range of the electromagnetic spectrum, in order to be able to estimate radiative
heat losses. Therefore the solar absorptivity and thermal emissivity of various materials can be defined as follows [1]:

- **Solar absorptivity:**

$$\alpha_s(T) = \frac{\int_{0.3\,\mu m}^{2.5\,\mu m} \alpha_\lambda(T) I_\lambda \, d\lambda}{\int_{0.3\,\mu m}^{2.5\,\mu m} I_\lambda \, d\lambda}$$

where T is the temperature of the absorbing surface and I_λ is the solar irradiance.

- **Thermal emissivity:**

$$\varepsilon_T(T) = \frac{\int_{1\,\mu m}^{100\,\mu m} \varepsilon_\lambda(T) M_\lambda^\circ(T)\,\mathrm{d}\lambda}{\int_{1\,\mu m}^{100\,\mu m} M_\lambda^\circ(T)\,\mathrm{d}\lambda}$$

where T is the temperature of the emitting surface and $M_\lambda^\circ(T)$ is the spectral hemispherical emissive power of the black body at temperature T, given by Planck's Law.

RADIATIVE HEAT EXCHANGE

Until now only the radiative properties and laws of one surface have been considered. This subsection will examine radiative heat exchange between two surfaces. There are two main characteristics that differentiate radiative heat exchange from the other two heat transfer mechanisms, conduction and convection:

- The first is that, because of the electromagnetic nature of thermal radiation, radiative heat exchange occurs even without the presence of a physical medium.
- The second is that, although conduction and convection stop when the thermodynamic systems between which the transfer takes place arrive at the same temperature, radiative heat exchange occurs even between identical surfaces at the same temperature. This means that radiative equilibrium is a dynamic phenomenon.

The net radiative power between two infinite, black parallel plates (see Figure 13.1) is calculated by applying the Stefan–Boltzmann Law.

The emissive power of the plate at temperature T_1 is

$$M_1^\circ = \sigma T_1^4$$

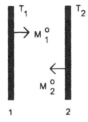

Figure 13.1 Radiative heat exchange between two infinite, black parallel plates

and the emissive power of the plate at temperature T_2 is:

$$M_2{}^\circ = \sigma T_2{}^4.$$

If $T_1 > T_2$, then the net radiative flux equals:

$$M_{1-2}^\circ = \sigma(T_1{}^4 - T_2{}^4).$$

The index of the net emissive power indicates the sense of the net radiative flow; this is from the hotter surface to the colder one. In this example this is from surface 1 to surface 2.

If the plates are not black, but their total hemispherical emissivities are ε_1 and ε_2 respectively, then the net radiative flux will be:

$$M_{1-2} = \sigma(\varepsilon_1 T_1{}^4 - \varepsilon_2 T_2{}^4).$$

In reality the surfaces exchanging radiation are not infinite but have finite dimensions. It is clear then that the radiative heat exchange between them depends strongly upon their geometry. To calculate the net radiative flux in this case, it is necessary to introduce the concept of the *configuration factor*. This is defined as the fraction of radiated energy leaving one surface and impinging on the second surface directly, if the two surfaces are diffuse in emission. The term 'directly' means that the energy impinging on the second surface does not come from reflection or re-radiation by other surfaces. Two important points can be made with reference to this definition:

- The configuration factor is a parameter without dimensions.
- Configuration factors depend only upon the geometry of the involved surfaces and not on any other physical property.

As an example we will give the configuration factor between two infinitesimal areas dA_1 and dA_2 (see Figure 13.2). The configuration factor between surface and surface is given by the following relation:

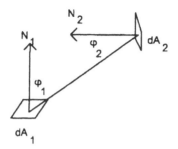

Figure 13.2 Radiative heat exchange
between two infinitesimal surfaces

$$F_{dA_1 \rightarrow dA_2} = \frac{\cos \varphi_1 \cos \varphi_2 \, dA_2}{\pi r^2}.$$

Calculation formulae or tabulated values of the configuration factor for other common configurations can be found in all books concerning radiative heat transfer. For more complicated problems, the reader can refer to radiative heat transfer handbooks.

The configuration factor can also be found in the literature under the names 'radiation shape factor', 'shape factor', 'view factor' and 'angle factor' [2].

Principles of radiative cooling

If two elements at different temperatures are facing one another, a net radiant heat flux from the hotter element will occur, even without any medium between them. If the colder element is kept at a fixed temperature, the other element will cool down to reach equilibrium with the colder element. This physical principle forms the basis of radiative cooling. The element at high temperature is the building envelope, or another radiator – attached to it or not – and the colder element is the sky vault.

Radiative cooling commonly occurs on the surface of the earth and it is the only mechanism that allows to the planet to dissipate the heat received from the sun, in order to maintain its thermal equilibrium.

Radiative cooling of a body, called a radiator, is based on the radiation exchange between it and the sky vault. Given that this radiator is a real body and not a black body, it is characterized by its emissivity.

SKY EMISSIVITY

Every object on the surface of the earth exchanges thermal radiation with all surrounding objects and therefore with the atmosphere. The atmosphere emits thermal radiation, except in the spectral region 8–13 μm, with a spectral distribution very close to that of a black body at a temperature equal to the dry-bulb temperature of the air close to the ground. The observed deviations from black-body radiation are attributed to the sky having its own wavelength-dependent emissivity, known as the sky emissivity, $\varepsilon_{sky}(\lambda)$. In practice, it is assumed that the sky emissivity does not depend upon wavelength for the long-wave part of the spectrum.

Atmospheric thermal emission is mainly due to the vibrational and rotational transitions of the asymmetrical molecules of the various constituents. These molecules are mainly water vapour, carbon dioxide and ozone. The symmetrical molecules O_2 and N_2, which compose 99% of the atmosphere, are transparent to infrared radiation (beyond 3 μm) and water vapour and carbon dioxide have few transitions in the spectral region of 8–13 μm. Consequently, in practice, the atmosphere can be considered transparent in this spectral region, which therefore is called the 'atmospheric window'.

The influence of the various atmospheric components on the atmospheric thermal radiation differs depending on the component:

- More than 90% of the total emitted radiation comes from the first 5 km in altitude. The contribution of each constituent to the total flux is 95.7% for H_2O + continuum, 2.8% for CO_2 (including CH_4 and N_2O) and 1.5% for O_3 (ozone).
- Ozone has a nearly constant peak of emission at 9.6 μm (near the centre of the atmospheric window), as it comes from absorption in the stratosphere, where its concentration is predominant.
- The atmospheric window is limited at about 14 μm because of the emission of CO_2. The carbon dioxide concentration is practically constant and no significant variation of the emitted thermal energy has been observed as a result of its variation, because the emission spectrum of CO_2 is superimposed on the emission spectrum of water vapour.
- The contribution of all other elements in the atmosphere is very small.

The emission spectrum of a clear-sky atmosphere during summer (dry-bulb temperature 21°C, wet-bulb temperature 16°C) is illustrated in Figure 13.3 [3]. The spectrum is given for the following zenith angles: 0°, 60°, 75° and 90°. The absorption band near the centre of the atmospheric window is due to ozone, independent of everyday weather. The relatively low emission within the range of the window at low zenith angles increases as the optical path through the low-altitude warm atmospheric layers also increases. When the zenith angle approaches 90°, then the emission spectrum of the atmosphere tends asymptotically to the black-body spectrum. This explains why the best position for placing a radiator is horizontal, since for low zenith angles the opening of the atmospheric window is higher.

The low emission within the atmospheric window is due to the continuum emission of water molecules. Therefore, the depth of the atmospheric window depends on the weather and thus it varies from day to day and even between two different periods of the same day, as well as varying from one geographical location to another. The effect of the water content of the atmosphere upon its spectral emission is illustrated in Figure 13.4 [3].

Since the water vapour content of the atmosphere is closely correlated with the absolute humidity or the dew-point temperature of the air close to the ground, these two physical quantities will be used later to estimate atmospheric emission.

Among the other atmospheric constituents that affect atmospheric thermal radiation are ozone and aerosols, but it has been shown that their effect is less important than that of water. Their emission spectra are illustrated in Figures 13.5 and 13.6 respectively [3]. From these figures it can be observed that only maritime aerosol significantly affects emission rates, while the effect of urban pollution is minor.

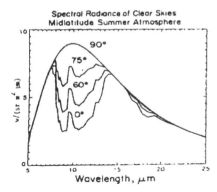

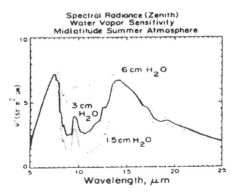

Figure 13.3 Emission spectrum of a clear atmosphere [3]

Figure 13.4 Effect of water vapour on the infrared emission spectrum depth of the atmosphere [3]

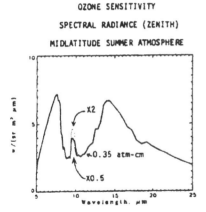

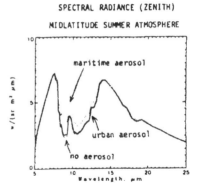

Figure 13.5 Sensitivity of the emission spectrum of the atmosphere to ozone variations [3]

Figure 13.6 Sensitivity of the emission spectrum of the atmosphere to the atmospheric aerosol [3]

If an object on the earth's surface emits thermal radiation within the atmospheric window and the atmospheric conditions are such that the atmospheric window is 'open', (i.e. low relative humidity and clear sky), then the object's temperature decreases. A radiator performs better under clear-sky conditions than under partly cloudy or average sky conditions. This was only to be expected since under clear-sky conditions sky thermal radiation is low, enabling the radiator to emit more energy towards the low-temperature heat sink than under average sky conditions. Increased amounts of clouds absorb and re-emit the infrared radiation. As a result, this slows the rate of radiative cooling from the radiating surface to the night sky.

The determinant parameter in evaluating the radiative cooling potential of a certain location is the sky temperature depression (DT_{sky}). This is defined as the difference between the ambient air temperature and the 'sky temperature' (i.e. the temperature of the black body having the same spectral distribution as the sky), which can be calculated as follows:

$$DT_{sky} = (1 - \varepsilon_{sky}^{1/4})T_a \tag{13.6}$$

where T_a is the ambient temperature and ε_{sky} is the sky emissivity.

Many correlations have been reported in the literature for calculating the sky emissivity (ε_{cs}). Here the Berdahl and Martin relation has been used [4]:

$$\varepsilon_{cs} = 0.711 + 0.56(T_{dp}/100) + 0.73(T_{dp}/100)^2 \tag{13.7}$$

where T_{dp} is dew-point temperature, defined as [5]

$$T_{dp} = C_3[\ln(RH) + C_1]/\{C_2 - [\ln(RH) + C_1]\} \tag{13.8}$$

with $C_1 = C_2 T_{dry}/(C_3 + T_{dry})$, $C_2 = 17.08085$, $C_3 = 234.175$, where T_{dry} is the ambient dry-bulb temperature (°C) and RH is the relative humidity (ranging from 0 to 1).

The instantaneous clear sky emissivity was estimated using the following expression, which takes into account the diurnal variation [4]:

$$\Delta\varepsilon_d = 0.013 \cos(2\pi t/24) \tag{13.9}$$

with t the hour of the day.

The values of sky emissivity obtained from equations (13.7) and (13.9) are valid for clear-sky conditions. Under cloudy sky the sky emissivity (ε_s) can be calculated from the following relationship [4]:

$$\varepsilon_s = \varepsilon_{cs}(1 + 0.0224n - 0.0035n^2 + 0.00028n^3) \tag{13.10}$$

where n is the total opaque cloud amount (0 for clear sky and 1 for overcast sky).

Measurements have shown that the long-wave radiation coming from the zenith of the sky vault is less than the radiation coming from the horizon. This is because the optical thickness of the atmosphere is smaller towards zenith than close to the horizon. Therefore the optimum position for a flat-plate radiative cooler is horizontal and does not depend, as in the case of the flat-plate solar collector, upon the latitude of the site.

RADIATIVE COOLING SYSTEMS

The simplest passive radiative cooling technique is to paint the roof white. White paint does not significantly affect the radiation rate at night, since white and black paints have almost the same emissivity in the long-wave range. The advantage of a white-painted roof is that as the roof absorbs less solar radiation during daytime, its temperature remains lower and therefore it can be easily cooled by radiation at night.

This technique has been applied in the traditional architecture of the Mediterranean basin, for example in the islands of the Cyclades in Greece. Measurements of this technique gave a cooling potential of 0.014 kWh m^{-2} day^{-1} [6]. This is a simple technique of rather poor performance and therefore applicable mainly in very hot countries.

Movable insulation

Movable insulation systems are applied on the roofs of buildings. They consist of an insulating material that can be moved over the roof of the building (Figure 13.7). These systems allow the exposure of the thermal mass of the roof (or of an additional sensible heat storage material placed on the roof) to the sky during the night. During the day the same mass is covered by an insulating layer to minimize heat storage in the thermal mass due to solar radiation. This layer can be operated manually or automatically. The advantage of such a system is that during winter its operation can be inverted: the mass is exposed to the sun and insulated at night to reduce radiative heat losses and channel the excess heat inside the building. There are two disadvantages: one is the additional cost of such a device, which is even higher if the option of automatic operation is chosen. The second is that, since the method is completely passive and no working fluids are involved, it is of minor interest for multi-storey buildings, since only the spaces directly under the roof will benefit from it.

Movable thermal mass

This technique is a variation of the previous one. Its cost is even higher, since it requires the construction of a thermally insulated pond on the roof of the building (Figure 13.8). It consists of a movable insulation device, underneath which there is a water pond. Between the pond and the roof of the building there is a gap in which the water from the pond can be canalized. This system is operated as follows: the pond is filled with water at night and exposed to the sky to be cooled by radiation. In

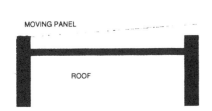

Figure 13.7 Principle of a moving insulation radiative system

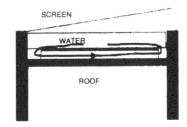

Figure 13.8 Principle of a movable thermal mass radiative system

the morning the pond is covered and the cooled water is drained from the pond and circulates between the insulation and the thermal mass of the roof, dissipating heat from the space below. The additional cost over that of the previous technique is related to the necessary reinforcements of the building structure to support the additional mass of water, to the extra attention required to avoid water absorption from the building elements and to the devices necessary for circulating the water. The advantage of this system, compared to movable insulation, is that it can cool other spaces and not only those directly in contact with the roof. This is because the cooling medium is water and it can be channelled to other spaces as well.

The flat-plate air cooler

This system has already been used for cooling water in a loop similar to that of a solar collector linked to a storage tank. This is a very simple device, looking almost like a flat-plate air solar collector without glazing. It consists of a horizontal rectangular duct. The top of the duct is the radiator, which is a metal plate. An example of such a cooler is illustrated in Figure 13.9 [7].

Attention should be paid to covering the metal plate with a material highly emissive in the long-wave part of the electromagnetic spectrum, since the emittance of metals decreases with wavelength. This is shown in Figure 13.10 [7], where the variation of the reflectivity of various metals is shown as a function of the wavelength.

The simplest solution is ordinary matt black paint. It is obvious that with such a coating, having almost opposite optical properties in the long-wave and in the visible ranges, there will be a higher efficiency, since the radiation losses of the cooler in the long-wave spectrum are higher. A radiator provided with a surface coating that approximates this behaviour is called a selective long-wave radiator. A survey of materials of high emissivity within the range of 8–14 μm has shown that the oxides and carbonates of titanium, aluminium, calcium and zinc are appropriate coatings,

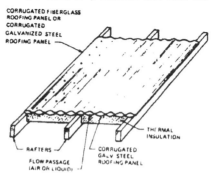

Figure 13.9 Example of flat-plate radiative cooler [7]

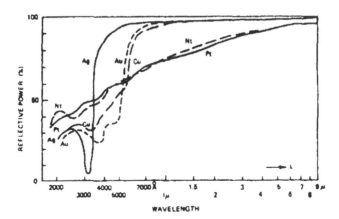

Figure 13.10 Reflectivity of various metals as a function of the wavelength [7]

since their reflectivity in the solar range of the spectrum is also high. White paints fabricated with the above substances have a long-wave emissivity (within the 8–14 μm part of the spectrum) ranging from 0.90 to 0.95 and an emissivity in the visible part of 0.05. Another type of selective radiator that could be used is evaporated aluminium covered by 12.5 μm of Tedlar (a polyvinyl-fluoride plastic) [7]. However a study of various experimental data obtained using selective coatings has demonstrated that the use of selective surfaces did not show any distinct advantage regarding their efficiency, compared to radiators with ordinary coatings [8].

The problems related to selective coatings are their cost and their reduced lifetimes. Therefore, in the current state of technology, the use of selective coatings is not always worth considering. However, a detailed discussion of the conditions (which rarely occur) under which the use of selective coatings for radiative cooling might be interesting is presented in [8].

The remaining part of the duct of the radiative cooler is insulated in the same way as the casing of flat-plate solar collectors.

The flat-plate cooler operates as follows: The radiator is cooled because of the night exposure to the sky vault. Air that circulates underneath the radiator by means of ventilators is cooled convectively and then injected into the space to be cooled. Such a system can be connected to the conventional air-conditioning system, in the same way as other hybrid cooling systems (see Chapter 11 on ground cooling; the reader should refer to that chapter for further information regarding linking and control problems).

What mainly differentiates the flat-plate air cooler from the solar air heater, except of course the different operating period during the day, is the absence of glazing, since ordinary glazing used in solar systems is not transparent to long-wave radiation. However some air coolers are covered with a so-called wind screen. The role of this wind screen is: When the operation of a radiator starts, its temperature is higher than or equal to the ambient temperature. In this case the radiator loses heat not only by radiation, but also convectively. The convective heat losses can be natural if there is no wind or forced convection. After some time the temperature of the radiator will become lower than the ambient temperature. At this moment, convection, up to now facilitating the operation of the cooler, will start opposing it. The wind screen is then applied, to reduce these convective heat gains from the warmer ambient air to the radiator, thus improving the efficiency of the system. Since windscreens must be transparent to infrared radiation, thin polyethylene films (60–100 μm), usually without ultraviolet inhibitors, can be used for this purpose. This is a commonly available material, which is not very expensive, and it has a long-wave transmittance of 70% [9]. It is reported however [8], that even a single layer of polyethylene reduces the emission of the radiator by 25%, because of its imperfect transmittance. Nonetheless, if the temperature of the radiator is below the ambient temperature, the convective heat gains are reduced to a much greater extent, especially if the wind velocity is high. Therefore the net cooling energy balance supports the use of the windscreen.

When a wind-screened radiator is used, attention should be paid to avoiding dew formation, since this will negatively affect the performance of the cooler. Experimental data have shown that dew formation raises the temperature of the radiator at night [9]. The temperature of the radiator rises from the moment that water droplets of dew start to form on the windscreen, which then starts to become opaque to long-wave radiation, since water is not transparent to this range of the electromagnetic spectrum. It is possible that after a certain time, which can be as much as several hours, the temperature of the radiator with a wind screen will approach the temperature of an exposed radiator, since the wind screen will become the radiating element. These observations indicate that the use of a wind screen is not effective in humid climates. To avoid dew formation, the difference between the air temperature and the dew-point temperature must be higher than the difference between the temperature of the radiator surface and the ambient temperature. When a wind screen is applied to a radiator, the film should be as tight as possible to avoid fluttering due to the wind. Such a fluttering creates convective currents in the still air between the wind screen and the radiator, thus reducing its insulating properties. Shading during daytime increases the lifetime of the wind screen by reducing the deterioration due to ultraviolet radiation. Wind screens also require also frequent cleaning since they attract dust particles which reduce their transparency. Experiments have shown that the use of a single windscreen gives a higher cooling load than the use of a double windscreen. The colour of the polyethylene was not found to influence the efficiency of the radiator [9]. In view of all the problems related to wind screens discussed above, serious consideration of all the parameters of a particular radiative system should be undertaken before they are used.

MODELLING THE FLAT-PLATE RADIATIVE COOLER

The useful cooling energy and the outlet temperature of the air provided by the radiator are calculated according to the methodology of Ito and Miura [10].

Cooling power of a radiator

The net heat flux (q_r) of a non-selective radiator at temperature (T_r) is calculated as a linear function of an effective heat transfer coefficient (h_e) and a minimum threshold temperature (T_{th}) (also referred to as stagnation temperature), as follows:

$$q_r = h_e(T_r - T_{th})$$ (13.11)

where h_e is defined by $h_e = h + 4\sigma T_a^3$ and T_{th} is the minimum temperature that the radiator can attain, defined by $T_{th} = T_a - q_0/h_e$. The convective heat transfer coefficient h is a function of the wind velocity V and is calculated from the following expressions [9].

- Radiator with no wind screen:

$$h = 5.7 + 3.8V \qquad V < 4 \text{ m s}^{-1}$$ (13.12)

$$h = 7.3V^{0.8} \qquad V > 4 \text{ m s}^{-1}.$$ (13.13)

- Radiator with wind screen:

$$h = 0.5 + 1.2V^{0.5}.$$ (13.14)

The net radiative power of a black body (q_0) at the ambient temperature (T_a) is given by:

$$q_0 = \sigma T_a^4 - q_s$$ (13.15)

where q_s is the sky irradiance, $q_s = \sigma \varepsilon_{sky} T_a^4$.

Fluid temperature

The problem of calculating the temperature of the heat transfer fluid flowing through a one-dimensional path in a radiator has been solved by Ito and Miura [10] in the same way as the case of a solar collector [11]:

$$T_{fo} - T_{th} = (T_{fi} - T_{th})\exp(-U_p A/mc_p)$$ (13.16)

with T_{fi} the inlet temperature of the heat transfer fluid, T_{fo} the outlet temperature of the heat transfer fluid and A the surface of the radiator. U_p is the overall heat transfer coefficient between the air circulating under the radiator and the ambient air and can be calculated from the following expression:

$$U_p = \frac{1}{(1/h_e) + (1/h) + (d_{rad}/k_{rad})}$$ (13.17)

with h_e the effective heat transfer coefficient previously defined, h the convective heat transfer coefficient between the radiator surface and the ambient air, calculated from equations (13.12)–(13.14), d_{rad} the thickness and k_{rad} the thermal conductivity of the radiator plate.

Equation (13.16) can be used to calculate the outlet temperature of the heat transfer fluid, given that the minimum threshold temperature T_{th} is known. One should note that the dependence of the thermal properties of the radiator and of the heat transfer fluid on temperature are taken into account in the numerical model.

POTENTIAL OF RADIATIVE COOLING

To estimate the potential of radiative cooling in various locations, the behaviour of a flat-plate cooler has been simulated. An open-loop radiative cooling system with an uncovered air collector, whose surface is exposed to the atmosphere at night, cools the air that circulates through the system. The air cooler was assumed to be a horizontal 2 m long rectangular air duct. The dimensions of the flow section were 1 m by 0.20 m. The radiator was considered to be a 0.003 m stainless steel plate, having an emittance of 0.90 in the infrared bandwidth. It was assumed that the cooler was functioning only during the night time with an air velocity through the radiator set at 2.5 m s^{-1}. The radiator was assumed to be horizontal. Simulations have been also carried out for the same system covered with a single wind screen.

The model presented in the previous section has been used to calculate the sky temperature depression and the air temperature at the outlet of the radiator. The useful cooling energy provided by the flat-plate radiative air cooler for various locations of southern Europe has been determined. This calculation was based on the temperature difference between the inlet and the outlet of the radiator. Two series of simulations have been performed for each location; one for optimum (clear) and one for average sky conditions.

Tables 13.1 to 13.5 give the total number of hours for a typical day in each month of the cooling season (May–September) for which the sky temperature depression reaches a given value. These data can also be presented in the form of histograms, which can be very useful for determining the feasibility of radiative cooling applications at a given location.

Table 13.1 Number of events for a given sky temperature depression in May [12]

Sky temperature depression (°C)

Location	Weather conditions	1	2	3	4	5	6	7	8	9	10	11	12	13	14	15	16	17	18	19	20	21	22
Ajaccio, Italy (Lat=41, 93n)	Optimum															1	2	2	2	2			
	Average										2	2	1	2	2								
Almeria, Spain (Lat=36, 85n)	Optimum														3		2	2	3	1			
	Average											4	2	2	3								
Ancona, Italy (Lat=43, 62n)	Optimum																3	3	2	1			
	Average								1	3	2	3											
Athens, Greece (Lat=37, 9n)	Optimum																	4	2	3			
	Average													3	2	2	2						
Atlanta, USA (Lat=33, 65n)	Optimum												1	2	1		1	1	1	1		1	2
	Average							2	1	1	1	1		1			1	1	2	1			
Barcelona, Spain (Lat=41, 38n)	Optimum																	4	3	2			
	Average													4	2	3							
Brindisi, Italy (Lat=40, 65n)	Optimum																3	2	2	2			
	Average												2	2	2	2	1						
Cagliari, Italy (Lat=39, 23n)	Optimum																	2	3	2	2		
	Average													2	3	2	2						
Catania, Spain (Lat=37, 47n)	Optimum																	4	2	3			
	Average													2	3	2	2						
Charleston, USA (Lat=32, 9n)	Optimum								1	1	1	1	1	1			1	1	1	2			
	Average				1	1	1	1	1		1	1	1		2	1							
Dubrovnic, Croatia (Lat=42, 63n)	Optimum																	4	3	2			
	Average												2	3	2	2							
Genoa, Italy (Lat=44, 40n)	Optimum																		6	3			
	Average													5	3	1							
Gibraltar (Lat=36, 15n)	Optimum														3	3	2	3					
	Average											4	2	2	3								
Ierapetra, Greece (Lat=35, 00)	Optimum														3	3	3	2					
	Average											4	3	2	3								
Leghorn, Italy (Lat=43, 55)	Optimum																3	2	3	1			
	Average												1	3	2	2	1						
Marseilles, France (Lat=43, 30n)	Optimum																2	2	2	1	3		
	Average											1	2	2	1	2	1						
Miami, USA (Lat=25, 8n)	Optimum										1	2	1	2	1	1	2	1					
	Average				1	2	1	1	1	1	2	2											
Milos, Greece (Lat=36, 44n)	Optimum																4	3	2	2			
	Average													4	3	2	2						
Naples, Italy (Lat=40, 68)	Optimum																	4	2	3			
	Average													2	3	2	2						
Nice, France (Lat=43, 68n)	Optimum																3	3	2	1			
	Average												4	2	2	1							
Nicosia, Cyprus (Lat=35, 15n)	Optimum																1	3	1	2	1	2	1
	Average														1	3	1	2	1	2	1		
Palma, Spain (Lat=39, 57n)	Optimum																	3	1	2	2	1	
	Average													2	2	2	1	2					
Parnos, Cyprus (Lat=34, 75n)	Optimum																	3	2	1	2	2	1
	Average												4	1	2	1	2						
Perpignan, France (Lat=42, 73n)	Optimum																	2	3	2	2		
	Average								1	3	2	1	2										
Raleigh, USA (Lat=35, 87n)	Optimum				2	1		1	1					1	1			1	1	2			
	Average				2	1		1	1					1			1	2	1				
Rome, Italy (Lat=41, 9n)	Optimum												1	3	2	2	1						
	Average													3	2	2	2						
Split, Croatia (Lat=43, 52n)	Optimum																3	3	3				
	Average											3	3	3									
Thesaloniki, Greece (Lat=40, 33n)	Optimum																3	2	2	2			
	Average												1	3	1	2	2						
Trieste, Italy (Lat=45, 65n)	Optimum																	2	3	3	1		
	Average													4	2	3							
Valletta, Malta (Lat=35, 90n)	Optimum																	9	2				
	Average													9	2								
Valencia, Spain (Lat=39, 47n)	Optimum																1	3	2	3			
	Average													2	3	2	2						
Venice, Italy (Lat=45, 43n)	Optimum																	2	3	2	2		
	Average												3	2	3	1							

Table 13.2 Number of events for a given sky temperature depression in June [12]

Location	Weather conditions	Sky temperature depression (°C)																					
		1	2	3	4	5	6	7	8	9	10	11	12	13	14	15	16	17	18	19	20	21	22
Ajaccio, Italy	Optimum															3	2	1	2	1			
(Lat=41, 93n)	Average								1	2	2	1	2	1									
Almeria, Spain	Optimum													3	2	2	2						
(Lat=36, 85n)	Average										3	2	2	2									
Ancona, Italy	Optimum																4	3	2				
(Lat=43, 62n)	Average										3	3	2	1									
Athens, Greece	Optimum																4	2	2	1			
(Lat=37, 9n)	Average											2	3	2	2								
Atlanta, USA	Optimum										1	1	1	1	1	1	1	2					
(Lat=33, 65n)	Average				1	1	1	1	1	1	1	1	1	1									
Barcelona, Spain	Optimum																4	2	3				
(Lat=41, 38n)	Average										3	3	2	1									
Brindisi, Italy	Optimum													3	2	2	2						
(Lat=40, 65n)	Average										3	2	2	2									
Cagliari, Italy	Optimum																2	2	2	2	1		
(Lat=39, 23n)	Average												3	2	2	1	1						
Catania, Spain	Optimum																2	2	2	3			
(Lat=37, 47n)	Average													4	2	1	2						
Charleston, USA	Optimum								2	1	1			1	1	1	1	1					
(Lat=32, 9n)	Average			1	1	1	1		1	1	1	1	1										
Dubrovnic, Croatia	Optimum															3	2	3	1				
(Lat=42, 63n)	Average												4	3	2								
Genoa, Italy	Optimum															3	3	2	1				
(Lat=44, 40n)	Average												4	2	3								
Gibraltar	Optimum													3	2	2	2						
(Lat=36, 15n)	Average									2	3	2	2										
Ierapetra, Greece	Optimum															5	2	2					
(Lat=35, 00)	Average												5	2	2								
Leghorn, Italy	Optimum																4	3	2				
(Lat=43, 55)	Average												4	2	2	1							
Marseilles, France	Optimum															2	2	1	2	2			
(Lat=43, 30n)	Average											1	2	2	1	2	1						
Miami, USA	Optimum								2	1	2	1	1	1	2	1							
(Lat=25, 8n)	Average				1	2	1	1	1	2	1	2											
Milos, Greece	Optimum														1	4	3	1					
(Lat=36, 44n)	Average													4	3	2							
Naples, Italy	Optimum															2	3	2	2				
(Lat=40, 68)	Average											1	4	2	2								
Nice, France	Optimum														3	3	2	1					
(Lat=43, 68n)	Average										2	3	2	2									
Nicosia, Cyprus	Optimum																2	2	1	2	1	1	1
(Lat=35, 15n)	Average												1	2	1	2	1	2					
Palma, Spain	Optimum															3	2	2	1	1			
(Lat=39, 57n)	Average											3	1	2	2	1							
Parnos, Cyprus	Optimum													1	2	2	1	2	1				
(Lat=34, 75n)	Average											1	2	2	1	2	1						
Perpignan, France	Optimum															3	2	2	2				
(Lat=42, 73n)	Average									3	2	2	2										
Raleigh, USA	Optimum										1	1	1		1	1	1		1	2			
(Lat=35, 87n)	Average				1	1			1	1	1	1	1	1	1								
Rome, Italy	Optimum																1	3	2	1	2		
(Lat=41, 9n)	Average												1	2	2	2	2						
Split, Croatia	Optimum																4	2	3				
(Lat=43, 52n)	Average											4	2	3									
Thesaloniki, Greece	Optimum															3	2	2	2				
(Lat=40, 33n)	Average											3	2	2	2								
Trieste, Italy	Optimum																4	3	2				
(Lat=45, 65n)	Average											4	3	2									
Valletta, Malta	Optimum															7	2						
(Lat=35, 90n)	Average											4	4	1									
Valencia, Spain	Optimum															2	3	2	2				
(Lat=39, 47n)	Average											1	3	2	2	1							
Venice, Italy	Optimum															3	2	3	1				
(Lat=45, 43n)	Average								2	3	2	2											

Table 13.3 Number of events for a given sky temperature depression in July [12]

Location	Weather conditions	Sky temperature depression (°C)																					
		1	2	3	4	5	6	7	8	9	10	11	12	13	14	15	16	17	18	19	20	21	22
Ajaccio, Italy	Optimum													3	2	1	2	1					
(Lat=41, 93n)	Average										3	2	1	2	1								
Almeria, Spain	Optimum													3	2	2	2						
(Lat=36, 85n)	Average									3	2	2	2										
Ancona, Italy	Optimum														4	2	2	1					
(Lat=43, 62n)	Average										4	2	2	1									
Athens, Greece	Optimum															4	2	2	1				
(Lat=37, 9n)	Average														4	2	2	1					
Atlanta, USA	Optimum								1	1	1	1	1	1	1	2							
(Lat=33, 65n)	Average				1	1	1	1	1	1	1	1	2										
Barcelona, Spain	Optimum														2	3	3	1					
(Lat=41, 38n)	Average											4	3	2									
Brindisi, Italy	Optimum													3	2	2	2						
(Lat=40, 65n)	Average										3	2	2	2									
Cagliari, Italy	Optimum														1	3	2	2	1				
(Lat=39, 23n)	Average												1	3	2	2	1						
Catania, Spain	Optimum																4	2	3				
(Lat=37, 47n)	Average														3	2	2	2					
Charleston, USA	Optimum							2	1	1	1	1	1	1	1	2							
(Lat=32, 9n)	Average		1	2	1	1		1	1	1	2	1											
Dubrovnic, Croatia	Optimum														3	3	2	1					
(Lat=42, 63n)	Average											4	2	3									
Genoa, Italy	Optimum															5	3	1					
(Lat=44, 40n)	Average											3	3	3									
Gibraltar	Optimum											2	3	2	2								
(Lat=36, 15n)	Average									1	3	2	2	1									
Ierapetra, Greece	Optimum															5	2	2					
(Lat=35, 00)	Average														4	2	3						
Leghorn, Italy	Optimum															4	2	3					
(Lat=43, 55)	Average											2	3	2	2								
Marseilles, France	Optimum														1	2	2	1	2	1			
(Lat=43, 30n)	Average												3	1	2	1	2						
Miami, USA	Optimum									2	2	1	1	1	2	2							
(Lat=25, 8n)	Average				3	1	1	1	2	1	2												
Milos, Greece	Optimum															5	3	2					
(Lat=36, 44n)	Average														4	3	2						
Naples, Italy	Optimum														1	4	2	2					
(Lat=40, 68)	Average										1	4	2	2									
Nice, France	Optimum													3	2	3	1						
(Lat=43, 68n)	Average										4	2	2	1									
Nicosia, Cyprus	Optimum															2	2	1	1	2	1		
(Lat=35, 15n)	Average													1	2	2	1	1	2				
Palma, Spain	Optimum													3	1	2	2	1					
(Lat=39, 57n)	Average											2	2	1	2	2							
Parnos, Cyprus	Optimum												2	2	1	2	2						
(Lat=34, 75n)	Average										2	2	1	2	1								
Perpignan, France	Optimum														3	2	1	2	1				
(Lat=42, 73n)	Average										1	2	2	2	2								
Raleigh, USA	Optimum									2	1	1	1	1	1	1	1						
(Lat=35, 87n)	Average			2		1	1	1	1			1	2										
Rome, Italy	Optimum																3	2	2	2			
(Lat=41, 9n)	Average												3	2	2	2							
Split, Croatia	Optimum															3	3	2	1				
(Lat=43, 52n)	Average											3	3	2	1								
Thesaloniki, Greece	Optimum														2	2	2	2	1				
(Lat=40, 33n)	Average												2	2	2	2	1						
Trieste, Italy	Optimum															5	3	1					
(Lat=45, 65n)	Average											1	4	3	1								
Valletta, Malta	Optimum														6	3							
(Lat=35, 90n)	Average											4	5										
Valencia, Spain	Optimum														3	3	2	1					
(Lat=39, 47n)	Average										2	3	2	2									
Venice, Italy	Optimum														3	2	2	2					
(Lat=45, 43n)	Average										2	3	2	2									

Table 13.4 Number of events for a given sky temperature depression in August [12]

Location	Weather conditions	1	2	3	4	5	6	7	8	9	10	11	12	13	14	15	16	17	18	19	20	21	22
Ajaccio, Italy	Optimum													2	2	2	1	2	2				
(Lat=41, 93n)	Average									2	2	2	1	2	2								
Almeria, Spain	Optimum											3	2	2	2	2							
(Lat=36, 85n)	Average							3	2	2	2	2											
Ancona, Italy	Optimum													3	2	3	2	1					
(Lat=43, 62n)	Average									5	2	2	2										
Athens, Greece	Optimum														3	2	2	3	1				
(Lat=37, 9n)	Average													3	3	2	2	1					
Atlanta, USA	Optimum								2	1	1	1	1	1	1	1	2						
(Lat=33, 65n)	Average				2	1	1	1	1	1		2	1	1									
Barcelona, Spain	Optimum													6	2	3							
(Lat=41, 38n)	Average									5	2	3	1										
Brindisi, Italy	Optimum												5	2	3	1							
(Lat=40, 65n)	Average									5	2	2	2										
Cagliari, Italy	Optimum														4	2	2	2	1				
(Lat=39, 23n)	Average											3	3	1	2	2							
Catania, Spain	Optimum															5	2	3	1				
(Lat=37, 47n)	Average														5	2	3	1					
Charleston, USA	Optimum							2	1	1	1	1	1	1	1	2							
(Lat=32, 9n)	Average	1	1	1	1	1	1	1	1	1	2												
Dubrovnic, Croatia	Optimum														4	2	3	2					
(Lat=42, 63n)	Average												4	3	2	2							
Genoa, Italy	Optimum														4	4	2	1					
(Lat=44, 40n)	Average											7	2	2									
Gibraltar	Optimum											3	2	2	2	2							
(Lat=36, 15n)	Average							2	3	1	2	3											
Ierapetra, Greece	Optimum														4	3	2	2					
(Lat=35, 00)	Average												4	3	2	2							
Leghorn, Italy	Optimum															5	3	2	1				
(Lat=43, 55)	Average											3	3	2	3								
Marseilles, France	Optimum														3	2	1	2	2	1			
(Lat=43, 30n)	Average											3	2	2	1	2	1						
Miami, USA	Optimum								2	2	1	1	1	2	2								
(Lat=25, 8n)	Average				3	1	1	1	1	2	2												
Milos, Greece	Optimum														3	3	3	2					
(Lat=36, 44n)	Average													3	3	2							
Naples, Italy	Optimum														3	3	2	2	1				
(Lat=40, 68)	Average											3	3	2	2	1							
Nice, France	Optimum													5	2	2	2						
(Lat=43, 68n)	Average									5	2	2	2										
Nicosia, Cyprus	Optimum														3	2	1	1	2	2			
(Lat=35, 15n)	Average											3	1	2	1	2	2						
Palma, Spain	Optimum														3	2	2	2	2				
(Lat=39, 57n)	Average									4	2	1	2	2									
Parnos, Cyprus	Optimum											2	2	2	1	2	2						
(Lat=34, 75n)	Average									3	1	2	2	1	2								
Perpignan, France	Optimum														2	3	2	2	2				
(Lat=42, 73n)	Average								2	2	2	2	3										
Raleigh, USA	Optimum							1	2			1	1	1	1	1	1	1	1				
(Lat=35, 87n)	Average	1	1	1	1	1	1				1	1	1	2									
Rome, Italy	Optimum														3	2	2	2	3				
(Lat=41, 9n)	Average											1	3	2	2	2	1						
Split, Croatia	Optimum														4	2	3	2					
(Lat=43, 52n)	Average											4	3	2	2								
Thesaloniki, Greece	Optimum														3	2	1	2	1	2			
(Lat=40, 33n)	Average										1	3	1	2	1	2	1						
Trieste, Italy	Optimum														3	4	2	2					
(Lat=45, 65n)	Average											6	2	3									
Valletta, Malta	Optimum													8	3								
(Lat=35, 90n)	Average								6	4	1												
Valencia, Spain	Optimum														4	3	2	2					
(Lat=39, 47n)	Average									3	3	2	3										
Venice, Italy	Optimum														3	3	2	2	1				
(Lat=45, 43n)	Average									3	2	2	2	2									

Table 13.5 Number of events for a given sky temperature depression in September [12]

Location	Weather conditions	Sky temperature depression (°C)																					
		1	2	3	4	5	6	7	8	9	10	11	12	13	14	15	16	17	18	19	20	21	22
Ajaccio, Italy (Lat=41, 93n)	Optimum													3	2	2	2	2					
	Average									4	1	2	2	2									
Almeria, Spain (Lat=36, 85n)	Optimum													3	2	2	3	1					
	Average									4	2	2	2	1									
Ancona, Italy (Lat=43, 62n)	Optimum																6	2	3				
	Average											5	3	3									
Athens, Greece (Lat=37, 9n)	Optimum																4	3	2	2			
	Average														5	2	3	1					
Atlanta, USA (Lat=33, 65n)	Optimum										2	2	1				1	1	2	1	1		
	Average					2	1	1	1	1	1	1	1	2								
Barcelona, Spain (Lat=41, 38n)	Optimum																6	2	3				
	Average											5	3	2	1								
Brindisi, Italy (Lat=40, 65n)	Optimum													4	2	3	2						
	Average									2	3	3	3	1									
Cagliari, Italy (Lat=39, 23n)	Optimum																3	3	2	2	1		
	Average											5	2	2	2								
Catania, Spain (Lat=37, 47n)	Optimum																3	3	2	3			
	Average											2	3	3	2	1							
Charleston, USA (Lat=32, 9n)	Optimum							2	1	1	1	1		1	1	1	1	2					
	Average		2	1	1	1	1					1	1	2	1								
Dubrovnic, Croatia (Lat=42, 63n)	Optimum																6	3	2				
	Average											6	2	3									
Genoa, Italy (Lat=44, 40n)	Optimum																6	3	1				
	Average									2	5	3	1										
Gibraltar (Lat=36, 15n)	Optimum													3	3	3	2						
	Average									2	3	2	2	2									
Ierapetra, Greece (Lat=35, 00)	Optimum																5	2	3	1			
	Average											4	3	2	2								
Leghorn, Italy (Lat=43, 55)	Optimum													3	2	2	1	3					
	Average											4	3	2	2								
Marseilles, France (Lat=43, 30n)	Optimum													2	2	2	1	2	2				
	Average									2	2	1	1	2	1	2							
Miami, USA (Lat=25, 8n)	Optimum							2	2	1	1	2	1	2									
	Average		1	2	1	2	1	1	2	1													
Milos, Greece (Lat=36, 44n)	Optimum																4	3	3	1			
	Average											5	2	3	1								
Naples, Italy (Lat=40, 68)	Optimum																5	2	2	2			
	Average													4	3	3	1						
Nice, France (Lat=43, 68n)	Optimum																5	2	2	2			
	Average													4	3	2	2						
Nicosia, Cyprus (Lat=35, 15n)	Optimum												3	1		2	1	1	2	1			
	Average													1	2	2	1	1	2	2			
Palma, Spain (Lat=39, 57n)	Optimum													3	2	2	2	1					
	Average											3	2	1	2	1							
Parnos, Cyprus (Lat=34, 75n)	Optimum														1	3	2	2	1				
	Average												1	3	2	2	2	1					
Perpignan, France (Lat=42, 73n)	Optimum													4	2	1	2	2	1				
	Average										2	3	1	2	2	1							
Raleigh, USA (Lat=35, 87n)	Optimum										2	1	1	1	1	1	1	1	1	1	1		
	Average						1	2		1	1	1		1	1	1	2						
Rome, Italy (Lat=41, 9n)	Optimum													4	2	2	2	1					
	Average											3	3	1	2	2							
Split, Croatia (Lat=43, 52n)	Optimum																6	3	2				
	Average											6	2	3									
Thesaloniki, Greece (Lat=40, 33n)	Optimum												3	2	2	2	2						
	Average											4	2	2	1	2							
Trieste, Italy (Lat=45, 65n)	Optimum													6	2	3							
	Average															7	4						
Valletta, Malta (Lat=35, 90n)	Optimum											2	8	1									
	Average													3	3	2	3						
Valencia, Spain (Lat=39, 47n)	Optimum											5	2	3	1								
	Average													3	3	2	3						
Venice, Italy (Lat=45, 43n)	Optimum											2	3	3	2	1							
	Average																						

An example is given in Figure 13.11, where the corresponding monthly values (May–September) of the distribution of the sky temperature depression is illustrated for Ajaccio (France), Raleigh (USA), Athens (Greece) and Nicosia (Cyprus). Athens is the most appropriate site for radiative cooling applications among the ones presented in this figure, because the sky temperature depression seldom drops below 14°C. It is important to note here the differences among the four cities regarding the number of continuous hours during which high temperature-depression values are obtained.

For Athens and Nicosia, the sky temperature depression values are at the same levels but Athens exhibits a longer number of continuous hours for a given temperature depression. The highest sky temperature depression values for Ajaccio are observed for almost the same number of hours as for Nicosia, but these values are lower than the values of sky temperature depression obtained in Nicosia. Finally Raleigh, which has the highest relative humidity during the cooling season among the four sites, has sky temperature depression values of less than 10°C for a significant number of hours.

Table 13.6 gives the mean daily useful cooling energy provided per square metre of radiating surface for each of the 28 southern European cities in Table 13.1 and Figure 13.12. The values given are those obtained for optimum (clear) sky conditions and average sky conditions. For each one of these conditions, there are two values. The first column corresponds to the case of an uncovered radiator and the second column to the case of a radiator covered with a wind screen. Each figure presents the results obtained for a typical day of each month in the cooling season, May through September.

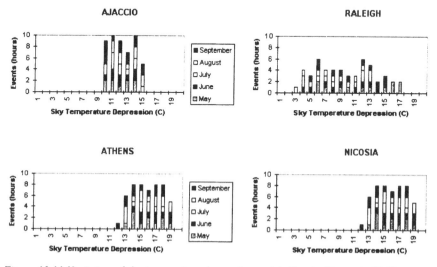

Figure 13.11 Variation of sky temperature depression at (a) Ajaccio (France), (b) Nicosia (Cyprus), (c) Athens (Greece) and (d) Raleigh, NC (USA)

Table 13.6 Mean daily useful cooling energy (Wh m⁻²) in the months of May to September with fluid velocity at 2.5 m s⁻¹. Information for each numbered location is given in Table 13.1 for May, Table 13.2 for June, Table 13.3 for July, Table 13.4 for August and Table 13.5 for September [12]. The locations are shown on the map in Figure 13.12

Location	May Clear sky No cover	May Clear sky Wind screen cover	May Average sky No cover	May Average sky Wind screen cover	June Clear sky No cover	June Clear sky Wind screen cover	June Average sky No cover	June Average sky Wind screen cover	July Clear sky No cover	July Clear sky Wind screen cover	July Average sky No cover	July Average sky Wind screen cover	August Clear sky No cover	August Clear sky Wind screen cover	August Average sky No cover	August Average sky Wind screen cover	September Clear sky No cover	September Clear sky Wind screen cover	September Average sky No cover	September Average sky Wind screen cover
1	141	188	100	133	126	176	97	135	149	190	122	155	184	233	151	191	163	217	122	162
2	121	196	96	156	93	151	74	120	89	143	71	114	99	166	78	131	111	179	96	139
3	107	167	74	115	109	168	79	122	105	162	80	124	136	202	106	156	89	179	64	128
4	118	178	96	137	110	172	95	149	100	167	95	159	124	204	116	191	134	210	117	184
5	120	177	94	139	113	166	87	128	110	162	86	126	129	189	98	143	136	199	103	152
6	74	144	59	114	72	138	60	115	65	129	57	114	83	160	72	138	93	165	71	125
7	100	171	80	136	99	167	81	137	97	163	86	145	117	195	104	175	115	194	91	154
8	109	177	91	148	115	178	99	153	123	179	112	164	150	220	142	208	141	218	120	184
9	102	166	78	127	105	163	83	129	101	155	86	133	107	189	93	164	114	192	95	159
10	106	172	79	128	95	162	73	123	94	159	75	127	113	191	58	148	114	193	85	144
11	127	199	103	161	89	151	79	134	75	147	72	140	87	173	82	163	96	178	82	154
12	98	174	78	139	68	131	55	106	78	132	63	106	90	159	73	129	91	160	71	126
13	97	166	76	129	100	162	78	127	98	157	78	125	129	197	103	158	115	195	90	152
14	79	154	61	118	94	159	74	125	86	152	73	128	103	182	83	146	100	177	78	139
15	118	178	96	146	95	159	85	144	75	144	71	138	91	176	85	165	92	179	81	184
16	123	181	94	138	127	178	99	153	123	172	100	140	152	211	124	172	129	199	98	152
17	116	171	89	131	116	163	89	124	112	157	87	122	124	181	98	144	130	191	98	144
18	136	220	115	187	110	177	103	166	108	174	104	167	125	201	119	191	133	205	124	190
19	119	176	94	139	115	169	89	131	112	164	94	138	130	191	104	152	128	189	96	141
20	122	199	100	162	94	151	81	131	93	144	82	126	118	182	100	154	124	192	109	168
21	99	168	65	110	102	165	68	111	100	161	74	120	118	191	84	135	129	199	103	152
22	118	183	91	141	116	179	92	143	114	175	96	147	137	211	117	180	127	206	100	162
23	89	157	63	112	96	162	74	126	94	158	78	131	109	190	88	155	107	189	83	146
24	124	179	95	137	125	183	103	151	114	169	99	147	144	210	124	181	150	215	121	174
25	83	160	59	114	88	155	63	110	79	151	60	116	96	184	75	143	98	190	73	141
26	89	180	69	140	84	149	68	119	84	141	73	124	95	161	76	128	95	167	72	126
27	121	178	94	138	127	177	99	139	119	166	92	128	147	204	112	156	148	206	114	160
28	94	160	67	114	91	154	65	110	88	149	66	112	117	185	89	143	111	188	80	135

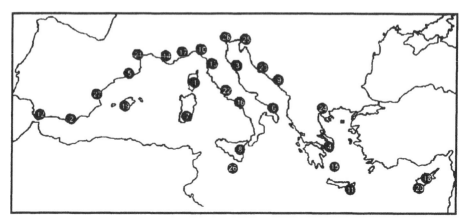

Figure 13.12 Map showing the positions of the 28 locations for which data is given in Tables 13.1 to 13.6

The mean daily useful cooling energy delivered by the flat-plate radiative cooler at the various southern European cities, ranges from 55 to 208 Wh m^{-2} for average sky conditions, and from 68 to 220 Wh m^{-2} for clear sky conditions. For the US cities the corresponding values range between 41 and 136 Wh m^{-2} for average sky conditions and between 69 and 182 Wh.m^{-2} for clear-sky conditions.

The influence of the wind screen can play an important role at some locations which are dominated by high wind speeds. For example, at Brindisi during the month of May under clear skies, the useful cooling energy of a covered radiator is 95% higher than the corresponding value of the uncovered radiator. A similar increase is also observed in August and September on the island of Milos, Greece, located in the Aegean Sea. During these months the area is dominated by strong northern winds that influence greatly the performance of the system. On the other hand, at some locations where the wind speed is relatively low, a wind screen is less effective. For example, at Ajaccio the corresponding values differ only by 33%.

The mathematical analysis presented above has been used for the development of a simplified methodology for radiative cooling calculations. The simplified calculation forms together with data necessary for these calculations are given in the Appendix at the end of this chapter. In the Appendix the reader can also find the calculation forms completed for a typical example.

PROBLEMS RELATED TO RADIATIVE COOLING

The main problem of radiative cooling techniques is that they cannot be applied in humid areas, since humidity absorbs long-wave radiation. An additional problem is related to the cost of these systems. Systems other than the simple radiator, incorporated into the structural elements of the building, increase the cost of the building, since sophisticated mechanisms for the movable parts are required, the structure of the building needs to be reinforced to support the weight of the additional thermal mass of the roof and additional treatment against humidity is required when a movable water mass is involved. For these reasons use of radiative cooling is not yet widespread and further research on practical applications is required.

CONCLUSIONS

Radiative cooling is an alternative natural cooling technique, based on radiative heat exchange. Because of the physics on which radiative cooling is based, it cannot be applied in areas having a significant relative humidity. An assessment of the radiative cooling potential for the northern Mediterranean countries gave satisfactory results. However, the related technology, not always simple, is expensive and therefore such systems cannot be applied massively in new constructions.

REFERENCES

1 Lampert, C.M. (1985). 'Fundamentals and properties of coatings for solar energy conversion', Lecture notes for the *Workshop on Physics of Non-Conventional Energy Sources and Material Science for Energy*, 2–20 September, International Centre for Theoretical Physics, Trieste, Italy.
2 Pitts, D. R. and L. E. Sissom (1977). *Theory and Problems of Heat Transfer*. McGraw-Hill, New York.
3 Clark, G. (1981). 'Passive/hybrid comfort cooling by thermal radiation', *Proceedings of the International Passive and Hybrid Cooling Conference*, Miami Beach, FL, eds A. Bowen, E. Clark and K. Labs, pp. 682–714.
4 Martin, M. (1989). 'Radiative cooling', *Passive Cooling*, ed. J. Cook, Ch. 4, pp. 138–196. MIT Press, Cambridge, MA.
5 *Aspirations – Psychrometer Tafel* (1976). *Deutscher Wetterdienst*, 5th ed. Vieweg Verlag, Braunschweig.
6 Santamouris, M. (1990). 'Natural cooling techniques', *Proceedings of Passive Cooling Workshop*, Ispra, Italy, eds. E. Aranovich, T. Steemers and O. Fernandes.
7 Pytlinski, J.T., G.R. Conrad and H.L. Connel (1986). 'Radiative cooling and solar heating potential using various roofing materials', *Proceedings of the International Conference on Renewable Energy Sources*, Miami, FL, ed. N. Veziroglu, pp. 101–144.
8 Givoni, B. (1982). 'Cooling by longwave radiation', *Passive Solar Journal*, Vol. 1, No 3, pp. 131–150.
9 Mostrel, M. and B. Givoni (1982). 'Windscreens in radiant cooling', *Passive Solar Journal*, Vol. 1, p. 229.
10 Ito, S. and N. Miura (1989). 'Studies of radiative cooling sytems for storing thermal energy', *Journal of Solar Energy Engineering*, Vol. 111, p. 251.
11 Duffie, J.A. and W.A. Beckman (1980). *Solar Engineering of Thermal Processes*. John Wiley & Sons, New York.
12 Argiriou, A., M. Santamouris, C. Balaras and S. Jeter (1993). 'Potential of radiative cooling in Southern Europe', *International Journal of Solar Energy*, Vol. 50, No. 3, pp. 197–204.

FURTHER READING

1 Addeo, A., E. Monza, M. Peraldo, B. Bartoli, B. Coluzzi, V. Silvestrini and G. Troise (1978). *Il Nuovo Cimento*, Vol. 1, p. 419.
2 Bartoli, B., B. Catalanoti, B. Coluzzi, B. Cuomo, V. Silvestrinni, and G. Troise (1977). *Applied Energy*, Vol. 3, p. 267.
3 Berger, X., B. Cubizolles and I. Donet (1988). *Solar & Wind Technology*, Vol. 5, p. 353.
4 Berger, X., C. Awanou and J. Bathiebo (1988). *International Journal of Ambient Energy*, Vol. 9, p. 155.
5 Catalanotti, S., V. Cuomo, G. Piro, D. Ruggi, V. Silvestrini and G. Troise (1975). *Solar Energy*, Vol. 17, p. 83.
6 *Climates of the States* (1975). *Climatography of the United States* No. 60-8, No. 60-9, No. 60-31 and No. 60-38.
7 Fragoudakis, A., G. Papadakis and S. Kyritsis (1989). *International Journal of Solar Energy*, Vol. 7, p. 73.
8 Golli, S. and Ph. Gremir (1981). *Journal de Physique, Colloque*, Supplement to No. 1, Vol. 42, C1.431.
9 Grenier, P. (1979). *Revue de Physique Appliquée*, Vol. 14, p. 87.
10 *Horizontal Study on Passive Cooling* (1990). C.E.C. – BUILDING 2000 Project, ed. M. Santamouris, Ch. 1, pp. 1–7. Organized by DG 12, EEC.
11 Steemers, T. C. (1991). *International Journal of Solar Energy*, Vol. 10, No 5, pp. 5–14.
12 *Weather in the Mediterranean* (1964). Vol. II. Her Majesty's Stationery Office, London.

APPENDIX A
SIMPLIFIED METHOD FOR EVALUATING
THE PERFORMANCE OF A RADIATIVE COOLER

Table 13.A1 Thermophysical properties of metals that can be used as radiator plates (necessary for calculations at STEP 2)

Metal type	Density (kg m^{-3})	Specific heat (J kg^{-1} K^{-1})	Conductivity (W m^{-1} K^{-1})
Copper	8 940	380	389
Aluminium	2 700	860	200
Stainless Steel	7 900	510	16

Table 13.A2 Heat transfer coefficient (W m^{-2} K^{-1}) between the cooled air and the radiator plate (necessary for calculations at STEP 6)

Air velocity (m s^{-1})	Hydraulic diameter (m)									
	0.020	0.139	0.058	0.077	0.095	0.113	0.131	0.148	0.165	0.182
1	6.764	3.416	2.299	1.741	1.406	1.183	1.024	0.904	0.811	3.845
2	6.764	3.416	2.299	1.741	7.614	7.337	7.113	6.927	6.767	6.629
3	6.764	3.416	11.634	10.961	10.471	10.090	9.782	9.526	9.307	9.117
4	6.764	3.416	14.585	13.741	13.127	12.650	12.264	11.942	11.668	11.430
5	6.764	18.197	17.380	16.375	15.643	15.075	14.615	14.231	13.904	13.620
6	6.764	21.831	20.057	18.898	18.053	17.397	16.866	16.423	16.046	15.718
7	6.764	24.643	22.640	21.332	20.378	19.637	19.038	18.538	18.112	17.743
8	6.764	27.369	25.145	23.692	22.632	21.810	21.144	20.589	20.116	19.706
9	34.754	30.023	27.584	25.989	24.827	23.925	23.195	22.586	22.067	21.617
10	37.754	32.615	29.965	28.233	26.970	25.990	25.197	24.535	23.972	23.483
11	40.691	35.151	32.295	30.428	29.068	28.012	27.157	26.444	25.836	25.309
12	43.570	37.639	34.581	32.582	31.125	29.994	29.078	28.315	27.664	27.100
13	46.399	40.083	36.826	34.697	33.146	31.941	30.966	30.153	29.460	28.859
14	49.181	42.486	39.034	36.778	35.133	33.857	32.823	31.961	31.227	30.590
15	51.921	44.853	41.209	38.827	37.091	35.743	34.652	33.742	32.967	32.294
16	54.623	47.187	43.353	40.847	39.020	37.603	36.455	35.498	34.682	33.974
17	57.822	49.489	45.468	42.840	40.924	39.437	38.233	37.230	36.374	35.632
18	59.920	51.673	47.557	44.808	42.805	41.249	39.990	38.940	38.045	37.269
19	65.520	54.009	49.621	46.753	44.662	43.039	41.726	40.630	39.696	38.887
20	65.092	56.231	51.662	48.676	46.500	44.810	43.442	42.301	41.329	40.486

Table 13.A3 Heat transfer coefficient (W m^{-2} K^{-1}) due to the wind, for a radiator without and with windscreen (necessary for the calculations at STEP 9)

Wind speed (m s^{-1})	No wind screen	With wind screen
0	5.70	0.50
1	9.50	1.70
2	13.30	2.20
3	17.10	2.58
4	20.90	2.90
5	26.45	3.18
6	30.61	3.44
7	34.63	3.67
8	38.53	3.89
9	42.34	4.10
10	46.06	4.29
11	49.71	4.48
12	53.29	4.66
13	56.82	4.83
14	60.29	4.99
15	63.71	5.15

Table 13.A4 Correction factor that gives the sky emissivity from the clear-sky emissivity as a function of cloudiness (necessary for the calculations at STEP 20)

Cloudiness factor	Correction factor
0	1.000
0.5	1.010
1	1.019
1.5	1.027
2	1.033
2.5	1.038
3	1.043
3.5	1.047
4	1.051
4.5	1.055
5	1.059
5.5	1.063
6	1.068
6.5	1.074
7	1.081
7.5	1.089
8	1.098
8.5	1.109
9	1.121
9.5	1.136
10	1.153

Form 1:   Simplified method for radiative cooling calculations

Radiator input data

1   Give the thickness of the radiator plate d (m):

2   Give the length (=dimension parallel to the air flow) of the radiator L (m):

3   Give the width of the radiator plate W (m): (Width = the dimension of the plate perpendicular to the flow; default value 1 m)

4   Give the height of the duct under the plate Z (m):
    (If unknown, a default value of 0.010 m can be used.)

5   Give the thermal conductivity of the radiator plate k (W m^{-1} K^{-1}):
    (Values are given in Table 13.A1)

6   Give the infrared emissivity of the radiator plate, ε:
    (If unknown, a default value of 0.9 can be assumed.)

7   Give the infrared transmittance of the wind screen τ:
    (1 if there is no wind screen; if unknown 0.75 can be used as a default value.)

8   Give the air velocity under the radiator plate u (m s^{-1}):
    (Default value 2.5 m s^{-1})

Weather input data

9   Give the ambient temperature T_A (°C):   _____

10   Give the ambient relative humidity RH (%):   _____

11   Give the wind velocity v (m s^{-1}):   _____

12   Give the average cloudiness c:
     (0 for clear sky, maximum 10 for overcast sky).   _____

13   Give the air temperature at the inlet of the radiator T_I (°C):
     (If equal to the ambient temperature, repeat value of STEP 9.)   _____

Calculations

14   Calculate the resistance of the radiator plate,
     $R_r = d/k$ [STEP1/STEP5]   _____

15   Calculate the hydraulic diameter $D_h = 2W[Z/(W + Z)]$
     [2 × STEP 3 × [(STEP 4)/(STEP 3 + STEP 4)]]   _____

16   Determine the heat transfer coefficient value between the
     circulated air and the radiator h (this value is given in Table 13.A2,
     as a function of D_h, (STEP 15) and u (STEP 8)   _____

17   Determine the heat loss coefficient due to wind h_w:
     (From Table 13.A3, as a function of the value of STEP 11).   _____

18   Calculate the absolute ambient temperature. $T_{abs} = T_A + 273$,
     [STEP 9+273]   _____

19   Calculate the third power of the absolute ambient temperature $T_A{}^3$:
     [STEP 18]^3 _____

20   Calculate the radiative heat transfer coefficient h_r from:
     $h_r = 0.000000227 \times \tau \times \varepsilon \times T_A{}^3$
     [0,000000227 × STEP 7 × STEP 6 × STEP 19]   _____

21   Calculate the effective heat transfer coefficient: $h_e = h_w + h_r$
     [STEP 17+STEP20]   _____

22   Calculate the effective resistance, $R_e = 1/h_e$ [1/(STEP 21)]   _____

23   Calculate the resistance between the circulating air and the
     radiator: $R_a = 1/h$ [1/(STEP 16)]   _____

24   Calculate the overall thermal resistance, R: $R = R_r + R_e + R_a$
     [STEP 14 + STEP 22 + STEP 23]   _____

25   Calculate the overall heat transfer coefficient I: $U = 1/R$ [1/(STEP 24)]   _____

26   Calculate the clear sky emissivity, ε_{cs}:
     $\varepsilon_{cs} = 0.71988 + 0.0041614T_A$ [0.71988 + 0.0041614 × STEP 9]   _____

27   Determine the correction factor for the sky emissivity c from the
     average cloudiness factor (STEP 12) and Table 13.A4   _____

28   Calculate the sky emissivity, ε_s: $\varepsilon_s = \varepsilon_{cs} \times c$ [STEP 26 × STEP 27]   _____

29   Calculate the fourth power of the absolute ambient temperature:
     $T_{abs}{}^4$ [(STEP 17)^4]   _____

30   Calculate the parameter $P_1 = T_{abs}{}^4/h_e$ [(STEP 29)/(STEP 21)]

31   Calculate the parameter $P_2 = 1 - \varepsilon_s$ [1 – STEP 28]

32   Calculate the parameter $P_3 = \varepsilon \times 0.000000057 \times P_2$
      [STEP 6 × 0.000000057 × STEP 31]

33   Calculate the threshold temperature: $T_{th} = T_A - P_3 \times P_1$
      [STEP 9 – STEP32 × STEP 30]

34   Calculate the parameter: $P_4 = U \times W$ [STEP 25 × STEP 3]

35   Calculate the parameter: $P_5 = Z \times u \times 1173$ [STEP 4 × STEP 8 × 1173]

36   Calculate the parameter: $P_6 = P_4/P_5$ [(STEP 34)/(STEP 35)]

37   Calculate the parameter: $P_7 = \exp(-P_6)$ [exp(–STEP 36)]

Final result

38   Calculate the air temperature at the outlet of the radiator:
      $T_{out} = T_{th} + (T' - T_{th}) \times P_7$ [STEP 33 + (STEP 13 – STEP 33) × P_7]

APPENDIX B
EXAMPLE OF APPLYING THE SIMPLIFIED METHODOLOGY
OF RADIATIVE COOLING

A radiator is given having the following characteristics:

Thickness of the metal plate:	2 mm
Length:	5 m
Width:	1 m
Height of the duct under the plate:	1 cm
Infrared emissivity of the radiator plate:	0.9
No wind screen	
Air velocity under the radiator plate:	2.5 m.s^{-1}

The following weather data are also given:

Ambient temperature:	27°C
Relative humidity:	40%
Wind velocity:	0 m.s^{-1}
Clear sky	

The temperature at the inlet of the radiator is 27°C. By applying the proposed methodology, the calculated outlet temperature is 26.2°C.

Form 2:   Simplified method for radiative cooling calculations

Radiator input data

1   Give the thickness of the radiator plate d (m):   0.002
2   Give the length (=dimension parallel to the air flow) of the
    radiator L (m):   10
3   Give the width of the radiator plate W (m): (Width = the
    dimension of the plate perpendicular to the flow;
    default value 1 m)   1
4   Give the height of the duct under the plate Z (m):
    (If unknown, a default value of 0.010 m can be used.)   0.01
5   Give the thermal conductivity of the radiator plate k (W m^{-1}
    K^{-1}): (Values are given in Table 13.A1)   16
6   Give the infrared emissivity of the radiator plate, ε:
    (If unknown, a default value of 0.9 can be assumed.)   0.9
7   Give the infrared transmittance of the wind screen τ:
    (1 if there is no wind screen; if unknown 0.75 can be used as
    a default value.)   1
8   Give the air velocity under the radiator plate u (m s^{-1}):
    (Default value 2.5 m s^{-1})   2.5

Weather input data

9   Give the ambient temperature T_A (°C):   27
10 Give the ambient relative humidity RH (%):   40
11 Give the wind velocity v (m s^{-1}):   0
12 Give the average cloudiness c:
    (0 for clear sky, maximum 10 for overcast sky).   0
13 Give the air temperature at the inlet of the radiator T_l (°C):
    (If equal to the ambient temperature, repeat value of STEP 9.)   27

Calculations

14 Calculate the resistance of the radiator plate,
    $R_r = d/k$ [STEP1/STEP5]   0.000125
15 Calculate the hydraulic diameter $D_h = 2W[Z/(W + Z)]$
    [2 × STEP 3 × [(STEP 4)/(STEP 3 + STEP 4)]   0.01980198
16 Determine the heat transfer coefficient value between the
    circulated air and the radiator h (this value is given in Table
    13.A2, as a function of D_h, (STEP 15) and u (STEP 8)   6.764
17 Determine the heat loss coefficient due to wind h_w:
    (From Table 13.A3, as a function of the value of STEP 11).   5.7

18 Calculate the absolute ambient temperature. $T_{abs}=T_A + 273$,

[STEP 9+273]   300

19 Calculate the third power of the absolute ambient
temperature T_A^3:

[STEP 18]^3 27000000

20 Calculate the radiative heat transfer coefficient h_r from:
$h_r = 0.000000227 \times \tau \times \varepsilon \times T_A^3$

[0,000000227 × STEP 7 × STEP 6 × STEP 19]   5.5161

21 Calculate the effective heat transfer coefficient: $h_e = h_w + h_r$

[STEP 17+STEP20]   11.2161

22 Calculate the effective resistance, $R_e = 1/ h_e$ [1/(STEP 21)]   0.08915755

23 Calculate the resistance between the circulating air and the
radiator: $R_a = 1/h$ [1/(STEP 16)]   0.147841514

24 Calculate the overall thermal resistance, R: $R = R_r + R_e + R_a$

[STEP 14 + STEP 22 + STEP 23]   0.237124064

25 Calculate the overall heat transfer coefficient I: $U = 1/R$

[1/(STEP 24)]   4.217201672

26 Calculate the clear sky emissivity, ε_{cs}:
$\varepsilon_{cs} = 0.71988 + 0.0041614 T_A$ [0.71988 + 0.0041614 × STEP 9]   0.8322378

27 Determine the correction factor for the sky emissivity c from   1
the average cloudiness factor (STEP 12) and Table 13.A4

28 Calculate the sky emissivity, ε_s: $\varepsilon_s = \varepsilon_{cs} \times c$ [STEP 26 × STEP 27]   0.8322378

29 Calculate the fourth power of the absolute ambient
temperature: T_{abs}^4 [(STEP 17)^4]   8100000000

30 Calculate the parameter $P_1 = T_{abs}^4/h_e$ [(STEP 29)/(STEP 21)]   722176157.5

31 Calculate the parameter $P_2 = 1 - \varepsilon_s$ [1 – STEP 28]   0.1677622

32 Calculate the parameter $P_3 = \varepsilon \times 0.000000057 \times P_2$

[STEP 6 × 0.000000057 × STEP 31]   8.6062E-09

33 Calculate the threshold temperature: $T_{th}=T_A – P_3 \times P_1$

[STEP 9 – STEP32 × STEP 30]   20.78480693

34 Calculate the parameter: $P_4 = U \times W$ [STEP 25 × STEP 3]   4.217201672

35 Calculate the parameter: $P_5 = Z \times u \times 1173$ [STEP 4 × STEP 8 ×
1173]   29.325

36 Calculate the parameter: $P_6 = P_4/ P_5$ [(STEP 34)/(STEP 35)]   0.143809094

37 Calculate the parameter: $P_7 = \exp(-P_6)$ [exp(-STEP 36)]   0.866053067

Final result

38 Calculate the air temperature at the outlet of the radiator:
$T_{out} = T_{th} + (T' - T_{th}) \times P_7$ [STEP 33 + (STEP 13 – STEP 33) × P_7]   26.16749395

14

Simplified methods for passive cooling applications

Calculation of the thermal performance of buildings using passive cooling systems and techniques requires the use of exact simulation codes such as TRNSYS [1], ESP [2], etc. Sometimes, also, calculation routines dealing with the performance of specific passive cooling components are not available, for example routines for buried pipes, and therefore it is impossible to perform calculations.

Various simplified manual calculation methods have been proposed in this book. All the methods deal with the performance of specific components and techniques like earth-to-air heat exchangers, radiative coolers, natural ventilation, etc. To calculate the impact of these systems and techniques on the global thermal performance of a building it is necessary to couple them thermally with the building operation.

NORMA is a simplified method developed to calculate the performance of buildings using passive cooling and techniques. The method is based on the well known principle of 'Balance Point Temperature' and it is validated against extensive and detailed simulation data obtained using TRNSYS. The basic elements of the method are given in the following sections.

COOLING REQUIREMENTS OF AIR-CONDITIONED (A/C) BUILDINGS

A full description of the calculation procedure to obtain the cooling load of an A/C building has been presented in Chapter 7 of the present book. Some basic principles are repeated in this section.

The instantaneous cooling load Q_c, for an A/C building can be written as :

$$Q_c = [k(T_o - T_i) + Q_s + Q_{in}]^+ \tag{14.1}$$

where k is the building load coefficient, (W °C^{-1}), T_o is the ambient (outdoor) temperature (°C), T_i is the indoor temperature (°C), Q_s are the solar 'gains' entering the building through transparent and opaque elements (W) and Q_{in} are the internal gains (W).

If $Q_T = Q_s + Q_{in}$ then

$$Q_c = [k(T_o - T_i) + Q_T]^+. \tag{14.2}$$

If a balance temperature T_b is used, where

$$T_b = T_i - Q_T / k \tag{14.3}$$

the instantaneous cooling load can be calculated as a linear function of the outdoor temperature T_o:

$$Q_c = [k(T_o - T_b)]^+. \qquad (14.4)$$

The monthly cooling load Q_{cm} can be calculated by integration of equation (14.4). Therefore

$$Q_{cm} = 3600 \, k \, \mathrm{CDD}(T_{bm}) \qquad (14.5)$$

where $\mathrm{CDD}(T_{bm})$ are the cooling degree hours based on the hourly value of the balance temperature T_b.

In order to verify the accuracy of the algorithms previously proposed to calculate the cooling load of A/C buildings, simulations of the monthly cooling load were performed for an extensive number of buildings. Calculations were performed using the TRNSYS computer programme. The results obtained for both monthly and annual cooling loads, as calculated by TRNSYS and the present method, are given in Figure 14.1. As shown, there is a very good agreement between the two sets of data. Regarding the annual cooling load, the absolute difference between the two sets of values is between 0.0 to 15%, with a mean value close to 6.3%. The absolute difference between the monthly predicted values is between 0.0 and 25%. Higher percentage differences are observed for months with a very low cooling load.

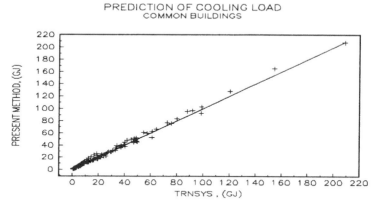

Figure 14.1 Calculated values of the cooling load using the present method and TRNSYS

COOLING REQUIREMENTS OF NATURALLY
VENTILATED (N/V) BUILDINGS

The balance temperature for buildings employing natural ventilation is different from standard air-conditioned buildings. If the mass of the air intake due to natural ventilation is denoted as m, the new balance temperature T_{bnv} under natural ventilation conditions can be computed from the following equation:

$$T_{bnv} = T_i + Q_T / K' \tag{14.6}$$

where

$$K' = k + mc. \tag{14.7}$$

Therefore, the cooling load is given by

$$Q_c = [K'(T_o - T_{bnv})]^+. \tag{14.8}$$

The monthly cooling load Q_{cm} can be calculated by integration of equation (14.8). Therefore

$$Q_{cm} = 3600\, K'\, \mathrm{CDD}(T_{bnv}) \tag{14.9}$$

where $\mathrm{CDD}(T_{bnv})$ are the modified cooling degree hours for natural ventilation calculated from the following expression:

$$\mathrm{CDD}(T_{bnv}) = \sum (T_o - T_b) S_j \tag{14.10}$$

$$S_j = 1 \text{ if } T_o > T_{bnv}$$

$$S_j = 0 \text{ if } T_o \leq T_{bnv}$$

where T_b is defined in equation (14.3).

COOLING REQUIREMENTS OF BUILDINGS EQUIPPED WITH
EARTH-TO-AIR HEAT EXCHANGERS (BURIED PIPES)

The instantaneous cooling load Q_c of a building equipped with buried pipes can be written as follows:

$$Q_c = [k(T_o - T_i) + Q_T - Q_{BP}]^+ \tag{14.11}$$

where Q_{BP} is the rate of energy produced from the buried pipes and k is the building load coefficient. Also

$$T_{pb} = T_i - (Q_T - Q_{BP})/k \tag{14.12}$$

where T_{pb} is the balance temperature for buildings equipped with buried pipes.

Therefore, the cooling load is given by

$$Q_c = [k(T_o - T_{pb})]^+. \tag{14.13}$$

The daily or monthly cooling load Q_{cbp} can be calculated by integration of equation (14.13). Therefore

$$Q_{cbp} = 3600\,k\,\text{CDD}(T_{pb}) \tag{14.14}$$

where $\text{CDD}(T_{pb})$ are the modified cooling degree hours for buildings equipped with buried pipes, calculated from the following expression :

$$\text{CDD}(T_{pb}) = \sum (T_o - T_{pb})S_j \tag{14.15}$$

$$S_j = 1 \text{ if } T_o > T_{pb}$$

$$S_j = 0 \text{ if } T_o \leq T_{pb}.$$

For the calculation of the rate of energy offered by the buried pipes Q_{BP} we use the following expression:

$$Q_{BP} = mc\,\text{DDBP}/t \tag{14.16}$$

where m and c are the mass rate and the specific heat of the circulated air. t is the daytime period in hours and DDBP are the degree hours for buried pipes defined as follows:

$$\text{DDBP} = \sum (T_o - T_{bpx})S_j \tag{14.17}$$

$$S_j = 1 \text{ if } T_o > T_{bpx} \text{ and } T_i > T_{bpx}$$

$$S_j = 0 \text{ if } T_o \leq T_{bpx}$$

$$S_j = 0 \text{ if } T_o > T_{bpx} \text{ and } T_i < T_{bpx}$$

where T_{bpx} is the exit temperature of the air from the pipes. The buried-pipes degree hours are calculated for the entire daytime period.

The methodology to calculate the exit air temperature from the buried pipes T_{bp} is given in Chapter 11. In order to validate the algorithms previously presented, a series of simulations was performed using the TRNSYS program coupled with routines which simulate dynamically the performance of earth-to-air heat exchangers. The routines are presented in [3]; they have been validated against extensive experimental data and found to be accurate. At the same time calculations were performed with the previously presented methodology.

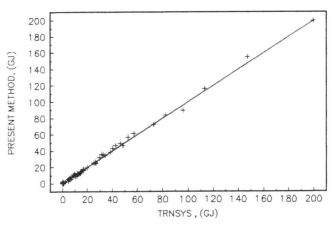

Figure 14.2 Comparison of the cooling load of buildings equipped with earth-to-air heat exchangers, as calculated using TRNSYS and the present method

The results obtained for the monthly and annual cooling loads, both from dynamic simulation and from the present method, are plotted in Figure 14.2. As shown, there is a very good agreement. For the estimated annual cooling loads, the differences obtained are between 1.0 and 8%. Higher values are presented for low cooling loads. For the monthly cooling load predictions, the differences are between 0.0 and 15%.

COOLING REQUIREMENTS OF BUILDINGS USING NIGHT-VENTILATION TECHNIQUES

The instantaneous cooling load Q_c for a building, where night-ventilation techniques are used, can be written as following :

$$Q_c = [k(T_o - T_i) + Q_T - Q_{NV}]^+ \qquad (14.18)$$

where Q_{NV} is the energy reduction due to the use of night ventilation and k is the building load coefficient of the daytime period. T_{bvn} is defined as

$$T_{bvn} = T_i - Q_T/k + Q_{NV}/k \qquad (14.19)$$

where T_{bvn} is the balance temperature for A/C buildings using night-ventilation techniques. Therefore, the cooling load is given by the following expression:

$$Q_c = [k(T_o - T_{bvn})]^+. \qquad (14.20)$$

The monthly cooling load Q_{cm} can be calculated by integration of equation (14.20). Therefore,

$$Q_{cm} = 3600\,k\,\mathrm{CDD}(T_{bvn})\qquad(14.21)$$

where $\mathrm{CDD}(T_{bvn})$ are the modified cooling degree hours for A/C buildings using night-ventilation techniques, calculated from the following expression:

$$\mathrm{CDD}(T_{bvn}) = \sum (T_o - T_{bvn})S_j\qquad(14.22)$$

$$S_j = 1 \text{ if } T_o > T_{bvn}$$

$$S_j = 0 \text{ if } T_o \leq T_{bvn}.$$

The energy reduction due to the night ventilation Q_{NV}, as introduced in equation (14.18), is calculated from the following algorithm:

$$Q_{NV} = (mc\,\mathrm{NDD})\,/\,\mathrm{DAY}\qquad(14.23)$$

where m is the mean air mass flow rate during the night period, c is the air specific heat, and DAY is the daytime period in hours. NDD are the night degree days calculated on a temperature base T_{ngh} defined as follows:

$$\mathrm{NDD} = \sum (T_{ngh} - T_o)S_j\qquad(14.24)$$

$$S_j = 1 \text{ if } T_{ngh} > T_o$$

$$S_j = 0 \text{ if } T_{ngh} < T_o$$

where T_{ngh} is the mean night-time indoor temperature of the building without night ventilation. It can be calculated from the following expression:

$$T_{ngh} = (h_{in}\,AT_k + m_a c T_{on})/(hA + m_a c)\qquad(14.25)$$

where

$$T_k = f_3(T_i + T_{on}).\qquad(14.26)$$

h_{in} is the internal heat transfer coefficient, A is the total internal surface area of the building, m_a is the night air flow rate of the building when no night-ventilation techniques are used, T_{on} is the mean night-time ambient temperature and f_3 is a coefficient defining the mean temperature of the mass.

In order to check if the total energy losses due to night ventilation, $mc\,\mathrm{NDD} \times 3600$, are higher than the maximum possible stored energy, we define a parameter, MCMAX, given by

$$\text{MCMAX} = \sum (M_i C_i)(T_{ngh} - T_{on}) > 0 \tag{14.27}$$

where $\sum (M_i C_i)$ is the effective thermal capacitance of the building. The index i indicates a material used in the building's structure. In the case where mc NDD $\times$ 3600 > MCMAX, the cooling degree hours should be appropriately adjusted and should be taken as equal to

$$\text{NDD} = f_1 \text{ MCMAX} / (3600mc). \tag{14.28}$$

where f_1 is a coefficient expressing the efficiency of heat transfer from the wall to the air and the degree to which the night-ventilation air is coupled to the thermal mass. This parameter is mainly a function of air-flow patterns inside the building and of the possible cover of the mass. A suggested value is $f_1 = 0.8$.

A second check should compare the energy losses due to night ventilation, mc NDD, with the cooling load of the building when no night-ventilation techniques are applied, Q_{cm}, where Q_{cm} is defined in equation (14.5). In the case when mc NDD > Q_{cm} then

$$\text{NDD} = f_2 \, Q_{cm} / (3600mc) \tag{14.29}$$

where f_2 is a coefficient expressing the ability of the building to carry over the 'coolth' into the occupied period on the following day. This coefficient is a function of the occupancy pattern and the thermal mass of the building. For heavyweight buildings f_2 can vary between 0.8 and 1 as a function of the occupancy pattern for buildings occupied at least ten hours a day.

In order to verify the accuracy of the algorithms previously proposed to calculate the cooling load of night-ventilated A/C buildings, simulations of the monthly cooling load were performed for various buildings. Calculations were performed using the TRNSYS computer program.

The results obtained for both monthly and annual cooling loads, as calculated with TRNSYS and the present method, are given in Figure 14.3. As shown, there is very good agreement between the two sets of data. For the annual cooling load, the absolute difference between the two sets of values is between 0.0 and 10%, with a mean value close to 4.6%. The absolute difference between the monthly predicted values is between 0.0 and 28%. Higher percentage differences are observed for months with a very low cooling load.

COOLING REQUIREMENTS OF BUILDINGS USING NIGHT-VENTILATION TECHNIQUES AND BURIED PIPES

The instantaneous cooling load Q_c for a building where night-ventilation techniques are used and earth-to-air heat exchangers operate can be written as follows:

PREDICTION OF COOLING LOAD
BUILDINGS WITH NIGHT VENTILATION

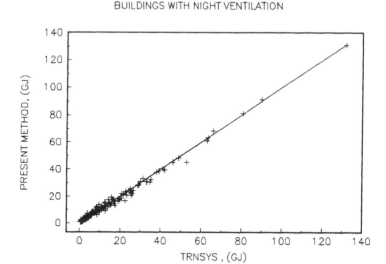

Figure 14.3 Comparison of the results obtained using TRNSYS and the present method

$$Q_c = [k(T_o - T_i) + Q_T - Q_{NVV} - Q_{BP}]^+ \qquad (14.30)$$

where Q_{NVV} is the energy reduction due to the use of night ventilation in buildings with buried pipes, k is the building load coefficient, as defined above and Q_{BP} is the rate of heat gain from the buried pipes defined in equation (14.16).

T_{bvpn} is defined as

$$T_{bvpn} = T_i - Q_T/k + (Q_{NVV} + Q_{BP})/k \qquad (14.31)$$

where T_{bvpn} is the balance temperature for buildings using night-ventilation techniques and equipped with buried pipes. Therefore, the cooling load is given by

$$Q_c = [k(T_o - T_{bvpn})]^+. \qquad (14.32)$$

The monthly cooling load Q_{cm} can be calculated by integration of equation (14.32). Therefore

$$Q_{cm} = 3600k\, CDD(T_{bvpn}) \qquad (14.33)$$

where $CDD(T_{bvpn})$ are the modified cooling degree hours for buildings using night-ventilation techniques and equipped with buried pipes, calculated from the following expression:

$$\mathrm{CDD}(T_{\mathrm{bvpn}}) = \sum (T_o - T_{\mathrm{bvpn}})S_j \tag{14.34}$$

$$S_j = 1 \text{ if } T_o > T_{\mathrm{bvpn}}$$

$$S_j = 0 \text{ if } T_o \le T_{\mathrm{bvpn}}.$$

The energy reduction due to the night ventilation, Q_{NVV}, as introduced in equation (14.30), is calculated from the following algorithm:

$$Q_{\mathrm{NVV}} = (mc\ \mathrm{NDD}) / \mathrm{DAY}. \tag{14.35}$$

In order to check if the total energy losses due to night ventilation $mc\ \mathrm{NDD} \times 3600$, are higher than the maximum possible stored energy, we define a parameter, MCMAX, as in equation (14.27). In the case when $mc\ \mathrm{NDD} \times 3600 > \mathrm{MCMAX}$ the cooling degree hours should be appropriately adjusted and should be taken as equal to

$$\mathrm{NDD} = f_1\ \mathrm{MCMAX} / (3600mc). \tag{14.36}$$

A second check should compare the energy losses due to night ventilation, $mc\ \mathrm{NDD}$, with the cooling load of the building when no night-ventilation techniques are applied, Q_{cm}, where Q_{cm} is defined in equation (14.5). When $mc\ \mathrm{NDD} > Q_{\mathrm{cm}}$

$$\mathrm{NDD} = f_2\ Q_{\mathrm{cm}} / (3600mc). \tag{14.37}$$

In order to validate the algorithms previously presented, a series of simulations were performed using the TRNSYS program coupled with routines dynamically simulating the performance of earth-to-air heat exchangers. The routines are presented in [3] and they have been validated against extensive experimental data and found to be accurate. At the same time, calculations were performed with the previously presented methodology.

The results obtained for both monthly and annual cooling loads, as calculated by TRNSYS and the present method, are given in Figure 14.4. As shown, there is very good agreement between the two sets of data. For the annual cooling load, the absolute difference between the two sets of values is between 0.0 and 16%, with a mean value close to 8.3%. The absolute difference between the monthly predicted values is between 0.0 and 24%. Higher percentage differences are observed for months with a very low cooling load.

HOW TO CALCULATE THE COOLING DEGREE HOURS

Various methods have been proposed to estimate the cooling degree days or degree hours [4–6]. The simplest method is that proposed in [5], where the cooling degree days can be calculated with the following expression:

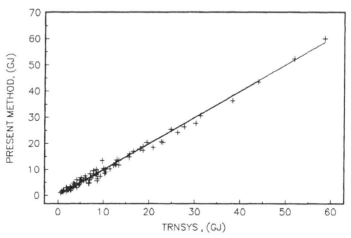

Figure 14.4 Calculated values of the cooling load using TRNSYS and the present method

$$D_c(T_b) = \sigma_m (N/24)^{1.5} [h/2 + (\ln(\cos h(1.698h)))/3.396 + 0.20441] \quad (14.38)$$

where T_b is the base temperature, σ_m is the standard deviation of the monthly average temperature T_o, N is the total number of hours in a month and the parameter h is calculated from the formula:

$$h = (T_o - T_b)/((N/24)\sigma_m)^{0.5}. \quad (14.39)$$

The following equation was fitted to calculate values of σ_m:

$$\sigma_m = 1.45 - 0.0290T_o + 0.0664\sigma_{yr} \quad (14.40)$$

where σ_{yr} is the standard deviation of the monthly average temperatures from the annual average temperature.

Equation (14.39) can be used to calculate the monthly cooling degree days for any base when σ_{yr} and the mean monthly temperature are known.

If σ_{yr} is not known, a more simple method is proposed. The monthly cooling degree hours in a temperature base T_b are calculated from the expression:

$$D_c(T_b) = N \sum_1^{24} (T_o(t) - T_b)^+ \quad (14.41)$$

where $T_o(t)$ are the mean monthly hourly values of the ambient temperature and N is the number of days in a month.

When the mean monthly hourly values of the ambient temperature are not known, the following expression, developed in [5], is proposed :

$$(T_o(t) - T_o) / A = 0.4632 \cos(t' - 3.805) + 0.0984 \cos(2t' - 0.36)$$

$$+ 0.0168 \cos(3t' - 0.822) + 0.0138 \cos(4t' - 3.513) \quad (14.42)$$

where A is the long-term monthly average amplitude of ambient temperature (°C) and

$$t' = 2(t - 1) / 24. \tag{14.43}$$

The parameter t is the hour of the day defined such that $t = 1$ at 1.00 a.m. and $t = 24$ at midnight. When the amplitude A is not known, the following expression is proposed:

$$A = 25.8 K_T - 5.21 \tag{14.44}$$

where K_T is the ratio of the total solar radiation striking a horizontal surface in a month to the monthly extraterrestrial radiation on a horizontal surface.

REFERENCES

1 TRNSYS 13.1 (1990). A Transient System Simulation program. Solar Energy Laboratory, University of Wisconsin-Madison, WI 53706, USA.
2 ESP-r (1993). A program for Building Energy Simulation. Energy Simulation Research Unit, Department of Mechanical Engineering, University of Strathclyde, UK.
3 Mihalakakou, G., M. Santamouris and D. Asimakopoulos (1994). 'On the use of ground for heat dissipation', *Journal of Energy*, Vol. 19, pp. 17–25.
4 Thom, H.C.S. (1954). 'Normal degree days below any base', *Monthly Weather Review*, Vol. 82, p. 5.
5 Erbs, D.G., S.A.Klein and W.A. Beckman (1983). 'Estimation of degree days and ambient bin data from monthly average temperatures', *ASHRAE Journal*, Vol. 60, pp. 1124–1130.
6 Schoenau, G. J. and R. A. Kehrig (1990). 'A method for calculating degree days to any base temperature', *Energy and Buildings*, Vol. 14, pp. 299–302.

Biographies of the authors

Professor Matheos Santamouris was born in Athens in 1956. He gained a diploma in Physics at the University of Patras in 1979, a D.E.A. Degree in Energy Physics at the Institut National Polytechnique de Grenoble in 1981 and a Doctorate in Solar Energy at the University of Patras in 1986. His main research direction is energy and buildings, passive cooling, thermal comfort and energy efficiency. He was the Co-ordinator of the PASCOOL research programme of the European Commission which is to aiming to develop methods, tools and design guidelines in the field of passive cooling of buildings. He also coordinates the OFFICE, AIOLOS and BUILT projects of the Commission, dealing with energy efficiency of buildings and the use of passive cooling techniques. He has participated in numerous international and national research projects.

He is the author of three books in the field of passive solar energy for buildings and has published over 200 scientific papers in international journals and in refereed conference proceedings. He is a member of the Editorial Boards of, and acts as a referee for, various international scientific journals. He is currently Assistant Professor at the Department of Applied Physics, University of Athens, Greece.

Professor Demosthenes Asimakopoulos qualified as an electronic engineer. but now applies his expertise in the field of the environment. He holds a PhD degree from University College London, UK. His main research area is instrumentation for environmental applications.

He has directed a large number of international experimental campaigns targeting the study of the indoor and outdoor environment and of the climate and was responsible for the design of the first high-resolution acoustic sounder, which is currently used in a number of atmospheric applications all over the world.

He is the founder of the European Energy Efficiency Education Center in Athens, Greece. He has published over 300 scientific papers in international journals and in refereed conference proceedings and these have been cited by more than 700 authors.

He is a Fellow of the British Meteorological Society, a Chartered Engineer of the Institute of Electrical Engineers, a Member of the Technical Chamber of Greece, an international journal referee and an Expert of the EUC. He has recently been elected Vice-Chairman of the European Environment Agency in Copenhagen, Denmark, and is a NATO expert on global environmental changes.

He is currently a Professor and Head of the Department of Applied Physics of the University of Athens and Director of the Institute of Meteorology and Physics of the Atmospheric Environment of the National Observatory of Athens. He is also a visiting Professor for the European Association for Environmental Management Education at the Polytechnic School of Torino, Italy, and the University of Trier, Germany.

Arthanassios A. Argiriou was born in Athens in 1960. He has a physics degree (University of Patras, 1983), D.E.A. in Energy Physics (Institute National Politechnique de Grenoble, 1984) and Doctarat Transferts Thermiques (Université de Provence, 1987). He is a researcher at the Institute of Meteorology and Physics of the Atmospheric Environment, National Observatory of Athens. Dr Argiriou is the author of about 35 papers in international scientific journals and his current fields of research are building physics, active and passive solar cooling and rational use of energy in the building sector.

Argiro Dimoudi, PhD, MSc, Dipl of Civil-Structural Engineer, is an Associate Researcher with CRES – Centre for Renewable Energy Sources in Athens, Greece. She obtained her Diploma in Civil-Structural Engineering from the Aristotle University of Thessaloniki (Greece), her MSc on Energy and Buildings from Cranfield Institute of Technology (UK) and her PhD with emphasis on passive cooling of buildings from Bath University (UK). Ms Dimoudi worked for four years in collaboration with CIENE and was involved in EC and national projects on energy efficiency, passive solar design and energy education in building.

Constantinos A. Balaras, PhD, a Mechanical Engineer, is a researcher at NOA-IMPAE. He was awarded a PhD and MSME from Georgia Institute of Technology, USA, and a BSME from Michigan Technological University. His research activities are in the areas of energy conservation, thermal and solar building applications, renewable energy sources, analysis and numerical modelling of thermal energy systems and HVAC systems. He is a registered engineer and a member of ASME, ASHRAE, ISES and IIR. He has published over 50 papers in technical journals and conferences and in nine books.

Dimitrios Mandas, Mechanical Engineer, MSc was awarded the Diploma of Mechanical Engineer in 1991 from the National Technical University of Athens, Greece (NTUA) and an MSc in Energy Systems in 1994 from the University of Strathclyde, UK. Since graduation from NTUA, he has been working in collaboration with CIENE and the National Observatory of Athens, within the framework of various projects related to energy conservation, including the creation of educational material and software development.

Elena Dascalaki, Energy Physicist, holds a Diploma in Physics (1987) and an MSc in Atmospheric Physics (1990), both from the University of Athens. Her main area of research is natural ventilation for cooling purposes. She is a PhD candidate in the Department of Applied Physics, University of Athens. Since 1992 she has been working as a researcher under contract for this Department and also works in the field of passive cooling, energy conservation and indoor air quality in the building sector, for national and European research projects.

Iro Livada-Tselepidaki, Assistant Professor, was born in Egypt in 1944. She was awarded a Diploma in Mathematics, University of Athens in 1971, Master of Meteorology, Meteorological Institute, University of Athens in 1974 and a PhD in Physics, University of Athens, 1980. She has worked in the fields of statistical climatology and hydrometeorology since 1975 and on renewable energy sources and heat island effects since 1985.

Index